第四次气候变化国家评估报告

第四次气候变化国家评估报告特别报告

——港澳特别行政区气候变化评估报告——

《第四次气候变化国家评估报告》编写委员会 编著

图书在版编目（CIP）数据

第四次气候变化国家评估报告特别报告. 港澳特别行政区气候变化评估报告/《第四次气候变化国家评估报告》编写委员会编著. —北京：商务印书馆，2023
（第四次气候变化国家评估报告）
ISBN 978 - 7 - 100 - 22064 - 4

Ⅰ. ①第… Ⅱ. ①第… Ⅲ. ①气候变化—评估—研究报告—香港、澳门 Ⅳ. ①P467

中国国家版本馆 CIP 数据核字（2023）第 056648 号

第四次气候变化国家评估报告
第四次气候变化国家评估报告特别报告
港澳特别行政区气候变化评估报告
《第四次气候变化国家评估报告》编写委员会 编著

商 务 印 书 馆 出 版
（北京王府井大街 36 号邮政编码 100710）
商 务 印 书 馆 发 行
北 京 冠 中 印 刷 厂 印 刷
ISBN 978 - 7 - 100 - 22064 - 4

2023 年 11 月第 1 版　　开本 710×1000 1/16
2023 年 11 月北京第 1 次印刷　　印张 32 3/4

定价：198.00 元

《第四次气候变化国家评估报告》编写委员会

编写领导小组

组　长	张雨东	科学技术部
副组长	宇如聪	中国气象局
	张　涛	中国科学院
	陈左宁	中国工程院
成　员	孙　劲	外交部条约法律司
	张国辉	教育部科学技术与信息化司
	祝学华	科学技术部社会发展科技司
	尤　勇	工业和信息化部节能与综合利用司
	何凯涛	自然资源部科技发展司
	陆新明	生态环境部应对气候变化司
	岑晏青	交通运输部科技司
	高敏凤	水利部规划计划司
	李　波	农业农村部科技教育司
	厉建祝	国家林业和草原局科技司
	张鸿翔	中国科学院科技促进发展局
	唐海英	中国工程院一局
	袁佳双	中国气象局科技与气候变化司
	张朝林	国家自然科学基金委员会地学部

曾经是《第四次气候变化国家评估报告》编写领导小组成员，并为报告的编写做了大量工作和贡献，后因职务变动等原因不再作为成员的有徐南平、丁仲礼、刘旭、张亚平、苟海波、孙桢、高润生、吴远彬、杨铁生、文波、刘鸿志、庞松、杜纪山、赵千钧、王元晶、高云、王岐东、王孝强。

专家委员会

主　任	徐冠华	科学技术部
副主任	刘燕华	科学技术部
委　员	杜祥琬	中国工程院
	孙鸿烈	中国科学院地理科学与资源研究所
	秦大河	中国气象局
	张新时	北京师范大学
	吴国雄	中国科学技术大学
	符淙斌	南京大学
	丁一汇	中国气象局国家气候中心
	吕达仁	中国科学院大气物理研究所
	王　浩	中国水科院国家重点实验室
	方精云	北京大学/中国科学院植物研究所
	张建云	南京水利科学研究院
	何建坤	清华大学

周大地	国家发展和改革委员会能源研究所
林而达	中国农业科学院农业环境与可持续发展研究所
潘家华	中国社会科学院城市发展与环境研究所
翟盘茂	中国气象科学研究院

编写专家组

组　长	刘燕华		
副组长	何建坤	葛全胜	黄　晶
综合统稿组	孙　洪	魏一鸣	
第一部分	巢清尘		
第二部分	吴绍洪		
第三部分	陈文颖		
第四部分	朱松丽	范　英	

领导小组办公室

组　长	祝学华	科学技术部社会发展科技司
副组长	袁佳双	中国气象局科技与气候变化司
	傅小锋	科学技术部社会发展科技司
	徐　俊	科学技术部社会发展科技司

	陈其针	中国21世纪议程管理中心
成　员	易晨霞	外交部条约法律司应对气候变化办公室
	李人杰	教育部科学技术与信息化司
	康相武	科学技术部社会发展科技司
	郭丰源	工业和信息化部节能与综合利用司
	单卫东	自然资源部科技发展司
	刘　杨	生态环境部应对气候变化司
	汪水银	交通运输部科技司
	王　晶	水利部规划计划司
	付长亮	农业农村部科技教育司
	宋红竹	国家林业和草原局科技司
	任小波	中国科学院科技促进发展局
	王小文	中国工程院一局
	余建锐	中国气象局科技与气候变化司
	刘　哲	国家自然科学基金委员会地学部

曾经是《第四次气候变化国家评估报告》编写工作办公室成员，并为报告的编写做了大量工作和贡献，后因职务变动等原因不再作为成员的有吴远彬、高云、邓小明、孙成永、汪航、方圆、邹晖、王孝洋、赵财胜、宛悦、曹子祎、周桔、赵涛、张健、于晟、冯磊。

本书作者

指导委员	王　浩	院　士	中国水科院国家重点实验室
	翟盘茂	研究员	中国气象科学研究院
领衔专家	邵　敏	教　授	暨南大学

首席作者（按姓氏笔画排序）

第一章	王志石	教　授	澳门科技大学
	刘绍臣	教　授	暨南大学
	李细明	高级科学主任	香港天文台
第二章	李剑锋	副教授	香港浸会大学
	陈永勤	教　授	香港中文大学（深圳）
	陈英凝	教　授	香港中文大学
第三章	吴芷茵	主　席	香港合资格环保专业人员学会
	徐　袁	副教授	香港中文大学
第四章	王雪梅	教　授	暨南大学
	杜迪佳	助理教授	香港科技大学（广州）
	陆恭蕙	教　授	香港科技大学

主要作者（按姓氏笔画排序）

第一章	刘　润	副研究员	暨南大学

	何　超	研究员	暨南大学
	陈仲良	教　授	香港城市大学
	周　文	教　授	香港城市大学
	唐恒伟	科学主任	香港天文台
	游　燕	助理教授	澳门科技大学
第二章	王　琼	助理教授	中山大学
	甘剑平	教　授	香港科技大学
	冯志雄	教　授	香港科技大学
	刘启汉	教　授	香港科技大学
	严鸿霖	教　授	香港中文大学
	杨　军	副研究员	暨南大学
	吴宏伟	教　授	香港科技大学
	陆萌茜	助理教授	香港科技大学
	林靖宇	研究员	香港中文大学
	罗　明	副教授	中山大学
	黄存瑞	教　授	中山大学
	黄　喆	副研究员	香港中文大学
	黎育科	副教授	香港中文大学
第三章	卢永鸿	教　授	香港中文大学
	宋庆彬	助理教授	澳门科技大学
	郭美瑜	讲　师	香港浸会大学

序

气候变化不仅是人类可持续发展面临的严峻挑战，也是当前国际经济、政治、外交博弈中的重大全球性和热点问题。政府间气候变化专门委员会（IPCC）第六次评估结论显示，人类活动影响已造成大气、海洋和陆地变暖，大气圈、海洋、冰冻圈和生物圈发生了广泛而迅速的变化。气候变化引发全球范围内的干旱、洪涝、高温热浪等极端事件显著增加，对全球粮食、水、生态、能源、基础设施以及民众生命财产安全等构成长期重大影响。为有效应对气候变化，各国建立了以《联合国气候变化框架公约》及其《巴黎协定》为基础的国际气候治理体系。多国政府积极承诺国家自主贡献，出台了一系列面向《巴黎协定》目标的政策和行动。2021 年 11 月 13 日，联合国气候变化公约第 26 次缔约方大会（COP26）闭幕，来自近 200 个国家的代表在会期最后一刻就《巴黎协定》实施细则达成共识并通过《格拉斯哥气候公约》，开启了全球应对气候变化的新征程。

中国政府高度重视气候变化工作，将应对气候变化摆在国家治理更加突出的位置。特别是党的十八大以来，在习近平生态文明思想指导下，按照创新、协调、绿色、开放、共享的新发展理念，聚焦全球应对气候变化的长期目标，实施了一系列应对气候变化的战略、措施和行动，应对气候变化取得了积极成效，提前完成了中国对外承诺的 2020 年目标，扭转了二氧化碳排放快速增长的局面。2020 年 9 月 22 日，中国国家主席习近平在第七十五届联合国大会一般性辩论上郑重宣示：中国将提高国家自主贡献力度，采取更加有力的政策和措施，二氧化碳排放力争于2030年前达到峰值，努力争取2060年前实现碳中和。中国正在为实现这一目标积极行动。

科技进步与创新是应对气候变化的重要支撑。科学、客观的气候变化评估是应对气候变化的决策基础。2006年、2011年和2015年，科学技术部会同中国气象局、中国科学院和中国工程院先后发布了三次《气候变化国家评估报告》，为中国经济社会发展规划和应对气候变化的重要决策提供了依据，为推进全球应对气候变化提供了中国方案。

为更好满足新形势下中国应对气候变化的需要，继续为中国应对气候变化相关政策的制定提供坚实的科学依据和切实支撑，2018年，科学技术部、中国气象局、中国科学院、中国工程院会同外交部、国家发展改革委、教育部、工业和信息化部、自然资源部、生态环境部、交通运输部、水利部、农业农村部、国家林业和草原局、国家自然基金委员会等十五个部门共同组织专家启动了《第四次气候变化国家评估报告》的编制工作，力求全面、系统、客观评估总结中国应对气候变化的科技成果。经过四年多的不懈努力，形成了《第四次气候变化国家评估报告》。

这次评估报告全面、系统地评估了中国应对气候变化领域相关的科学、技术、经济和社会研究成果，准确、客观地反映了中国2015年以来气候变化领域研究的最新进展，而且对国际应对气候变化科技创新前沿和技术发展趋势进行了预判。相关结论将为中国应对气候变化科技创新工作部署提供科学依据，为中国制定碳达峰、碳中和目标规划提供决策支撑，为中国参与全球气候合作与气候治理体系构建提供科学数据支持。

中国是拥有14.1亿多人口的最大发展中国家，面临着经济发展、民生改善、污染治理、生态保护等一系列艰巨任务。我们对化石燃料的依赖程度还非常大，实现双碳目标的路径一定不是平坦的，推进绿色低碳技术攻关、加快先进适用技术研发和推广应用的过程也充满着各种艰难挑战和不确定性。

我们相信，在以习近平同志为核心的党中央坚强领导下，通过社会各界的共同努力，加快推进并引领绿色低碳科技革命，中国碳达峰、碳中和目标一定能够实现，中国的科技创新也必将为中国和全球应对气候变化做出新的更大贡献。

王志刚

科学技术部部长

2022年3月

前　　言

2018 年 1 月，科学技术部、中国气象局、中国科学院、中国工程院会同多部门共同启动了《第四次气候变化国家评估报告》的编制工作。四年多来，在专家委员会的精心指导下，在全国近 100 家单位 700 余位专家的共同努力下，在编写工作领导小组各成员单位的大力支持下，《第四次气候变化国家评估报告》正式出版。本次报告全面、系统地评估了中国应对气候变化领域相关的科学、技术、经济和社会研究成果，准确、客观地反映了中国 2015 年以来气候变化领域研究的最新进展。报告的重要结论和成果，将为中国应对气候变化科技创新工作部署提供科学依据，并为中国参与全球气候合作与气候治理体系构建提供科学数据支持，意义十分重大。

本次报告主要从“气候变化科学认识”“气候变化影响、风险与适应”“减缓气候变化”“应对气候变化政策与行动”四个部分对气候变化最新研究进行评估，同时出版了《第四次气候变化国家评估报告特别报告：方法卷》《第四次气候变化国家评估报告特别报告：科学数据集》《第四次气候变化国家评估报告特别报告：中国应对气候变化地方典型案例集》等八个特别报告。总体上看，《第四次气候变化国家评估报告》的编制工作有如下特点：

一是创新编制管理模式。本次报告充分借鉴联合国政府间气候变化专门委员会（IPCC）的工作模式，形成了较为完善的编制过程管理制度，推进工作机制创新，成立编写工作领导小组、专家委员会、编写专家组和编写工作办公室，坚持全面系统、深入评估、全球视野、中国特色、关注热点、支撑决策的原则，确保报告的高质量完成，力争评估结果的客观全面。

二是编制过程科学严谨。为保证评估质量，本次报告在出版前依次经历了内审专家、外审专家、专家委员会和部门评审“四重把关”，报告初稿、零稿、一稿、二稿、终稿“五上五下”，最终提交编写工作领导小组审议通过出版。在各部分作者撰写报告的同时，我们还建立了专家跟踪机制。专家委员会主任徐冠华院士和副主任刘燕华参事负责总体指导；专家委员会成员按照领域分工跟踪指导相关报告的编写；同时还借鉴 IPCC 评估报告以及学术期刊的审稿过程，开通专门线上系统开展报告审议。

三是报告成果丰富高质。本次报告充分体现了科学性、战略性、政策性和区域性等特点，积极面向气候变化科学研究的基础性工作、前沿问题以及中国应对气候变化方面的紧迫需求，深化了对中国气候变化现状、影响与应对的认知，较为全面、准确、客观、平衡地反映了中国在该领域的最新成果和进展情况。此外，此次评估报告特别报告也是历次《气候变化国家评估报告》编写工作中报告数量最多、学科跨度最大、质量要求最高的一次，充分体现出近年来气候变化研究工作不断增长的重要性、复杂性和紧迫性，同时特别报告聚焦各自主题，对国内现有的气候变化研究成果开展了深入的挖掘、梳理和集成，体现了中国在气候变化领域的系统规划部署和深厚科研积累。

本次报告得出了一系列重要评估结论，对支撑国家应对气候变化重大决策和相关政策、措施制定具有重要参考价值。一方面，明确中国是受全球气候变化影响最敏感的区域之一，升温速率高于全球平均。如中国降水时空分布差异大，强降水事件趋多趋强，面临洪涝和干旱双重影响；海平面上升和海洋热浪对沿海地区负面影响显著；增暖对陆地生态系统和农业生产正负效应兼有，中国北方适宜耕作区域有所扩大，但高温和干旱对粮食生产造成损失更为明显；静稳天气加重雾霾频率，暖湿气候与高温热浪增加心脑血管疾病发病与传染病传播；极端天气气候事件对重大工程运营产生显著影响，青藏铁路、南水北调、海洋工程等的长期稳定运行应予重视。另一方面，在碳达峰、碳中和目标牵引下，本次评估也为今后应对气候变化方面提供了重要参考。总而言之，无论是实施碳排放强度和总量双控、推进能源系统改革，还是加强气候变化风险防控及适应、产业结构调整，科技创新都是必由之路，更是重要依靠。

我们必须清醒地认识到，碳中和目标表面上是温室气体减排，实质是低碳技术实力和国际规则的竞争。当前，中国气候变化研究虽然取得了一定成绩，形成了以国家层面的科技战略规划为统领，各部门各地区的科技规划、政策和行动方案为支撑的应对气候变化科技政策体系，较好地支撑了国家应对气候变化目标实现，但也要看到不足，在研究方法和研究体系、研究深度和研究广度、科学数据的采集和运用，以及研究队伍的建设等方面还有提升空间。面对新形势、新挑战、新问题，我们要把思想和行动统一到习近平总书记和中央重要决策部署上来，进一步加强气候变化研究和评估工作，不断创新体制机制，提高科学化水平，强化成果推广应用，深化国际领域合作，尽科技工作者最大努力更好地为决策者提供全面、准确、客观的气候变化科学支撑。

本次报告凝聚了编写组各位专家的辛勤劳动以及富有创新和卓有成效的工作，同时也是领导小组和专家委员会各位委员集体智慧的集中体现，在此向大家表示衷心的感谢。也希望有关部门和单位要加强报告的宣传推广，提升国际知名度和影响力，使其为中国乃至全球应对气候变化工作提供更加有力的科学支撑。

徐冠华

中国科学院院士、科学技术部原部长

中 文 版

目　录

摘　要

香港和澳门作为我国的两个特别行政区，人口密度大、城市化水平高、经济发达。2017 年的《政府工作报告》明确了粤港澳大湾区发展战略，从全球视野出发，提出发挥港澳独特优势，推动内地与港澳深化合作。港澳地区位于亚热带季风气候带，水系发达、海岸线长，生态环境复杂多样，是我国气候变化最敏感的区域之一。在建设世界一流湾区的过程中，港澳地区的人口与经济聚集度将进一步提升，面临的气候风险也将不容小觑。在全球气候变化的背景下，系统评估港澳地区的气候变化尤为重要。本报告从港澳地区气候变化的事实，气候变化的影响、风险和适应，减缓气候变化，气候变化政策与地区合作四个方面评估了港澳地区的气候变化。

一、港澳地区气候变化的事实

港澳地区气候变化的事实评估了港澳地区气候变化的证据。数据显示在过去百年间，香港和澳门的气候变暖是明确的，随着极端天气气候事件趋频趋强，未来气候将继续呈加剧变化态势。

过去百年，港澳地区的气候呈变暖变湿趋势，海平面不断上升。1885～2018 年香港年平均温度的变暖趋势为每十年升高 0.13 摄氏度，而在 1989～2018 年达每十年升高 0.17 摄氏度。城市化的影响约占香港整体升温趋势的

40%～50%。1901～2018 年澳门年平均温度的变暖趋势为每十年升高 0.056 摄氏度。澳门面积小、临海、城市化对局地气候的影响较小，这些可能是澳门的平均升温速率低于香港的原因。港澳地区的降水量总体呈增加趋势，但年际波动较大。1885～2018 年，香港的年降水量呈微弱增加趋势，速率为每十年增加 22 毫米。1901～2018 年，澳门的年降水量也呈微弱的增加趋势，速率为每十年增加 41.4 毫米。1954～2018 年香港维多利亚港海平面上升速率为每年 3.1 毫米。1925～2010 年澳门沿海海平面上升速率为每年 1.35 毫米，特别是 1970 年后，上升速率达到每年 4.2 毫米。

过去百年，港澳地区极端热天显著增多，极端冷天显著减少，极端降水呈总体增加趋势。香港每年酷热天气[①]日数和热夜[②]日数呈明显的上升趋势，寒冷[③]天气日数则呈减少趋势。1901～2018 年，澳门日最高气温超过 33 摄氏度的天数虽未明显变化（–0.02 天/十年），但年低温（低于 13 摄氏度）天数呈下降趋势（–0.75 天/十年）。1947～2018 年，香港大雨[④]日数增加趋势为每十年 0.2 天，最高单小时降雨量纪录频繁被刷新。进入港澳 500 千米范围的热带气旋数目有长期减少趋势，但并不显著。1971～2010 年导致澳门需要发出八号风球的热带气旋个数有下降趋势，但影响其趋势的因素较多，并不一定由气候变化造成。

未来，港澳地区气候变化整体上也存在变暖变湿趋势，海平面继续上升，极端天气事件增多。在 RCP8.5[⑤]情景下，预计到 21 世纪末（2091～2100 年），香港年平均温度较 1986～2005 年将升高 3～6 摄氏度，年降水量较 1986～2005

①日最高气温在 33 摄氏度或以上。

②日最低气温在 28 摄氏度或以上。

③日最低气温在 12 摄氏度或以下。

④雨量大于 30 毫米/小时。

⑤RCP 表示典型浓度路径，用相对于 1750 年的 2100 年近似总辐射强迫水平来表示，在 RCP8.5 情景下为 8.5 瓦/平方米，该情景是温室气体排放非常高的情景，到 2100 年 CO_2 浓度大约为 936 ppm。

年增加约 180 毫米，香港及邻近水域的平均海平面高度较 1986～2005 年提高 0.73～1.28 米[①]。随着未来海平面继续上升和热带气旋强度的增加，风暴潮的威胁也会相应增加。在 RCP8.5 情景下，预计到 21 世纪末，香港平均年酷热天气日数将超过 100 天，而平均年热夜日数约为 150 天，寒冷天气日数显著减少；极端降雨日[②]会由 1986～2005 年实况观测的平均每年 4.2 天增加至约 5.1 天。在不同温室气体排放情景（排放情景由低至高排列：B1、A1B、A2）下，21 世纪末（2091～2100 年）澳门的平均气温很有可能较 1971～2000 年升高 1.9～3.4 摄氏度，多模式多情景的平均升温为 2.7 摄氏度。据预测，未来澳门海平面的上升速率将比全球平均快 20%。

二、港澳地区气候变化的影响、风险和适应

气候变化对港澳地区的洪水灾害、城市安全、水资源、海平面、人体健康等相关敏感领域产生了十分明显的影响，总体为“弊大于利”。未来的气候变化将进一步加剧风险。因此，港澳地区因地制宜，在适应气候变化领域采取了具有特色的预防策略和措施。

港澳地区面临巨大的城市泄洪和山体滑坡的压力，城市安全受到威胁。气候变化导致港澳地区的极端降水增加，且降水时间分布不均匀。海平面上升和风暴潮的影响，导致港澳地区的城市排水泄洪压力较大，对城市安全构成威胁。香港以山地为主的地形使其容易发生由暴雨引发的山体滑坡和泥石流等地质灾害，造成基础设施破坏以及生命和财产损失。为此，港澳地区通过建立完善的雨水排放系统、提升基础设施的防洪标准、采纳“蓝绿建设”理念进行规划、采用斜坡安全系统等方式来防范上述风险。

---

①包含本地地壳垂直运动的影响。

②极端降雨日：指日降水量在 100 毫米或以上。

气候变化背景下，港澳地区未来热带气旋的强度、降水强度及沿岸水淹风险将会增加。虽然影响港澳地区的热带气旋个数总体上呈下降趋势，但影响港澳及周边地区的强台风个数明显增多；气候变暖导致海平面上升、海岸侵蚀、极端风暴潮重现期缩短。港澳地区现有的热带气旋预警系统可以有效减少台风灾害带来的可能损失。未来需要加强更有针对性的观测，提高热带气旋预警能力，形成有效的区域联防机制，进一步减少台风灾害带来的各类损失。

气候变化带来更频繁的高温和热浪，改变舒适度，影响城市宜居性，同时给空气质量和人体健康带来负面影响。气候变暖与城市化导致的高温和低风速等气象条件的常态化，成为港澳地区夏秋季臭氧浓度上升和秋冬季 $PM_{2.5}$ 污染加重的重要因素。气候变化也间接影响自然环境中传染病的病原体、宿主和传播媒介，影响人体呼吸、免疫和消化等系统，对人类造成健康损害。气候变暖及其导致的极端天气气候事件增多还容易引发心理精神疾病。为此，港澳地区把城市气候纳入了发展考虑因素内，积极改善城市街道布局，推行高温热浪、低温寒潮的健康防护措施，加强公众对气候变化影响人体健康的认知。

三、港澳地区减缓气候变化

一方面，能源结构的调整及节能要求推动香港减排取得了显著成效。香港范畴一[①]排放量在 2014 年已达峰值，与 1990 年相比，2017 年范畴一排放

---

①参照 *Global Protocol for Community-Scale Greenhouse Gas Emission Inventories*，基于不同核算方式，规定了不同的碳排放范畴。范畴一采用基于生产的核算方法，指的是城市边界内排放源的温室气体排放。范畴二采用基于消费的核算方法，指在城市边界内使用电网供电、热能、蒸汽和/或冷却产生的温室气体排放。范畴三采用基于消费的核算方法，指城市边界内活动导致在城市边界以外发生的所有其他温室气体排放。

量增加了 15.6%，人均温室气体排放量和单位生产总值排放量分别大幅下降了 11%和 58%。香港的燃料组合从煤炭为主转型为天然气和核电（从内地购买），结合去工业化、能效提高，是其碳排放显著下降的主要原因。香港范畴二排放量于 2014 年达到峰值，但范畴三排放量仍然没有达峰迹象。1990～2017 年，香港范畴二排放量仅增加 19.0%，相较于 2014 年的排放水平，2017 年减少了 9.3%的排放量。然而在 1990～2017 年，香港的范畴三排放量增加了 142.3%。香港的大部分石油产品消耗是用于跨境运输，特别是香港国际机场及世界级的海港。因此，石油产品的净进口及其隐含的二氧化碳排放量持续快速增长，并没有趋于平缓的迹象。

另一方面，澳门的范畴一排放量仍未达峰，能否尽快达峰的关键是澳门的电力发展政策。从 2000～2016 年的行业平均水平来看，本地发电是澳门温室气体排放的最大来源，占 38.6%。天然气发电在澳门本地发电中使用比例增加、加大从南方电网购入电力，是导致澳门 2005～2012 年范畴一排放量下降的因素。如果新增电力需求用外购电来满足而本地发电保持较低比例，结合节能减排等其他政策，可能使范畴一排放尽快达峰，否则排放还将持续增加。澳门的范畴二和范畴三排放还没有出现达峰迹象。随着内地的减排努力，特别是南方电网温室气体排放因子的不断下降，外购电所带来的温室气体排放预期将在未来出现显著下降，从而推动范畴二的达峰。范畴三为范畴二加上跨境水运和航空能源使用所隐含的温室气体排放。由于外购电的主导效应以及澳门并非重要的航运中心，范畴三的排放趋势和范畴二类似，也没有出现达峰的迹象。

为达到本报告未来情景分析中在 2040 年实现香港范畴一排放减少 90%的目标，可以采用六大技术措施，包括停止化石燃料的发电出口、能源消费全面电气化、节能、燃料组合转型带来的低排放系数、碳捕获和储存以及废弃物焚烧发电。为实现澳门范畴一的减排目标，可以采用能源全面电气化、节能减废以及增加外购电等手段。

四、港澳地区气候变化的政策与地区合作

港澳地区现行的减缓、适应和应对气候变化政策在两地回归前就已实施，其中部分政策走在世界前列；回归后，港澳地区的气候政策持续探索，不断增加的区域合作机会也给气候变化的减缓、适应和应对带来新的机遇。

港澳地区建立了应对气候变化管理机制并开展合作，出台了应对气候变化规划、行动方案等政策文件。香港成立了气候变化督导委员会，澳门也组建了气候变化跨部门工作小组，分别统筹本地各部门应对气候变化和节能减排相关工作。香港特区与广东省政府于 2011 年签署了粤港应对气候变化合作协议，并成立粤港应对气候变化联络协调小组（即现时粤港环保及应对气候变化合作小组）。此外，香港出台了《香港应对气候变化策略及行动纲领——咨询文件》《香港都市节能蓝图 2005～2015+》《香港气候变化报告 2015》《香港气候行动蓝图 2030+》等系列文件，澳门也在《环境法纲要》中明确了应对气候变化的总体纲要和原则，并发布了《澳门特别行政区政府应对气候变化方案》《澳门环境保护规划（2010—2020）》等重要文件。

受国家乃至全球在应对气候变化行动中努力的影响，以及地方政府在改善空气质量方面采取的积极举措，港澳地区实施了一系列减缓气候变化的政策。港澳地区减缓气候变化的政策包含两个重点：①能源结构调整，以减少空气污染物和碳排放。②节约能源。通过提高本地天然气发电比例和开发可再生能源以改善发电燃料组合；在交通运输领域和消费领域也都制定相应措施鼓励清洁能源的使用；通过建筑物节能、加强和推广良好实践、提高强制性能源效益标准、激励社区行动等方式来促进节能。

港澳地区在适应气候变化方面，明确了适应气候变化的重点领域，因地制宜制定了具有独特性的政策。主要体现在：①气象灾害风险预警。香港通过一系列的恶劣天气警告系统及对应的天灾应变计划，为公众、航海航空业

及其他特殊用户提供不同的警告服务，并列明特区政府和社会各界应采取的措施和行动，为城市应对不同紧急情况提供了工作范本。澳门也仿照香港采用“挂风球”的方法来发布热带气旋信号。②提升城市防洪能力。香港通过提升防洪标准、雨水排放隧道、地下蓄洪池等雨洪防治措施缓解排水系统的防洪压力，并减低低洼地区的内涝威胁，兼顾了人口稠密地区的洪灾预防及短期、长期需求。③开展斜坡安全管理。香港按风险等级对人造斜坡进行分类，定期维修管理，同时对天然山坡作滑坡风险排序，按序优化防治工程，该方案被联合国认可为管理滑坡风险的良好做法。香港“长远防治山泥倾泻计划”被 C40 城市气候领导联盟选为 2017 年全球 100 个应对气候变化的良好措施。

科学研究是港澳地区气候变化政策的重要支撑，可对决策的方向和内容提供参考。香港天文台和澳门地球物理暨气象局提供的气象服务及研究结果奠定了港澳地区气候变化决策的基础，并逐渐发展成为政策依据的一部分。政府设立的专门环保基金是促进本地气候变化科研转化的重要支撑。香港特区政府设有环境及自然保育基金，支持“环保研究、技术示范和会议项目”，重点主题包括“气候变化——减缓、适应及应变”“生物多样性、保育及地质保护”，等协助应对气候变化项目的开发。香港特区政府拨款 2 亿元港币成立的“低碳绿色科研基金”于 2020 年 12 月开始接受申请，为有助于减碳和加强环保的科研项目提供更充裕的资助。澳门设有环保与节能基金，包含“环保、节能产品和设备资助计划”，旨在资助购买或更换能有助改善环境质量、更具能源效益或节水效果的产品和设备。此外，港澳地区一直大力推动与国际、国家气候变化科学研究服务机构接轨，互通有无，反哺气候决策。香港和澳门分别于 1948 年和 1996 年加入了世界气象组织，并由香港天文台和澳门地球物理暨气象局代表。香港天文台和澳门地球物理暨气象局与中国气象局、国家海洋局和广东省气象局等部门有合作项目，并保持着良好的工作关系。

在未来政策需求方面，由于港澳地区的管理体制不同于内地，提供了试验政策与实践的灵活性和机会。例如：①加快绿色金融的发展。香港已经具备成为中国绿色金融中心的能力，欢迎更多机构善用香港的资本市场、金融资源和专业服务为其绿色项目作投融资，内地和香港特区政府可考虑香港如何为内地的碳排放提供国际市场。②发挥人才优势，促政策革新。面对日益复杂的气候变化和环境问题，进一步发挥专才的优势，促进政府政策的前沿发展和科学制定，这对香港自身变革、保持竞争力意义重大。③加强跨学科跨部门的开放和协作。应对气候变化的复杂性需要采取多学科的方法，并需要加强政府、大学、行业和专业人士之间的协作。④加强气候抗御合作。进一步完善大湾区联合应对气候变化的工作机制，确保信息对称和联防联动渠道畅通，提升气候灾害预警预报和巨灾防范能力，同时开展关于恶劣天气事件的公众教育及宣传活动方面的合作。这些举措将增强社区对恶劣天气的抵御能力和准备能力，也将进一步促进大湾区社区的绿色和低碳生活方式。

第一章　气候变化的事实

自工业革命以来，随着人类社会发展的突飞猛进，生产效率快速提高，人类赖以生存的地球环境正遭受着前所未有的污染与破坏，全球气候也随之发生变化，并进一步加剧了环境的恶化。人类活动导致的碳排放及污染，对气候变化起着至关重要的影响（陈隆勋等，2004），同时也影响人类健康。尤其是在近几十年经济贸易全球化的背景下，温室气体排放和人为污染物的迅速增多，导致越来越多的气候及环境问题出现（如冰雪消融、海平面升高、极端天气事件频发、城市空气污染等），正威胁着人类的生存。面对如此艰巨的气候及环境问题，减缓气候变化已迫在眉睫。对全球变暖有显著影响的人为排放污染物可以分为两类：第一类是长寿命的温室气体，如二氧化碳等；第二类是短寿命气候污染物，通常包括对流层臭氧、甲烷、黑碳气溶胶和氟利昂及替代物（谢冰，2016）。这些污染物的大量存在不仅会使全球温度升高，还会对区域气候产生至关重要的影响，如加速冰川和雪盖的融化、影响太阳辐射到达地表的辐射通量，甚至改变季风系统和台风降水。

本章将介绍在全球变暖的大背景下，香港和澳门在各自的环境及社会发展过程中所经历的气候变化及未来气候预测。

第一节　观测到的变化

一、基本气候要素的变化

（一）香港

位于九龙尖沙咀的香港天文台总部是世界气象组织认可的长期观测站，自 1884 年起，香港天文台已经有逐日的气象观测记录，仅在 1940～1946 年因为第二次世界大战中断。超过 130 年的观测记录为地区的气候变化研究提供了珍贵的参考数据。已有研究利用香港天文台数据分析了香港多个基本气候要素的长期变化（Ginn *et al.*, 2010; Chan *et al.*, 2012; He *et al.*, 2016），本研究将把相关分析延伸至 2018 年。

1. 温度的变化

图 1–1 展示了香港天文台总部过去一百多年的年平均温度记录，清晰可见年平均气温有明显的上升趋势，升温速率为每十年 0.13 摄氏度，这一部分受全球变化的影响，另一部分反映了本地城市化的影响。甄荣磊等（Ginn *et al.*, 2010）通过对温度的时间序列的分析，指出香港的增温在 20 世纪 70 年代和 80 年代有加快迹象，原因除了全球暖化外，也与当时急速的城市化有关。香港天文台利用两种不同方法评估本地城市化影响，一是利用澳门作为郊区站，分析香港与澳门两地的升温趋势差异（Lee *et al.*, 2011; Chan *et al.*, 2012）；二是利用京士柏气象站探空数据中 850 hPa 条件下的温度与美国国家环境预报中心再分析数据作比较，得出背景变暖趋势（Chan *et al.*, 2012）。结合两种不同方法，结论是城市化影响约占整体升温原因的 40%～50%（Chan *et al.*, 2012）。

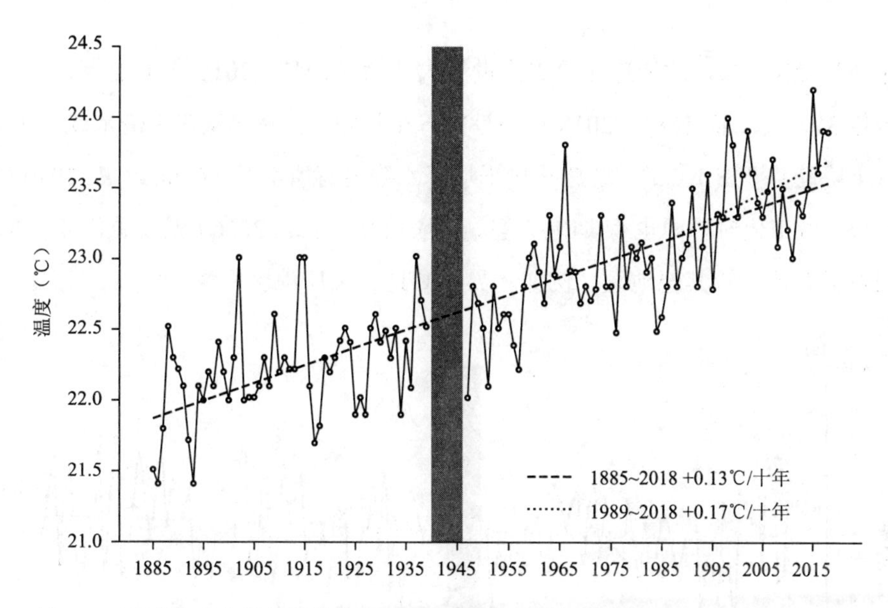

图 1–1　香港天文台总部录得的 1885～2018 年平均气温（1940～1946 年无数据）

香港各个季节都有显著的升温趋势（表1–1），其中以春季的升温速率最大，达每十年 0.16 摄氏度，夏季和秋季较小。按陈浩新等（Chan *et al.*, 2012）的分析方法，将 1981～2010 年日平均温度的第一四分位数及第三四分位数（即 19.1 摄氏度和 27.7 摄氏度）分别定义为凉日和暖日的阈值，可以发现凉日的数目越来越少，而暖日的数目越来越多，且有更早开始和更迟结束的趋势。

表 1–1　1885～2018 年香港各季节的平均温度上升速率（摄氏度/十年）

	冬季（12～2 月）	春季（3～5 月）	夏季（6～8 月）	秋季（9～11 月）
升温速率	0.12	0.16	0.11	0.11

2. 降水的变化

1885～2018 年，香港的年降水量有轻微上升趋势（图 1–2a），平均每十年增加 22 毫米，此增速在第二次世界大战后（1947～2018 年）更高，达每十年 37 毫米。而 1885～2018 年香港的降雨日数（日降水量在 1 毫米或以上）有下降趋势（图 1–2b）。这意味着平均降水强度有所上升（Ginn *et al.*, 2010）。此外，城市化对降雨也有影响，莫庆炎等（Mok *et al.*, 2006）发现香港市区的雨量上升速度较其他地区（例如新界、高地、离岸地区）快。

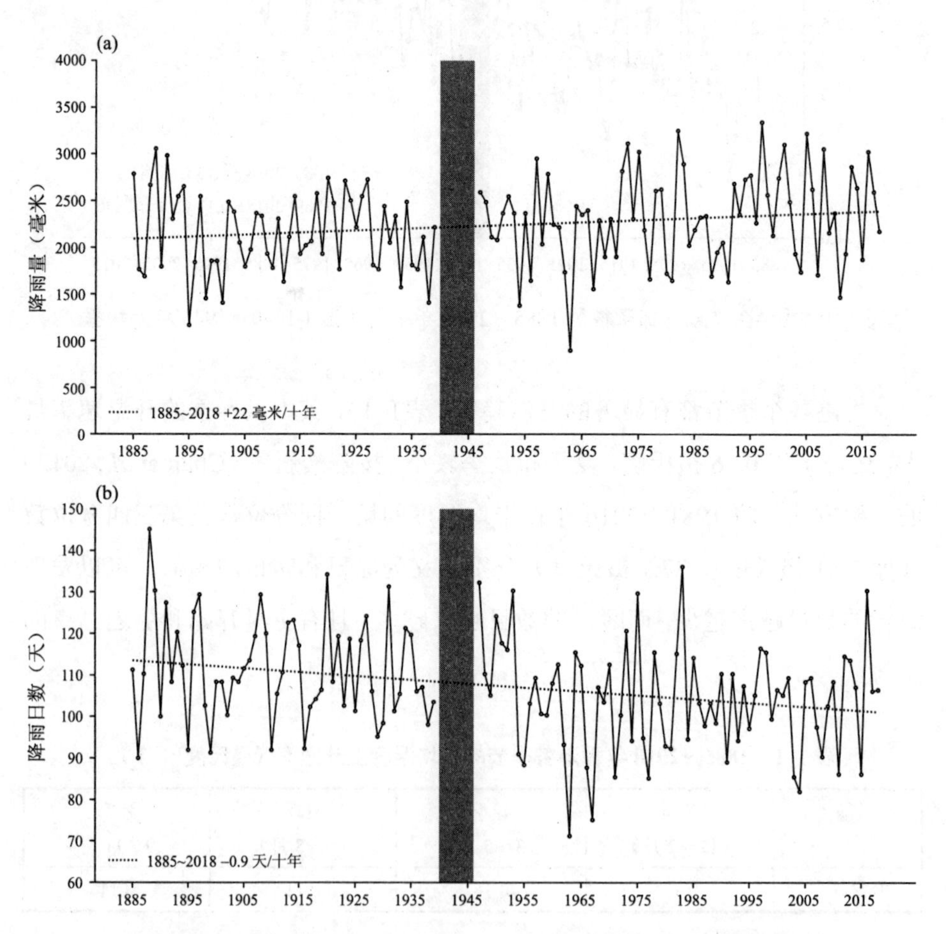

图 1–2 （a）香港天文台总部录得的 1885～2018 年降雨量（1940～1946 年无数据）；（b）香港天文台总部录得的 1885～2018 年降雨日数（1940～1946 年无数据）

3. 海平面的变化

在全球变暖背景下，海水受热膨胀、陆地冰雪融化，导致了全球海平面的上升。香港也不能幸免，香港自 1954 年开始在维多利亚港内（鲗鱼涌/北角）设置仪器观测海平面高度，结果显示年平均海平面高度有明显上升趋势（图 1–3a），上升趋势达每十年 31 毫米，与香港其他地区、澳门的测潮站及南海北部的卫星观测数据吻合（Wong *et al.*, 2003; He *et al.*, 2016）。

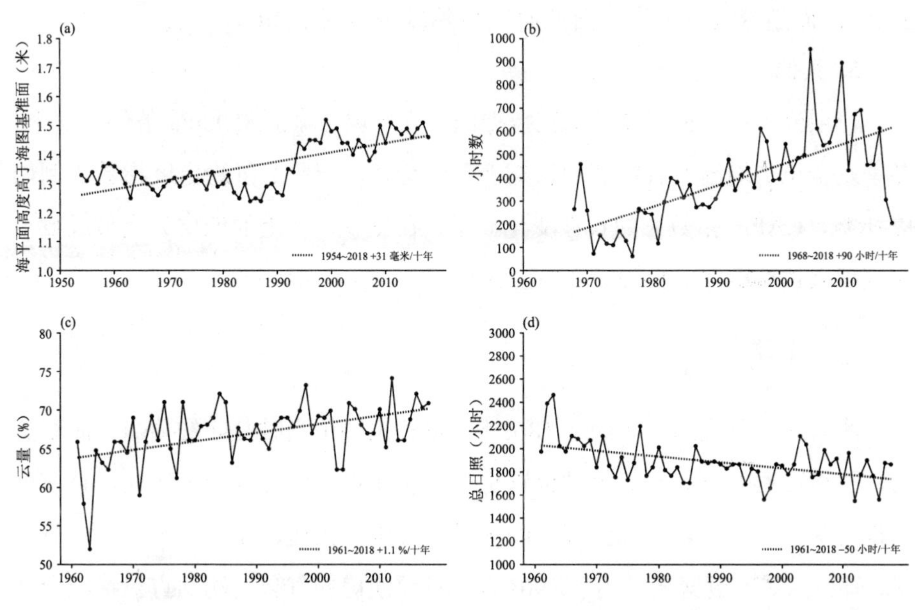

图 1–3　（a）香港维多利亚港年平均海平面高度（1954～2018 年）；
（b）香港天文台总部能见度低于五千米（不论天气状况）的时数（1968～2018 年）；
（c）香港天文台总部录得每年平均云量（1961～2018 年）；
（d）京士柏气象站录得每年总日照时数（1961～2018 年）

4. 能见度、云量和日照时数的变化

香港天文台的能见度数据由经训练的观测员每小时观测获得。统计显示，1968～2018 年香港的能见度低于 5 千米的时数（不论天气状况）由 20 世纪 70 年代至 21 世纪 10 年代初呈现明显上升趋势（图 1–3b），但近几年开始出

现下降迹象。梁延刚等（2008）指出香港低能见度时数的变化可能跟 $PM_{2.5}$ 的浓度有关，而 $PM_{2.5}$ 的来源主要是建筑、汽车废气排放和燃煤发电等人类活动。

香港云量的数据也由经训练的观测员每小时以八分法观测所得。1961～2018 年年平均云量每十年上升了 1.1%（图 1–3c）。云量上升的一个潜在原因是城市活动产生的大气凝结核浓度上升，这有利于云的形成（Ginn *et al.*, 2010）。而随着云量增多，日照时间自然减少（图 1–3d）。

5. 其他

一项分析京士柏气象站探空数据的研究发现，除了 850 hPa 条件下秋季、冬季及全年平均温度有显著的长期上升趋势外，一些大气稳定度指数（例如 K 指数，CAPE 指数）的长期变化显示，在夏季，香港的低层大气最近数十年也越来越不稳定（Tong and Lee, 2014）。

（二）澳门

澳门的气象观测工作始于 1861 年，最初主要由当时驻守的葡萄牙海军负责，虽然观察并不固定，测站位置也曾多次迁移，但可说历史悠久。目前可查询的基本气象要素（如温度、降水量、气压、风速、风向、蒸发量、日照量、辐射量等）数据最早始于 1901 年。目前澳门的气象监测站点包括纪念孙中山市政公园站、外港客运码头站、海事博物馆站、大炮台站、西湾大桥站、嘉乐庇总督大桥站、友谊大桥站、大潭山站、九澳站和路环分站。

1. 温度的变化

本研究根据澳门地球物理暨气象局提供的数据绘制了澳门年平均气温距平（取 1901～2018 年平均）的时间变化曲线图（图 1–4）。1901～2007 年，澳门年平均气温升高了约 0.71 摄氏度，升温率约为每十年 0.066 摄氏度。而 1901～2018 年升温率约为每十年 0.056 摄氏度，即在更长时间尺度上升温速率减慢。特别是 1998～2018 年，年平均气温数据出现了每十年–0.14 摄氏度的降温速率。除此之外，澳门面积小且临海，城市化进程对局地气候的影响较小，可能导致澳门的平均升温率低于全球及邻近地区。

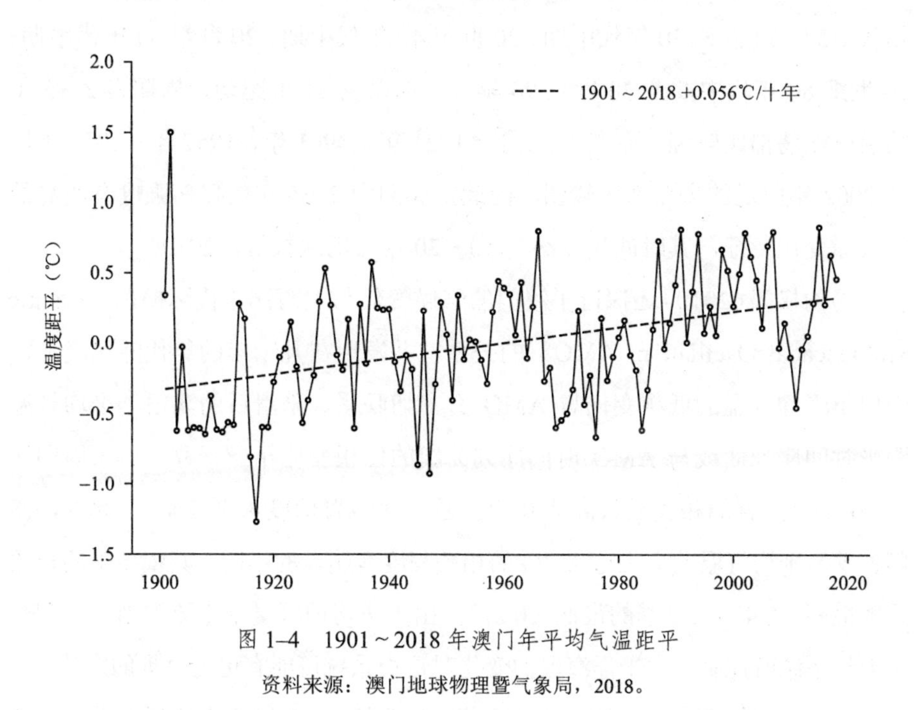

图 1–4　1901～2018 年澳门年平均气温距平

资料来源：澳门地球物理暨气象局，2018。

冯瑞权等（2010）通过对澳门年均气温的分析发现澳门地区气温的年代际变化相当显著：两个明显的偏冷期分别出现在 20 世纪头 20 年以及 20 世纪 60 年代末至 80 年代初；而在 20 世纪 30 年代至 40 年代，以及 50 年代末至 60 年代中期，气温明显变暖；另一个偏暖期则出现在 20 世纪 80 年代中期以后。因此澳门年平均气温变化呈现出时间尺度几十年的振荡，在此缓慢气候波动的背景下，叠加有明显的年际变化特征。图 1–4 可见，1903～1928 年、1940～1957 年、1967～1985 年，澳门处于温度较低的时期，2008～2014 年也出现了温度降低的现象；而在这四个时期之后，气温都有较明显的升温现象，1985 年之后气温升高幅度最大，持续时间较长。

有研究通过小波段分析，得到澳门的年平均温度变化存在比较显著的周期为 2～5 年的振荡，并有很强的时域局部性，即在 20 世纪初、20 世纪 10 年代中期、20 世纪 30 年代中期、20 世纪 40 年代中期、20 世纪 60 年代中期、20 世纪 80 年代中期至 20 世纪 90 年代中期以及 21 世纪初，周期为 2～5 年的温度振荡都比较强，而在 1910 年、1921 年、1941 年、1957 年、1981 年以及 2002 年附近的振荡都比较弱。因此，周期为 2～5 年的振荡强度有明显的年代际变化特征，其时间尺度约为 10～20 年（冯瑞权等，2010）。

冯瑞权等（2010）还探讨了澳门的气候变化与大西洋年代际振荡（Atlantic Multidecadal Oscillation, AMO）的关系，并发现在几十年的年代际尺度上，澳门年平均气温的低频变化与 AMO 有密切联系。平滑后的澳门年平均气温距平时间序列曲线与 AMO 时间序列曲线的位相变化相当一致，振荡周期约为 60 年；两者的相关系数高达 0.67（通过 99%置信度水平检验）。此外，澳门各季节平均气温的变化与 AMO 的相关程度有明显的不同，其中冬季的相关系数最高（0.47），夏季的最低（0.21）。由于华南地区夏季主要被副热带和热带天气系统所控制，太平洋海温的变化对这类系统可能有更为直接的影响。

本研究对 1901～2018 年澳门日最高气温超过 33 摄氏度的天数进行了统计，发现 118 年间高温天气天数并未增加（–0.02 天/十年）。同期，年低温（低于 13 摄氏度）的天数则呈下降趋势，下降速率为–0.75 天/十年。

钱诚等（Qian *et al.*, 2015）以 1912～2012 年连续日气温观测资料为基础，对澳门近百年来酷热天气日数和热夜日数这两个指标进行分析，发现这两个极端指标都与厄尔尼诺-南方振荡（El Niño-Southern Oscillation, ENSO）——东亚夏季风耦合系统的年际和年代际变化有关。

2. 降水的变化

1901～2018 年澳门的年降水量平均为 1 873 毫米。图 1–5 是年降水量的时间曲线，可以看出 1901～1982 年降水量的整体趋势是逐渐增加的，这 118 年中 1982 年的降水量达到了峰值，全年降水总量为 3 041 毫米。1982～1991 年的距平值下降幅度较大，近 20 年的距平值也是呈整体下降趋势。1901～2018 年澳门的年降水增加率为 41.4 毫米/十年，略低于全球平均水平，且近 20 年，澳门的降水量呈下降趋势。

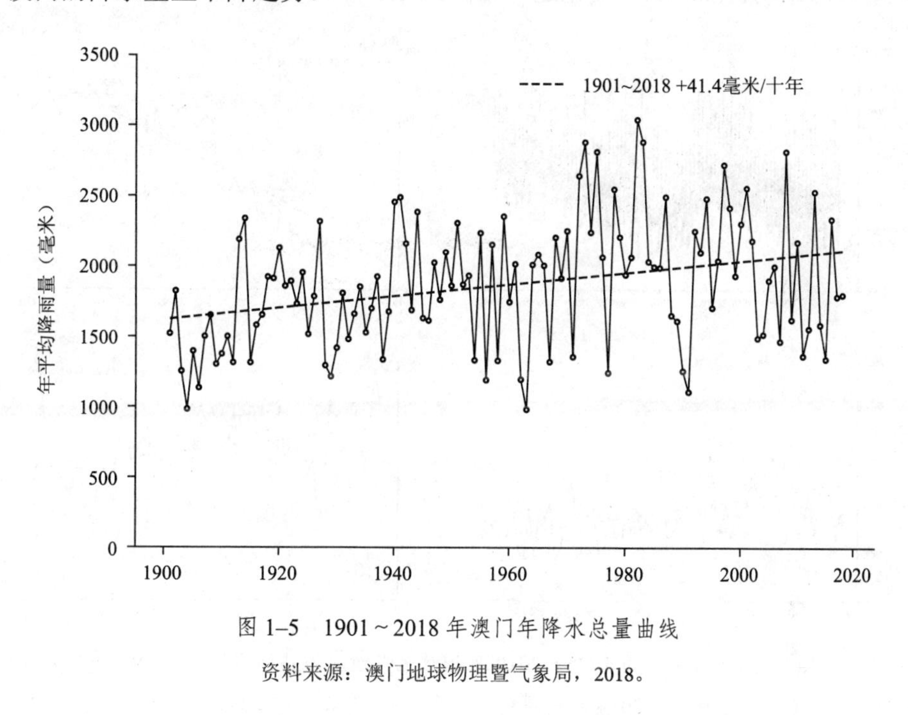

图 1–5　1901～2018 年澳门年降水总量曲线

资料来源：澳门地球物理暨气象局，2018。

3. 海平面的变化

1925～2010 年澳门海平面表现为上升趋势，上升速率为 1.35 毫米/年；而 1970 年后，海平面则以 4.2 毫米/年的速度加速上升。澳门在 1993～2012 年的海平面上升速率是全球平均的 1.1 倍（Wang L. *et al.*, 2016）。图 1–6a 为澳门 1925～2010 年的平均海平面变化。

4. 能见度、云量和日照时数的变化

根据澳门 1952～2018 年的年日照时长图（图 1–6c），1952～2018 年澳门年日照平均为 1 861.5 小时。1952～1963 年的年日照时长变化趋势是增加的，且在 1963 年日照时长达到峰值，峰值的日照时长为 7.2 小时。然而近 68 年，澳门的平均日照时长为 5.1 小时，1952～2018 年的日照时长线性变化率为–51.4 小时/十年。从 20 世纪 60 年代初以来，澳门的年日照时长数显著减少，这与全球平均日照时长的变化趋势相似。

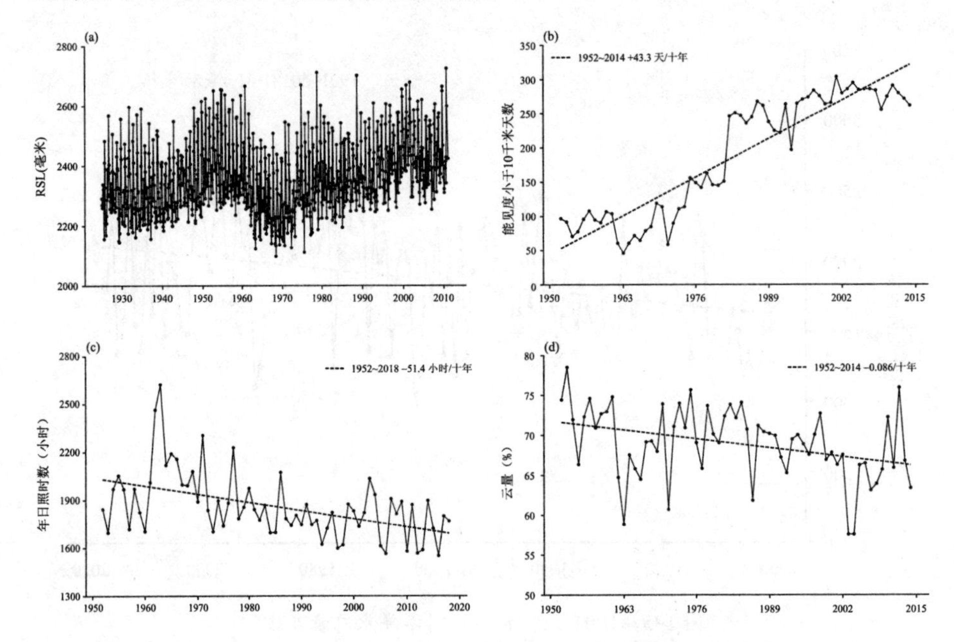

图 1–6 （a）澳门相对海平面 1925～2010 年相对于当地海图基准面的演变（Wang L. *et al.*, 2016）；（b）澳门 1952～2014 年出现能见度小于 10 千米的天数统计；（c）澳门 1952～2018 年的年总日照时长；（d）1952～2014 年澳门年平均云量曲线

资料来源：澳门地球物理暨气象局，2018。

1952～2014 年，澳门的年平均云量每十年减少了 0.86%（图 1–6d），与毗邻的广东省的云量变化趋势不同（伍红雨等，2011）。1952～2014 年，澳门能见度小于 10 千米的天数在增加，增速为 43.3 天/十年（图 1–6b）。

5. 其他

米金套(2010)指出澳门在 1998 年和 2004 年的热岛效应面积分别为 4.07 平方千米和 2.83 平方千米，表明在此期间，澳门的热岛效应有所减弱，但整体来说，澳门区域面积较小，其热岛效应程度还是比较强烈。

二、极端天气气候事件的变化

（一）香港

气候平均值的改变，同时也会改变极端天气发生的频率。以气温为例，若平均温度升高，出现极端高温的频率会升高，而出现极端低温的频率则会降低。香港每年的酷热天气日数和热夜日数皆呈明显上升趋势，而寒冷天气日数则呈减少趋势（Wong *et al.*, 2011）。香港天文台总部报告的 1885～1914 年与 1989～2018 年的平均年热夜数、酷热天气日数与寒冷天气日数的比较呈现强烈对比（图 1–7a），热夜数更是增长了数十倍。其他极端温度的指标例如每年的绝对最高和最低温度，也都呈明显上升趋势（Wong *et al.*, 2011）。

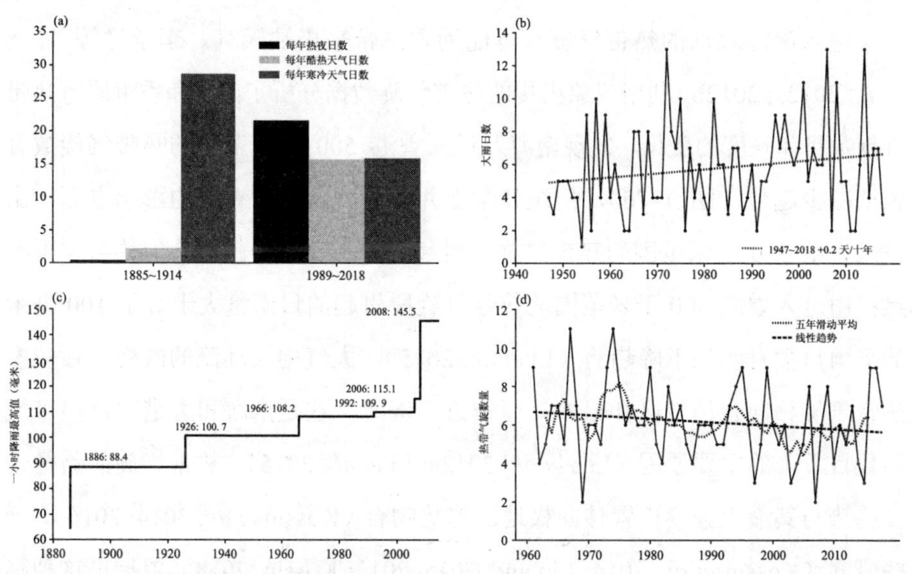

图 1–7　（a）香港天文台总部报告的 1885～1914 年与 1989～2018 年的平均年热夜数、酷热天气日数与寒冷天气日数的比较；（b）1947～2018 年香港天文台总部报告的大雨（雨量高于 30 毫米/小时）日数；（c）香港天文台总部报告的小时雨量最高纪录；（d）1961～2018 年进入香港 500 千米范围的热带气旋数

王伟文等（Wang W. *et al.*, 2016）基于极值理论（Extreme Value Theory, EVT）定义了香港的极端城市热岛，计算出冬季和夏季的阈值分别为 7.8 摄氏度和 4.8 摄氏度，并通过在 EVT 模型中引入参数变化，指出香港极端城市热岛在夏季夜间有显著的上升趋势。

甄荣磊等（Ginn *et al.*, 2010）指出香港出现大雨（雨量高于 30 毫米/小时）的日数有上升趋势（图 1–7b），利用非平稳时间相关极端值模型分析极端降水事件变化发现，小时雨量等于或超过 100 毫米的极端降水事件的回归期由 1900 年的 37 年缩短到 2000 年的 18 年，代表极端降水事件越来越频繁（Wong *et al.*, 2011）。香港天文台报告的最高小时雨量纪录也刷新得越来越密（图 1–7c），一定程度上反映了极端降水事件越来越多。同时，香港 4～9 月最长连续干日数也有明显上升趋势，反映降水有越来越不平均的迹象。

进入南海北部的热带气旋有可能为香港带来极端天气。李子祥等（Lee *et al.*, 2012a; 2012b）利用气象机构的热带气旋数据分析了南海和香港附近热带气旋数目和强度的变化，发现南海及进入香港 500 千米范围的热带气旋数有长期减少趋势（图 1–7d），但统计学上并不显著。在对香港的影响方面，由进入香港 500 千米范围的热带气旋所引起的降水有下降趋势，但趋势并不明显；由进入香港 300 千米范围的热带气旋所引起的日雨量大于等于 100 毫米的降雨日数有明显下降趋势（Li *et al.*, 2015）。大气中层环流的改变导致较多热带气旋移向台湾地区和日本，而非进入南海，这是南海和香港附近热带气旋数目下降的主要原因（Lee *et al.*, 2012a; Li *et al.*, 2015）。热带气旋的路径转变趋势与其最大强度位置移向极地的趋势吻合（Kossin *et al.*, 2014; 2016），一些研究（Kossin *et al.*, 2014; Liu and Chan, 2017; Kossin, 2018）更指出这种趋势在西北太平洋相当稳定，不太可能完全是自然变化所致。根据世界气象组织（WMO）的热带气旋与气候变化任务组与联合国亚太经济与社会理事会/世界气象组织（UNESCAP/WMO）台风委员会专家组的最新评估报告，人为气候变化影响 20 世纪 40 年代以来西北太平洋热带气旋最大强度位置北移趋势的信度水平属“低至中级”（Knutson *et al.*, 2019; Lee *et al.*, 2020）。

与进入香港 500 千米范围的热带气旋相关的最高十分钟平均风速和每秒钟阵风在横澜岛站点有轻微下降趋势，但在统计学上并不显著；而市区站点的风速有明显下降趋势，这很有可能跟本地城市化有关（Lee *et al.*, 2012a; Peng *et al.*, 2018）。热带气旋除了带来强风和暴雨外，也会为香港带来风暴潮。在过去一百多年，数个台风为香港带来过严重的风暴潮，造成了重大伤亡及破坏，例如 1962 年的超强台风“温黛”袭港期间，维多利亚港内录得的最高潮位为 3.96 米（表 1–2），而吐露港内的大埔滘录得的最高潮位更达 5.03 米（表 1–3）。李本滢等（Lee *et al.*, 2010）利用当时的观测数据得出维多利亚港 50 年一遇高潮位事件的回归值是 3.5 米，但 2017 年“天鸽”和 2018 年“山竹”令维多利亚港短短两年内就出现了两次超过 3.5 米的高潮位事件（表 1–2），其中“山竹”更为香港带来破纪录的风暴潮，“山竹”过后，维多利亚港内出现 3.5 米潮位事件的回归期已缩短一半（少于 25 年）。

表 1–2　维多利亚港内录得的热带气旋影响香港期间最高潮位及最大风暴潮（1954～2018 年）

排名	年份	热带气旋名称	维多利亚港内录得最高潮位（米）	年份	热带气旋名称	维多利亚港内录得最大风暴潮（米，最大风暴增水）
1	1962	“温黛”	3.96	2018	“山竹”	2.35
2	2018	“山竹”	3.88	1962	“温黛”	1.77
3	2017	“天鸽”	3.57	1954	“艾黛”	1.68
4	2008	“黑格比”	3.53	1964	“露比”	1.49
5	2001	“尤特”	3.38	1979	“荷贝”	1.45

表 1–3　大埔滘录得的热带气旋影响香港期间最高潮位及最大风暴潮（1962～2018 年）

排名	年份	热带气旋名称	在大埔滘录得最高潮位（米）	年份	热带气旋名称	在大埔滘录得最大风暴潮（米，最大风暴增水）
1	1962	“温黛”	5.03	2018	“山竹”	3.40
2	2018	“山竹”	4.71	1979	“荷贝”	3.23
3	1979	“荷贝”	4.33	1962	“温黛”	3.20
4	2017	“天鸽”	4.09	1964	“露比”	2.96
5	2008	“黑格比”	3.77	1964	“艾黛”	2.16

（二）澳门

王伟文等（Wang *et al.*, 2015）对澳门的长期观测资料进行了均质化处理，结果表明，澳门观测记录中极端降雨的频率显著增加，而极端降雨强度的增长趋势则不显著。澳门地区夏季热浪的强度与持续时间的增长趋势不显著。

在 1971～2010 年，导致澳门需要悬挂八号台风讯号的台风个数有下降的趋势，但并不显著。这个结论与登陆广东省热带气旋个数的变化是一致的（Li and Duan, 2010; Zhang *et al.*, 2011）。

三、温室气体和气溶胶的变化

（一）香港

1. 温室气体

香港天文台自 2009 年 5 月起开始在京士柏气象站设置仪器，测量市区的户外二氧化碳浓度。此外，香港天文台与香港理工大学（理大）合作，自 2010 年 10 月起在香港岛东南端鹤咀的理大本底大气监测站测量郊区二氧化碳浓度（冯颖怡等，2011）。2020 年 1 月，香港天文台更新了测量仪器，并将仪器迁到香港环境保护署位于鹤咀的超级空气监测站。由于位于鹤咀的监测站远离市区，较少受到本地二氧化碳源的直接影响，其测量的二氧化碳浓度大部分时间可代表香港的本底二氧化碳浓度。图 1–8 展示了香港京士柏站和鹤咀站的二氧化碳浓度变化，虽然观测历史只有十年左右，但仍然可以看出二氧化碳浓度在市区和郊区都有上升趋势，这与全球平均二氧化碳浓度的上升趋势吻合。

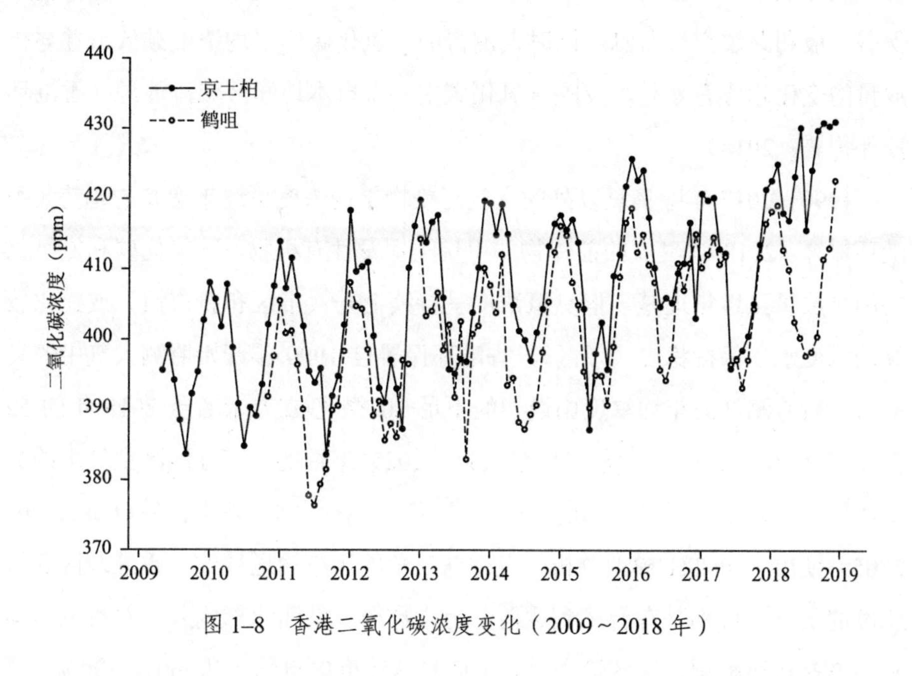

图 1–8　香港二氧化碳浓度变化（2009～2018 年）

2. 大气气溶胶

根据 2015 年的排放清单，香港特区的主要空气污染物排放量较 2010 年下降了 14%～45%。其间：①二氧化硫的排放量减少了 78%，主要原因是公用发电的排放量大幅减少。这与环保署一般空气监测站测得的二氧化硫年平均浓度与排放量的变化趋势大致相同，反映了除受气象因素影响之外，空气中的二氧化硫主要来自本地排放源。②氮氧化物的排放量减少了 39%，这与环保署路边空气监测站测得的氮氧化物年平均浓度与排放量的变化趋势大致相同，反映出路边的氮氧化物主要来自本地排放源。③PM_{10}和 $PM_{2.5}$的排放量均减少了 69%，主要是因为道路运输及公用发电的排放量大幅减少。这同环保署一般空气监测站同时期测得的年平均浓度与排放量的变化趋势有差异，反映了空气中的颗粒物主要来自本地和区域排放源。④挥发性有机化合物的排放量减少了 65%，主要因为非燃烧源及道路运输的排放量大幅减少。⑤一氧化碳的排放量减少了 37%，主要因为道路运输的排放量显著减少。环保署一般和路边空气监测站同时期测得的一氧化碳年平均浓度颇低，并与排放量的变化趋势有差异，反映一氧化碳主要来自本地和区域排放源（香港环境保护署，2018）。

1990～2017 年除臭氧以外的空气污染物年均浓度呈逐年递减的趋势（香港环保署，2018）。臭氧的前体物非甲烷类挥发性有机物、一氧化碳和氮氧化物的排放量虽逐年递减，但臭氧浓度呈持续上升（市区和新市镇）或在比较高的浓度水平保持稳定（郊区）。香港环保署自 1990 年开始监测大气中的臭氧，一般监测站的年均浓度由最初的不足 20 微克/立方米增至 2018 年的 52 微克/立方米，创多年新高。环境局预计 2025 年香港大部分地区空气中的臭氧浓度会超出空气质素指标的有关标准，高于现行水平。林嘉仕等（Lam *et al.*, 2005）使用了 PATH 模型分析，发现珠三角的污染输送导致了香港地区臭氧浓度的上升，特别是在台风到达前，污染物经香港西北部输入，对香港的臭氧污染有重要影响，区域输送对香港的臭氧浓度的贡献可达 60%～90%。黄建平等（Huang *et al.*, 2005）分析认为香港大约 30%的臭氧是由本地光化学反应生成，而约 70%来源于华南向香港的输送。

对 1995～2017 年香港大气 PM_{10}的化学组分观测数据进行分析，可以发现与一次污染物排放清单及环境浓度不同的变化趋势，且二次无机离子在 PM_{10}中所占百分比在上升，这也反映出大气中二次反应的复杂性以及二次反应对于空气质量的重要影响。

（二）澳门

1. 温室气体

2007～2014 年澳门温室气体的估算排放量总体呈下降趋势，但在 2015 和 2016 年有较大幅的回升。主要的温室气体排放源为本地发电、陆海空交通运输和废弃物焚化（澳门环境保护局，2018）。如澳门的国内生产总值（Gross Domestic Product, GDP）、人口、旅客量等持续增加，预计澳门的温室气体排放仍会继续上升。

2. 大气气溶胶

纵观 2007～2016 年澳门大气各污染物估算排放量的整体趋势，SOx 总体呈下降趋势，氨气上升趋势平缓，而一氧化碳、氮氧化物、非甲烷类挥发性有机物的估算排放量则呈明显上升趋势。PM_{10} 及 $PM_{2.5}$ 在 2007～2010 年呈下降趋势，2011 年之后又有所回升。石化燃料的使用是澳门大气污染物和温室气体产生的主要来源。

在空气污染物年平均浓度值整体的趋势方面，二氧化硫、PM_{10}、$PM_{2.5}$ 呈下降趋势，这些污染物的大气颗粒物浓度在 2014 年之后都已经低于标准值。其中二氧化硫年平均浓度值处于较低水平。近 5 年 $PM_{2.5}$ 的年平均浓度值更呈显著下降趋势，近 10 年二氧化氮及一氧化碳的年平均浓度值整体变化趋势较平缓，其中二氧化氮处于相对较高水平，局部地区浓度超标。值得注意的是近 10 年臭氧年平均浓度值呈上升趋势，臭氧是澳门 2017 年的主要空气污染物，在 9 月浓度较高。

汪琼琼等（Wang *et al.*, 2019）在澳门外港码头和大潭山采集滤膜样品进行化学分析并应用 PMF 对澳门的 $PM_{2.5}$ 来源进行解析，结果表明硫酸盐和有机碳是澳门 $PM_{2.5}$ 来源最多的两个部分，它们的元素碳浓度和百分比也较高（2.20 和 4.43 微克碳/立方米，8%与 15%），比 2011～2012 年（Wang. *et al.*, 2016）在佛山（2.95 微克碳/立方米，5%）和广州（1.92 微克碳/立方米，4%）

所测得的元素碳浓度都要高。同香港的监测结果相似，这些结果均反映出二次反应对澳门大气中细粒子的浓度有重要影响。

第二节　大气环流和气候现象的变化

一、气候系统内部变率

气候系统内部变率和人为强迫都可能对港澳地区的年代际气候变化有影响。年代际气候变化的归因研究非常依赖于可靠的观测数据和数值模式。气候模式的空间分辨率不足以准确刻画小范围内的气候变化，因此单站气候年代际变化的归因研究远比大范围区域气候变化的归因研究更困难。目前尚没有专门针对港澳地区气候变化的归因研究，但有大量的研究工作探讨了包含港澳地区在内的大范围区域的年代际气候变化归因和机制。结合这些工作，本节拟探讨主要的气候系统内部变率模态和人为强迫对港澳地区年代际气候变化可能产生的影响。

（一）年代际振荡

在年代际尺度上，最为著名的内部变率模态是太平洋年代际振荡（Pacific Decadal Oscillation, PDO）和 AMO。PDO 主要表现为北太平洋中纬度地区海温的年代际变率，PDO 正位相时，热带太平洋海表面温度（SST）偏高而北太平洋中纬度地区 SST 偏低；PDO 负位相时，热带太平洋 SST 偏低而北太平洋中纬度地区 SST 偏高。20 世纪 70 年代末期，PDO 由负位相进入正位相，直到 20 世纪 90 年代末期，PDO 再度回到负位相。与 PDO 负位相期间相比，PDO 正位相期间华南地区的夏季气温偏高、降水偏少，冬季气温偏低、降水偏多（卢楚翰等，2013；徐忆菲等，2017），东亚夏季风减弱（程乘等，2017）。目前，PDO

可能正在发生由负位相到正位相的转变，需要加强监测并做好相应的防范。

AMO 是北大西洋 SST 年代际变率的主导模态。AMO 的正位相表现为北大西洋 SST 偏高，负位相表现为北大西洋 SST 偏低。近 100 年来 AMO 的位相转换主要发生在：20 世纪 20 年代末期 AMO 由负位相转为正位相，20 世纪 60 年代 AMO 由正位相转为负位相，20 世纪 90 年代 AMO 由负位相转入正位相。AMO 正位相期间，包括粤港澳在内的东亚地区季节气温出现年代际整体偏高（Wang *et al.*, 2013; Han *et al.*, 2016）。另外，南海夏季风的爆发时间在 20 世纪 90 年代初期出现了年代际提前（Kajikawa and Wang, 2012; 张莉萍等, 2014; 简茂球等, 2018），与 AMO 的位相转换几乎同时发生，但尚不清楚南海夏季风爆发时间的年代际提前是不是 AMO 位相转变的结果。

（二）季风

东亚地区盛行季风气候，盛行风由冬季风（寒冷干燥的偏北风）转为夏季风（温暖湿润的偏南风）的时间在初夏前后。对我国大部分地区而言，夏季是强降水和主要气象灾害高发的季节，因此夏季风对经济社会的影响比冬季风更大。一般来说，南海和华南地区的夏季风建立时间在 5 月中旬，明显早于东亚季风区的其他区域，而南海夏季风的建立呈现出几天之内盛行风向迅速反转的“爆发”特征，因此南海夏季风的爆发时间被视为东亚夏季风全面建立的先兆信号，而南海夏季风的爆发时间也成为我国汛期短期气候的重要预测对象。一般而言，南海夏季风爆发后，来自南海的水汽能够直接向华南地区输送，南海和华南地区的降水量将猛增，华南前汛期进入盛期，暴雨集中发生。每年春季到初夏，国家气候中心和广东省气象局都要针对南海夏季风的爆发时间进行多次讨论，并实时监测、预测南海夏季风的爆发情况。南海夏季风的爆发时间与其强度有密切关系，南海夏季风爆发越早则其强度越弱，爆发越晚则其强度越强（张莉萍等，2014）。

南海夏季风爆发日期的年际变率受到多尺度气候变率的调控。先行研究显示，ENSO 等年际尺度海温异常信号对南海夏季风爆发的早晚具有调控作用（林爱兰等, 2013；邵勰等, 2015; He and Zhu, 2015；谷德军等, 2018），但仅根据海温异常信号，研究者难以预测南海夏季风爆发的具体日期，而是只能针对南海夏季风爆发会偏早或偏晚进行定性预测（邵勰等，2015）。热带季节内振荡也强烈地调控着南海夏季风的爆发时间（邵勰等，2014；李春晖等，2017）。南海夏季风的爆发时间最早为 4 月底，最晚在 6 月底，存在着强烈的年际变率。观测资料表明，南海夏季风的建立日期在 1993 年前后发生了一次突变，1994 年之后南海夏季风的平均爆发日期比 1979～1993 年发生了显著的年代际提前（Kajikawa and Wang, 2012；张莉萍, 2014；简茂球等, 2018）。但是，从更长的时间周期来看，1945 年以来南海夏季风爆发时间主要呈现波动特征，没有显著的线性趋势（Kajikawa and Wang, 2012）。

（三）华南寒潮

华南寒潮主要从西伯利亚和蒙古地区南下进入我国北方，再经我国东部、中部或西南部抵达华南地区。寒潮对华南的影响主要受到东亚大槽、贝加尔湖附近的短波槽和印缅槽调控。寒潮爆发前乌拉尔山附近常有阻塞高压（Zhou *et al.*, 2009; Park *et al.*, 2011; Cheung H. *et al.*, 2015），这与北大西洋往东传的罗斯贝波（Rossby wave）有关之外（Takaya and Nakamura, 2005），也与对流层与平流层的相互作用有关（Cheung *et al.*, 2016）。东亚大槽的强度和位置主要与赤道太平洋的讯号（如 ENSO）有关（Leung and Zhou, 2016; Leung *et al.*, 2017）。当东亚大槽加深，寒潮的路径就向南偏移，对华南地区的影响就加大（Leung *et al.*, 2015）。另外，近年研究也发现热带地区的 MJO 低频讯号通过会影响海洋性大陆的对流活动，从而进一步影响南海的偏北风和寒潮南下的强度（Jeong *et al.*, 2005; Hong and Li, 2009）。

从 20 世纪 80 年末至 21 世纪初，乌拉尔山阻塞高压的出现减少，西伯利亚冷高压偏弱，华南寒潮也随之减弱（Wu and Leung, 2009; Wang and Chen, 2014）。而从 21 世纪初起，乌拉尔山阻塞高压的出现偏多，西伯利亚冷高压增强，华南开始出现数十年一遇的强寒潮，如 2008 年 1 月和 2016 年 1 月（Zhou *et al.*, 2009; Cheung *et al.*, 2016）。近年华南寒潮的变化除了受到 PDO 和大西洋涛动等低频振荡的调控（Zhou *et al.*, 2007; Hong *et al.*, 2008; Luo *et al.*, 2016）以外，可能也受到了北极海冰减少、北极增暖和平流层极涡减弱的影响（Wu B. *et al.*, 2011; Wu Z. *et al.*, 2011; Kim *et al.*, 2014; Mori *et al.*, 2014; Luo *et al.*, 2018）。有学者利用 EOF 分析我国 1957～2010 年冬季寒潮长度的主模态，研究发现 1980 年前主模态的中心在北方，1980 年后主模态的中心移往华南地区，原因或与东亚地区的气旋轴活动向南偏移有关（Ma *et al.*, 2012）。根据气候模式的预测，寒潮长度的主模态也有向南移往华南的现象，这表示气候暖化或会加强华南寒潮的强度和持续性（Ma *et al.*, 2012）。另外，乌拉尔山阻塞高压未来的频率没有明显变化，但在年际尺度上与西伯利亚冷高压南伸的关系变强，这表示未来阻塞高压对华南寒潮的年际变化可能会有更强的影响（Cheung and Zhou, 2015）。

（四）台风

热带气旋的形成、强度、结构变化及移动，基本上受它外围环境的控制。影响港澳地区的热带气旋有两个主要发源地，一是在菲律宾以东海域，另一个是南海。菲律宾以东形成的热带气旋一般会穿过菲律宾后进入南海，登陆菲律宾后就会减弱，因此即使在进入南海后再度增强，大部分气旋也不会变得很强。然而，一些穿越吕宋海峡直接进入南海的热带气旋不会减弱，因此有机会发展成超强台风。此外，广东沿岸的水温颇高，加之大陆架较浅，在热带气旋经过时很少冷水上翻，所以假如大气条件（譬如垂直风切变）配合的话，影响港澳地区的热带气旋可以在登陆前继续增强。总体来说，登陆港澳地区前的热带气旋，增强的个数比减弱的多（图 1–9）。

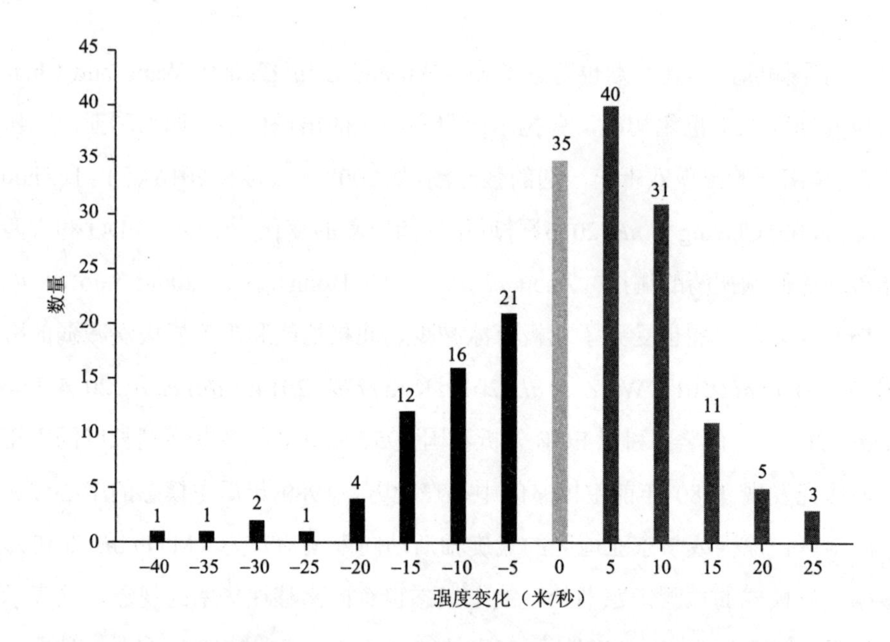

图 1–9　登陆港澳地区的热带气旋强度和 48 小时前强度的比较

资料来源：联合台风警报中心（Joint Typhoon Warning Center）1960～2013 年最佳路径数据库。

热带气旋的移动主要是受它背景的气流（引导气流）所推动（Chan and Gray, 1982）；而引导气流方向的变化与位于西北太平洋的副热带高压的变化有关。副高的位置、西伸点和强弱有所改变，导致影响广东省的热带气旋个数的年代际变化，而这些改变和 ENSO 及 PDO 有关（Liu and Chan, 2017）。另外一个因素是热带气旋形成的位置。在个数偏低的年代，热带气旋的平均形成位置较为偏北，因此气旋形成后，背景的引导气流会把它推往较北的地方，而不会进入南海。此外，假如整个西北太平洋热带气旋的个数减少，也会导致影响广东省的热带气旋个数减少（Zhang *et al.*, 2011）。

二、人类活动的影响

（一）大尺度环流

西北太平洋副热带高压（简称西太副高）是影响东亚夏季气候最重要的大尺度大气环流系统。西太副高的强度变异直接影响着东亚雨带位置（Zhu *et al.*, 2003; Ren *et al.*, 2013）和台风路径（Wu *et al.*, 2005; Kim *et al.*, 2012）在季节内和年际尺度上的变率。鉴于西太副高对华南气候的重要影响，西太副高的范围、位置、强度的年代际变化一直是学术界广泛关注的话题。

气候业务中通常采用位势高度监测西太副高的季节内变率和年际变率。然而，近年来的最新研究表明，气候变暖背景下全球位势高度会由于空气热力膨胀而整体升高（He *et al.*, 2015; 2018），可能无法真实反映西太副高反气旋性环流的年代际变化特征，因此有必要结合大气环流的变化并通过多种指标的横向比较来综合判定西太副高的年代际以上尺度的变化。观测资料显示，对流层低层西太副高的西边界位置在 20 世纪 70 年代末以后比此前更偏东，这显示西太副高发生了年代际减弱并向东撤退（Huang *et al.*, 2015）。若采用涡动位势高度（扣除了纬向平均值之后的位势高度）作为度量指标，对流层中层的西太副高也在 20 世纪 70 年代末期表现出年代际减弱（Wu and Wang, 2015）。

西北太平洋热带气旋的移动受到西太副高引导气流的强烈调控。西太副高的年代际减弱和东退可能对西北太平洋热带气旋的年代际变化有一定影响（Lee *et al.*, 2020）。登陆我国的热带气旋的登陆地点表现出北移趋势（Liu and Chan, 2017），具体表现为：登陆地点位于厦门以南的热带气旋频数呈现年代际减少的趋势，而登陆地点位于厦门以北的热带气旋频数呈现年代际增多的趋势（王磊等，2009）。根据杨玉华等（2009）的研究，虽然 1949～2006 年登陆中国的热带气旋频数呈年代际减少趋势，但达到台风级别的较强热带气旋频数并无显著的年代际变化，反而登陆的强台风频数、平均强度有增加趋势，特别是登陆华南的热带气旋强度的增强趋势更为明显（杨玉华等, 2009; Mei and Xie, 2016; Li *et al.*, 2017）。UNESCAP/WMO 台风委员会专家组的最新评估指出，在 1949～2017 年登陆中国的热带气旋及台风数据没有显著变化趋势，过去几十年影响香港（1961～2018 年）及澳门（1953～2017 年）的热带气旋数目也没有显著变化趋势（Lee *et al.*, 2020）。由于早期资料可靠性差（Moon *et al.*, 2019），目前还难以断定西北太平洋热带气旋的平均移动速度是否具有年代际减速趋势（Kossin, 2018; Moon *et al.*, 2019; Lai *et al.*, 2020）。

过去几十年的年代际变化可能兼有气候系统内部变率或者人为温室气体外强迫等不同因素的影响。基于三十多个气候系统模式的集合平均，能够抑制气候系统内部变率和模式随机误差的干扰，提取温室气体强迫的大气环流响应。虽然西北太平洋位势高度随着气候变暖而显著上升，但西太副高区位势高度的上升量小于纬向平均，西北太平洋大气环流的变化呈现异常气旋环流，显示西太副高强度减弱。相对于 20 世纪后半叶而言，在 RCP8.5 情景下，21 世纪后半叶西太副高的范围缩小，西脊点向东撤退了大约 10 度。由于 RCP8.5 情景下 21 世纪后半叶相对于 20 世纪后半叶升温幅度约为 2.8 摄氏度，因此全球气温每升温 1 摄氏度，西太副高的西边界就会向东撤退大约 3 度（He *et al.*, 2015）。

尽管人为气候变暖导致夏季西太副高的强度减弱，但诸多证据表明，气候变暖很可能导致东亚夏季风的强度增强。学术界一般采用东亚对流层低层的南风强度来度量东亚夏季风的环流强度。耦合模式预估试验的结果一致显示，在人为温室气体强迫下，整个中国东部地区对流层低层的南风从华南到华北一致增强；从更大的范围来看，东亚夏季风的增强与一个围绕青藏高原的异常气旋环流有关。种种证据表明，青藏高原热力强迫的增强对预估东亚夏季风环流增强具有关键作用（He *et al.*, 2019; He and Zhou, 2020）。参考 CMIP3 模式、CMIP5 模式以及其他多数气候系统模式的预估结果，东亚夏季风环流的增强方面表现非常一致（丁一汇等, 2013; Jiang and Tian, 2013; Kamae *et al*., 2014; Kitoh, 2017），这说明东亚夏季风环流随着气候变暖而增强的结论是不依赖于模式的。东亚夏季风强度偏强意味着东亚雨带偏北，南方雨量偏少而北方雨量偏多；反之亦然。古气候记录也显示，与较冷的时期相比，暖期的东亚夏季风偏强，雨带位置偏北（Man *et al*., 2012; Yang *et al*., 2015）。在人为温室气体强迫下，东亚夏季风环流增强但西太副高强度减弱，两者并不矛盾，这是因为东亚夏季风受多种不同因素的调控，而西太副高的大尺度环流只是这些众多因素之一。

（二）强降水

城市热岛效应会导致广州市雷雨增加（Meng *et al*., 2007）。在珠三角地区，由于人类活动，污染物浓度增加，导致降水率上升，大到暴雨的区域面积也有所增加（Wang Y. *et al*., 2011）。霍尔斯特等（Holst *et al*., 2016）在利用WRF模式模拟一个珠三角地区的强降水过程时，把人为加热也加进地表感热通量函数内，他发现当人为加热达到一定大的值时，城市的降水，尤其是强降水会增加，而一个面积很大的城市，其强降水的增加会更明显（Holst *et al*., 2016）。

第三节 未来气候变化预估

一、基本气候要素预估

（一）香港

香港天文台主要利用降尺度统计法来推算未来气候变化（海平面高度推算除外）。有关气候推算方法的细节可以参考其他研究（Lee *et al*., 2011; Chan *et al.*, 2014; Cheung M. *et al.*, 2015; Chan *et al.*, 2016; Tong *et al.*, 2017）。

图 1–10a 给出了在中低温室气体浓度情景（RCP4.5）和高温室气体浓度情景（RCP8.5）下香港年平均气温的推算。在 RCP8.5 情景下，预计到 21 世纪末（2091～2100 年），香港年平均温度较 1986～2005 年将平均升高 3～6 摄氏度，年平均温度的升幅在 RCP4.5 和 RCP6.0 情景下较小（Chan *et al.*, 2014）。

图 1–10b 给出了在 RCP4.5 和 RCP8.5 情景下香港年降雨量的未来推算，从推算的可能范围可见，各个模式的推算存在颇大差异，可见雨量推算的不确定性较温度推算的不确定性高。总体来说，在 RCP8.5 情景下，20 世纪末香港年降雨量会较 1986～2005 年平均多约 180 毫米（Cheung M. *et al.*, 2015）。图 1–10c 和图 1–10d 分别展示了在 RCP4.5 和 RCP8.5 情景下香港极端多雨（年降雨量多于 3 168 毫米）和极端少雨（年雨量少于 1 289 毫米）年数的未来推算。与 1885～2005 年的实况观测相比，预计 21 世纪出现极端多雨的年数会增加，而出现极端少雨的年数则大致维持不变（Cheung M. *et al.*, 2015）。

何宇恒等（He *et al.*, 2016）利用 CMIP5 全球气候模式输出的海平面高度数据、政府间气候变化专门委员会（Intergovernmental Panel on Climate Change, IPCC）第五次评估报告（*Fifth Assessment Report*, AR5）给出的全球陆地储冰量和储水量数据，结合全球定位系统量度的香港土地沉降数据推算香港及其邻近海域未来的海平面高度变化。结果显示，不论温室气体浓度情景如何，21 世纪香港及其邻近海域的海平面会继续上升。IPCC 于 2019 年 9 月发表《气候变化中的海洋和冰冻圈特别报告》（*Special Report on the Ocean and Cryosphere in a Changing Climate*, SROCC），上调了全球平均海平面上升的推算。基于 SROCC 提供的最新数据，香港及邻近水域的平均海平面上升推算也已更新。在 RCP8.5 情景下，2091～2100 年香港及邻近水域的平均海平面高度（包含本地地壳垂直运动的影响）会较 1986～2005 年的平均值高 0.73～1.28 米（图 1–10e）。随着平均海平面上升，热带气旋带来的风暴潮威胁也会相应增加。

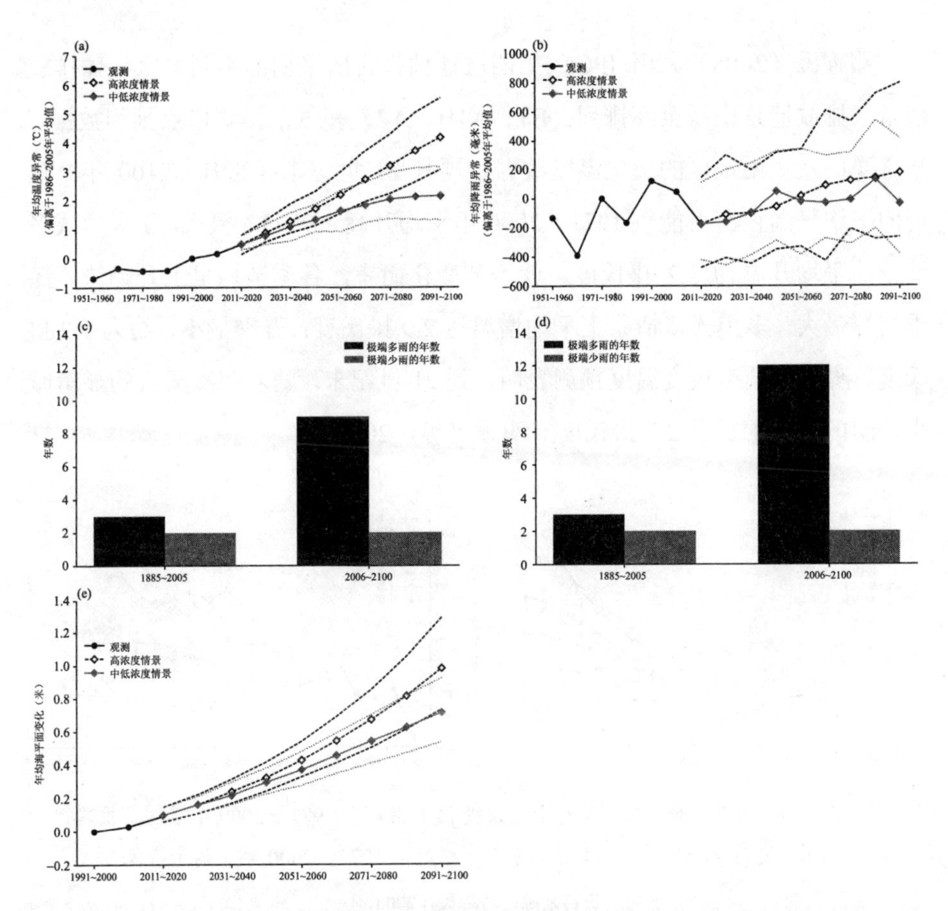

图 1–10　（a）在 RCP8.5 和 RCP4.5 情景下香港年平均气温距平（相对于 1986 ~ 2005 年平均）预测（符号连线是平均值，虚线是推算结果的可能范围）。历史观测以黑实线及圆点表示；（b）在 RCP8.5 和 RCP4.5 情景下香港年雨量距平（相对于 1986 ~ 2005 年平均）的预测；（c）在 RCP4.5 和（d）RCP8.5 情景下，香港极端多雨年数和极端少雨年数的预测；（e）在 RCP8.5 和 RCP4.5 情景下香港及邻近水域的平均海平面高度变化（相对于 1986 ~ 2005 年平均）的预测

（二）澳门

邓耀民（2008）利用 IPCC 第四次评估报告所采用的不同温室气体排放情景（排放情景由低至高排列：B1、A1B、A2）和气候模式格点预测数据，结合澳门及华南地区的历史温度数据，预测 21 世纪末（2091～2100 年）澳门的平均气温很有可能较 1971～2000 年平均升高 1.9～3.4 摄氏度，多模式多情景的平均升温为 2.7 摄氏度。就季节变化而言，各季节均呈上升趋势。冬季增幅最大，多模式多情景下平均增温达 2.9 摄氏度；夏季最小，约为 1.9 摄氏度。澳门的城市尺度温度预测表明，到 21 世纪末，随着寒冷天气的显著减少，澳门将大约提升 2.7 摄氏度（邓耀民等，2008）。

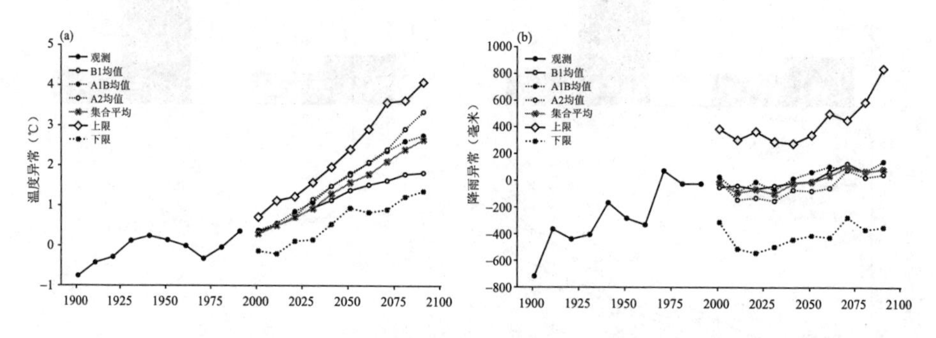

图 1–11　（a）澳门 21 世纪年均地面气温变化（相对于 1971～2000 年平均）推算；（b）澳门 21 世纪年均降雨量变化（相对于 1971～2000 年平均）推算

资料来源：邓耀民，2008。

邓耀民（2008）也预测了澳门的年降水（多模式多情景平均，排放情景由低至高排列：B1、A1B、A2）在 2040 年前将呈减少趋势，2040 年后则呈增加趋势。但不同模式推算结果差异较大，不确定性高。从多模式多情景的平均来看，未来冬春降水量较 1971～2000 年平均将减少，其中冬季降水偏少的情况尤为明显。夏秋两季的降水则在 2040、2050 年后呈增多趋势。

据预测，未来澳门海平面的增加速率将比全球平均快 20%，主要原因是澳门附近的海洋增温快于全球平均，且南海北部的南风增强导致海水堆积。到2020年、2060年、2100年，海平面升高的幅度分别可达8～12厘米、22～51厘米、35～118厘米，具体量值取决于温室气体排放情景和气候敏感度（Wang L. *et al.*, 2016）。高海平面加剧了风暴潮、洪涝、海岸侵蚀及海水入侵等灾害，将给澳门地区居民的生产生活和社会经济发展造成一定影响。

二、极端天气气候事件预估

极端气温方面，香港地区酷热天气日数和热夜日数预计将显著增多。在RCP8.5情景下，21世纪末（2091～2100年）平均每年酷热天气日数将超过100天，平均每年热夜日数约为150天，而寒冷天气日数则会显著减少（图1–12a和图1–12b）。人体舒适度不仅与温度相关，也与湿度有关，湿球温度是暑热压力的一个基本指标，高湿球温度表示环境既暖又湿，人体散热较困难。日最高湿球温度推算显示，不论在任何温室气体浓度情景下，21世纪香港每年极端“暖湿”天气（日最高湿球温度在28.2摄氏度或以上）日数和每年最长连续极端“暖湿”天气日数都会增加（图1–12c和图1–12d），增加幅度在RCP8.5情景下最为显著（Tong *et al.*, 2017）。

极端降水方面，在RCP8.5情景下，极端降雨（日雨量在100毫米或以上）日数会由1986～2005年实况观测的平均每年4.2日增加至21世纪末（2091～2100年）约5.1日。平均降雨强度（即年雨量除以降雨日数）、每年最高日雨量、每年最高连续3日雨量和每年最长连续干日数目会增加（表1–4、图1–13），但每年降雨日数会减少（Chan *et al.*, 2016）。

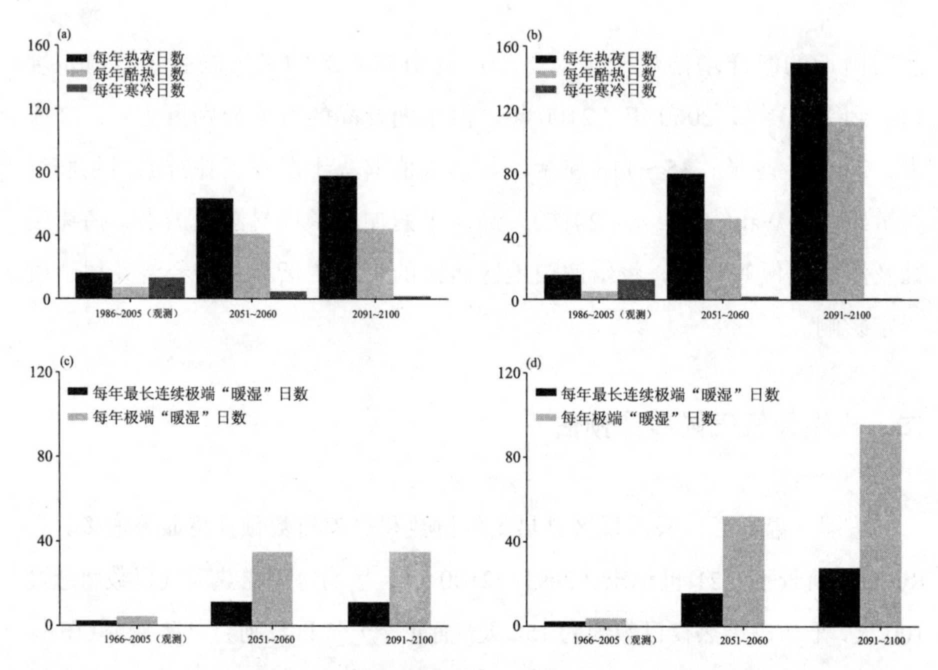

图 1–12 （a）在 RCP4.5 情景下，香港每年热夜数目、酷热日数和寒冷天气日数的未来变化；（b）在 RCP8.5 情景下，香港每年热夜数目、酷热日数和寒冷天气日数的未来变化；（c）在 RCP4.5 情景下，香港每年极端“暖湿”天气日数和最长连续极端“暖湿”天气日数的未来推算；（d）在 RCP8.5 情景下，香港每年极端“暖湿”天气日数和最长连续极端“暖湿”天气日数的未来推算

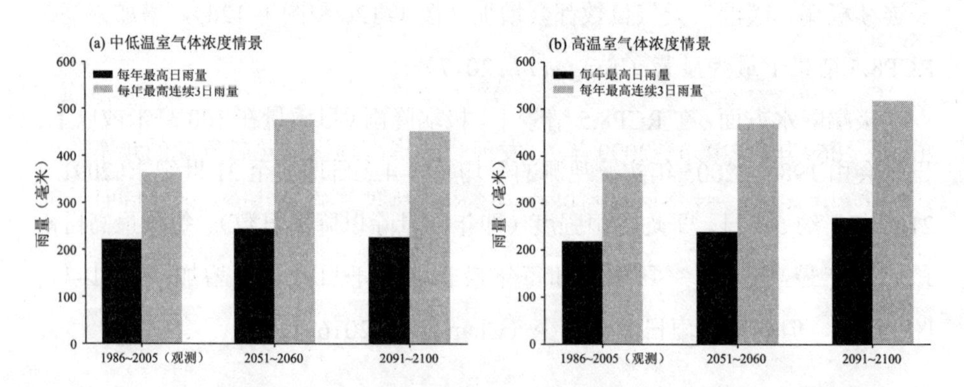

图 1–13 （a）RCP4.5 情景下香港每年最高日雨量和每年最高连续 3 日雨量的未来推算；（b）RCP8.5 情景下香港每年最高日雨量和每年最高连续 3 日雨量的未来推算

表 1–4　香港每年极端降雨日数、平均降雨强度、每年最高日雨量、每年最高连续 3 日雨量、每年最长连续干日数目和每年降雨日数的未来推算

	1986～2005 年观测	2051～2060 年推算	2091～2100 年推算	2051～2060 年推算	2091～2100 年推算
温室气体浓度情景	—	RCP4.5	RCP4.5	RCP8.5	RCP8.5
每年极端降雨日数	4.2	4.5	4.4	5.0	5.1
平均降雨强度（毫米/日）	23.4	25.0	24.0	25.4	26.7
每年最高日雨量（毫米）	221	246	228	243	273
每年最高连续 3 日雨量（毫米）	367	482	454	476	523
每年最长连续干日数目	46	49	52	54	59
每年降雨日数	102	103	102	100	97

绝大多气候模式的分辨率不足以预估未来影响港澳地区热带气旋的个数或强度。UNESCAP/WMO 台风委员会专家组的最新评估指出，未来西北太平洋（包括南海）的热带气旋强度、强热带气旋比率及相关降水率将会增加（Cha *et al.*, 2020）。横井觉和高薮缘（Yokoi and Takayabu, 2009）分析了五个 CMIP3 模式，发现在不同的情景下，南海台风发生的频率会有明显的下降。其他研究也有类似的结果（Wang R. *et al.*, 2011）。

一项研究通过降尺度模拟系统做了三个十年时段的预估（2030～2039 年、2060～2069 年和 2090～2099 年），发现登陆华南地区的热带气旋的频率有所减少（图 1–14a），这个结果与以前关于南海热带气旋的结论一致。同时，强度（以 APDI 代表）分布却显示，越靠近 21 世纪末，热带气旋越有更大的可能变强（图 1–14b）（Lok and Chan, 2018）。随着热带气旋强度增加及未来海平面上升，热带气旋带来的风暴潮及沿岸水淹风险将会增加（Cha *et al.*, 2020; Chen *et al.*, 2020）。

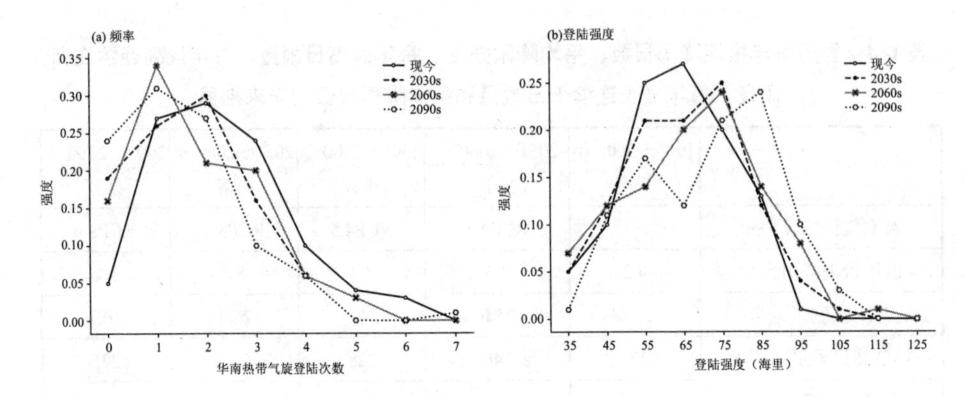

图 1–14 （a）登陆华南热带气旋十年（2030～2039 年、2060～2069 年、2090～2099 年）间预估次数；（b）登陆华南热带气旋十年（2030～2039 年、2060～2069 年、2090～2099 年）间强度的概率密度函数（probability density function, PDF）

资料来源：Lok and Chan，2018。

参考文献

Cha, E. J., T. R. Knutson, T. C. Lee, *et al.*, 2020. Third Assessment on Impacts of Climate Change on Tropical Cyclones in The Typhoon Committee Region—Part II: Future Projections. *Tropical Cyclone Research and Review*, 9(2).

Chan, H. S., M. H. Kok, T. C. Lee, 2012. Temperature Trends in Hong Kong from a Seasonal Perspective. *Climate Research*, 55(1).

Chan, H. S., H. W. Tong, S. M. Lee, 2014. Temperature Projection for Hong Kong in the 21st Century Using CMIP5 Models. *Hong Kong Meteorological Society Bulletin*, 24.

Chan, H. S., H. W. Tong, S. M. Lee, 2016. Extreme Rainfall Projection for Hong Kong in the 21st Century Using CMIP5 Models. 30th Guangdong-Hong Kong-Macao Seminar on Meteorological Science and Technology, Guangzhou, April 20-22, HKO Reprint No. 1222.

Chan, J. C. L., W. M. Gray, 1982. Tropical Cyclone Movement and Surrounding Flow Relationships. *Monthly Weather Review*, 110(10).

Chen, J. L., Z. Q. Wang, C. Y. Tam, *et al.*, 2020. Investigating Climate Change on Tropical Cyclones and Induced Storm Surges in the Pearl River Delta Region Using Pseudo-Global Warming Method. *Scientific Reports*, 10.

Cheung, H. H. N., W. Zhou, 2015. Implications of Ural Blocking for East Asian Winter Climate in CMIP5 GCMs. Part II: Projection and Uncertainty in Future Climate Conditions. *Journal of Climate*, 28(6).

Cheung, H. H. N., W. Zhou, S. M. Lee, *et al.*, 2015. Interannual and Interdecadal Variability of the Number of Cold Days in Hong Kong and Their Relationship with Large-Scale Circulation. *Monthly Weather Review*, 143(4).

Cheung, H. H. N., W. Zhou, M. Y. T. Leung, *et al.*, 2016. A Strong Phase Reversal of the Arctic Oscillation in Midwinter 2015/2016: Role of the Stratospheric Polar Vortex and Tropospheric Blocking. *Journal of Geophysical Research-Atmospheres*, 121(22).

Cheung, M. S., H. S. Chan, H. W. Tong, 2015. Rainfall Projection for Southern China in the 21st Century Using CMIP5 Models. 29th Guangdong-Hong Kong-Macao Seminar on Meteorological Science and Technology, Macao, January 20-22, HKO Reprint No. 1165.

Ginn, W. L., T. C. Lee, K. Y. Chan, 2010. Past and Future Changes in the Climate of Hong Kong. *Acta Meteorologica Sinica*, 24(2).

Han, Z., F. F. Luo, S. L. Li, *et al.*, 2016. Simulation by CMIP5 Models of the Atlantic Multidecadal Oscillation and Its Climate Impacts. *Advances in Atmospheric Sciences*, 33(12).

He, C., A. L. Lin, D. J. Gu, *et al.*, 2018. Using Eddy Geopotential Height to Measure the Western North Pacific Subtropical High in a Warming Climate. *Theoretical and Applied Climatology*, 131.

He, C., T. J. Zhou, A. L. Lin, *et al.*, 2015. Enhanced or Weakened Western North Pacific Subtropical High under Global Warming? *Scientific Reports*, 5.

He, C., Z. Q. Wang, T. J. Zhou, *et al.*, 2019. Enhanced Latent Heating over the Tibetan Plateau as a Key to the Enhanced East Asian Summer Monsoon Circulation under a Warming Climate. *Journal of Climate*, 32(11).

He, C., W. Zhou, 2020. Different Enhancement of the East Asian Summer Monsoon under Global Warming and Interglacial Epochs Simulated by CMIP6 Models: Role of the Subtropical High. *Journal of Climate*, 33(22).

He, J. H., Z. W. Zhu, 2015. The Relation of South China Sea Monsoon Onset with the Subsequent Rainfall over the Subtropical East Asia. *International Journal of Climatology*, 35(15).

He, Y. H., H. Y. Mok, E. S. T. Lai, 2016. Projection of Sea-Level Change in the Vicinity of Hong Kong in the 21st Century. *International Journal of Climatology*, 36(9).

Holst, C. C., J. C. L. Chan, C. Y. Tam, 2017. Sensitivity of Precipitation Statistics to Urban Growth in a Subtropical Coastal Megacity Cluster. *Journal of Environmental Sciences*, 59.

Holst, C. C., C. Y. Tam, J. C. L. Chan, 2016. Sensitivity of Urban Rainfall to Anthropogenic

Heat Flux: A Numerical Experiment. *Geophysical Research Letters*, 43(5).

Hong, C. C., H. H. Hsu, H. H. Chia, *et al.*, 2008. Decadal Relationship Between the North Atlantic Oscillation and Cold Surge Frequency in Taiwan. *Geophysical Research Letters*, 35.

Hong, C. C., T. Li, 2009. The Extreme Cold Anomaly over Southeast Asia in February 2008: Roles of ISO and ENSO. *Journal of Climate*, 22(13).

Huang, J. P., J. C. H. Fung, A. K. H. Lau, *et al.*, 2005. Numerical Simulation and Process Analysis of Typhoon-Related Ozone Episodes in Hong Kong. *Journal of Geophysical Research-Atmospheres*, 110.

Huang, Y. Y., H. J. Wang, K. Fan, *et al.*, 2015. The Western Pacific Subtropical High after the 1970s: Westward or Eastward Shift? *Climate Dynamics*, 44.

Jeong, J. H., C. H. Ho, B. M. Kim, *et al.*, 2005. Influence of the Madden-Julian Oscillation on Wintertime Surface Air Temperature and Cold Surges in East Asia. *Journal of Geophysical Research-Atmospheres*, 110.

Jiang, D. B., Z. P. Tian, 2013. East Asian Monsoon Change for the 21st Century: Results of CMIP3 and CMIP5 Models. *Chinese Science Bulletin*, 58(12).

Kajikawa, Y., B. Wang, 2012. Interdecadal Change of the South China Sea Summer Monsoon Onset. *Journal of Climate*, 25(9).

Kamae, Y., M. Watanabe, M. Kimoto, *et al.*, 2014. Summertime Land-Sea Thermal Contrast and Atmospheric Circulation over East Asia in a Warming Climate. Part I: Past Changes and Future Projections. *Climate Dynamics*, 43(9).

Kim, B. M., S. W. Son, S. K. Min, *et al.*, 2014. Weakening of the Stratospheric Polar Vortex by Arctic Sea-Ice Loss. *Nature Communications*, 5.

Kim, J. H., C. H. Ho, H. S. Kim, *et al.*, 2012. 2010 Western North Pacific Typhoon Season: Seasonal Overview and Forecast Using a Track-Pattern-Based Model. *Weather and Forecasting*, 27(3).

Kitoh, A., 2017. The Asian Monsoon and Its Future Change in Climate Models: A Review. *Journal of the Meteorological Society of Japan*, 95(1).

Knutson, T., S. J. Camargo, J. C. L. Chan, *et al.*, 2019. Tropical Cyclones and Climate Change Assessment. Part I: Detection and Attribution. *Bulletin of the American Meteorological Society*, 100(10).

Kossin, J. P., 2018. A Global Slowdown of Tropical-Cyclone Translation Speed. *Nature*, 558(7708).

Kossin, J. P., K. A. Emanuel, S. J. Camargo, 2016. Past and Projected Changes in Western North Pacific Tropical Cyclone Exposure. *Journal of Climate*, 29(16).

Kossin, J. P., K. A. Emanuel, G. A. Vecchi, 2014. The Poleward Migration of the Location of Tropical Cyclone Maximum Intensity. *Nature*, 509(7500).

Lai, Y. C., J. F. Li, X. Gu, *et al.*, 2020. Greater Flood Risks in Response to Slowdown of Tropical Cyclones over the Coast of China. *Proceedings of the National Academy of Sciences of the United States of America*, 117(26).

Lam, K. S., T. J. Wang, C. L. Wu, *et al.*, 2005. Study on an Ozone Episode in Hot Season in Hong Kong and Transboundary Air Pollution over Pearl River Delta Region of China. *Atmospheric Environment*, 39(11).

Lee, B. Y., W. T. Wong, W. C. Woo, 2010. Sea-level Rise and Storm Surge: Impacts of Climate Change on Hong Kong. HKIE Civil Division Conference, Hong Kong, April 12-14, HKO Reprint No. 915.

Lee, T. C., K. Y. Chan, W. L. Gin, 2011. Projection of Extreme Temperature in Hong Kong in the 21st Century. *Acta Meteorologica Sinica*, 25(1).

Lee, T. C., T. R. Knutson, H. Kamahori, *et al.*, 2012a. Impacts of Climate Change on Tropical Cyclones in the Western North Pacific Basin. Part I: Past Observations. *Tropical Cyclone Research and Review*, 1(2).

Lee, T. C., T. R. Knutson, T. Nakaegawa, *et al.*, 2020. Third Assessment on Impacts of Climate Change on Tropical Cyclones in the Typhoon Committee Region. Part I: Observed Changes, Detection and Attribution. *Tropical Cyclone Research and Review*, 9(1).

Lee, T. C., C. Y. Y. Leung, M. H. Kok, *et al.*, 2012b. The Long Term Variations of Tropical Cyclone Activity in the South China Sea and the Vicinity of Hong Kong. *Tropical Cyclone Research and Review*, 1(3).

Leung, M. Y. T., H. H. N. Cheung, W. Zhou, 2015. Energetics and Dynamics Associated with Two Typical Mobile Trough Pathways over East Asia in Boreal Winter. *Climate Dynamics*, 44.

Leung, M. Y. T., H. H. N. Cheung, W. Zhou, 2017. Meridional Displacement of the East Asian Trough and Its Response to the ENSO Forcing. *Climate Dynamics*, 48.

Leung, M. Y. T., W. Zhou, 2016. Direct and Indirect ENSO Modulation of Winter Temperature over the Asian-Pacific-American Region. *Scientific Reports*, 6.

Li, Q. Q., Y. H. Duan, 2010. Tropical Cyclone Strikes at the Coastal Cities of China from 1949 to 2008. *Meteorology and Atmospheric Physics*, 107.

Li, R. C. Y., W. Zhou, T. C. Lee, 2015. Climatological Characteristics and Observed Trends of Tropical Cyclone-Induced Rainfall and Their Influences on Long-Term Rainfall Variations in Hong Kong. *Monthly Weather Review*, 143(6).

Li, R. C. Y., W. Zhou, C. M. Shun, *et al.*, 2017. Change in Destructiveness of Landfalling Tropical Cyclones over China in Recent Decades. *Journal of Climate*, 30(9).

Liu, K. S., J. C. L. Chan, 2017. Variations in the Power Dissipation Index in the East Asia Region. *Climate Dynamics*, 48.

Lok, C. C. F., J. C. L. Chan, 2018. Changes of Tropical Cyclone Landfalls in South China Throughout the Twenty-First Century. *Climate Dynamics*, 51.

Luo, D. H., X. D. Chen, A. G. Dai, *et al.*, 2018. Changes in Atmospheric Blocking Circulations Linked with Winter Arctic Warming: A New Perspective. *Journal of Climate*, 31(18).

Luo, D. H., Y. Q. Xiao, Y. N. Diao, *et al.*, 2016. Impact of Ural Blocking on Winter Warm Arctic-Cold Eurasian Anomalies. Part II: The Link to the North Atlantic Oscillation. *Journal of Climate*, 29(11).

Ma, T. T., Z. W. Wu, Z. H. Jiang, 2012. How Does Coldwave Frequency in China Respond to a Warming Climate? *Climate Dynamics*, 39.

Macao Environmental Protection Bureau, 2018. Report on the State of the Environment of Macao 2017. http://www.dspa.gov.mo/richtext_report2017.aspx.

Macao Meteorological and Geophysical Bureau, 2018. Macao Annual Air Quality Reports 2018. http://www.smg.gov.mo/smg/airQuality/pdf/IQA_2018.pdf.

Man, W. M., T. J. Zhou, J. H. Jungclaus, 2012. Simulation of the East Asian Summer Monsoon During the Last Millennium with the MPI Earth System Model. *Journal of Climate*, 25(22).

Mei, W., S. P. Xie, 2016. Intensification of Landfalling Typhoons over the Northwest Pacific Since the Late 1970s. *Nature Geoscience*, 9.

Meng, W. G., J. H. Yan, H. B. Hu, 2007. Urban Effects and Summer Thunderstorms in a Tropical Cyclone Affected Situation over Guangzhou City. *Science in China Series D-Earth Sciences*, 50(12).

Mok, H. Y., Y. K. Leung, T. C. Lee, *et al.*, 2006. Regional Rainfall Characteristics of Hong Kong over the Past 50 Years. Conference on "Changing Geography in a Diversified World", Hong Kong, June 1-3, HKO Reprint No. 646.

Moon, I. J., S. H. Kim, J. C. L. Chan, 2019. Climate Change and Tropical Cyclone Trend. *Nature*, 570 (7759).

Mori, M., M. Watanabe, H. Shiogama, *et al.*, 2014. Robust Arctic Sea-Ice Influence on the Frequent Eurasian Cold Winters in Past Decades. *Nature Geoscience*, 7(12).

Park, T. W., C. H. Ho, S. Yang, 2011. Relationship Between the Arctic Oscillation and Cold Surges over East Asia. *Journal of Climate*, 24(1).

Peng, L., J. P. Liu, Y. Wang, *et al.*, 2018. Wind Weakening in a Dense High-Rise City due to Over Nearly Five Decades of Urbanization. *Building and Environment*, 138.

Qian, C., W. Zhou, S. K. Fong, *et al.*, 2015. Two Approaches for Statistical Prediction of Non-Gaussian Climate Extremes: A Case Study of Macao Hot Extremes During 1912-2012. *Journal of Climate*, 28(2).

Ren, X. J., X. Q. Yang, X. G. Sun, 2013. Zonal Oscillation of Western Pacific Subtropical High and Subseasonal SST Variations During Yangtze Persistent Heavy Rainfall Events.

Journal of Climate, 26(22).

Takaya, K., H. Nakamura, 2005. Geographical Dependence of Upper-Level Blocking Formation Associated with Intraseasonal Amplification of the Siberian High. *Journal of the Atmospheric Sciences*, 62(12).

Tong, H. W., W. C. Lee, 2014. Observed Long-Term Trend in the Upper Air over Hong Kong. 28th Guangdong-Hong Kong-Macao Seminar on Meteorological Science and Technology, Hong Kong, January 13-15, HKO Reprint No. 1122.

Tong, H. W., C. P. Wong, S. M. Lee, 2017. Projection of Wet-Bulb Temperature for Hong Kong in the 21st Century using CMIP5 Data. The 31st Guangdong-Hong Kong-Macao Seminar on Meteorological Science and Technology and the 22nd Guangdong-Hong Kong-Macao Meeting on Cooperation in Meteorological Operations, Hong Kong, February 27-March 1, HKO Reprint No. 1276.

Wang, J. L., B. Yang, F. C. Ljungqvist, *et al.*, 2013. The Relationship Between the Atlantic Multidecadal Oscillation and Temperature Variability in China During the Last Millennium. *Journal of Quaternary Science*, 28(7).

Wang, L., W. Chen, 2014. The East Asian Winter Monsoon: Re-Amplification in the Mid-2000s. *Chinese Science Bulletin*, 59(4).

Wang, L., G. Huang, W. Zhou, *et al.*, 2016. Historical Change and Future Scenarios of Sea Level Rise in Macau and Adjacent Waters. *Advances in Atmospheric Sciences*, 33(4).

Wang, Q. Q., Y. M. Feng, X. H. H. Huang, *et al.*, 2016. Nonpolar Organic Compounds as $PM_{2.5}$ Source Tracers: Investigation of Their Sources and Degradation in the Pearl River Delta, China. *Journal of Geophysical Research-Atmospheres*, 121(19).

Wang, Q. Q., X. H. H. Huang, F. C. V. Tam, *et al.*, 2019. Source Apportionment of Fine Particulate Matter in Macao, China with and without Organic Tracers: A Comparative Study Using Positive Matrix Factorization. *Atmospheric Environment*, 198.

Wang, R. F., L. G. Wu, C. Wang, 2011. Typhoon Track Changes Associated with Global Warming. *Journal of Climate*, 24(14).

Wang, W. W., W. Zhou, S. K. Fong, *et al.*, 2015. Extreme Rainfall and Summer Heat Waves in Macau Based on Statistical Theory of Extreme Values. *Climate Research*, 66(1).

Wang, W. W., W. Zhou, E. Y. Y. Ng, *et al.*, 2016. Urban Heat Islands in Hong Kong: Statistical Modeling and Trend Detection. *Natural Hazards*, 83(2).

Wang, Y., Q. Wan, W. Meng, *et al.*, 2011. Long-Term Impacts of Aerosols on Precipitation and Lightning over the Pearl River Delta Megacity Area in China. *Atmospheric Chemistry and Physics*, 11(23).

Wong, M. C., H. Y. Mok, T. C. Lee, 2011. Observed Changes in Extreme Weather Indices in Hong Kong. *International Journal of Climatology*, 31(15).

Wong, W. T., K. W. Li, K. H. Yeung, 2003. Long Term Sea Level Change in Hong Kong. *Hong Kong Meteorological Society Bulletin*, 13.

Wu, B. Y., J. Z. Su, R. H. Zhang, 2011. Effects of Autumn-Winter Arctic Sea Ice on Winter Siberian High. *Chinese Science Bulletin*, 56(30).

Wu, L. G., B. Wang, S. Q. Geng, 2005. Growing Typhoon Influence on East Asia. *Geophysical Research Letters*, 32(18).

Wu, L. G., C. Wang, 2015. Has the Western Pacific Subtropical High Extended Westward Since the Late 1970s? *Journal of Climate*, 28(13).

Wu, M. C., W. H. Leung, 2009. Effect of ENSO on the Hong Kong Winter Season. *Atmospheric Science Letters*, 10(2).

Wu, Z. W., J. P. Li, Z. H. Zhang, *et al.*, 2011. Predictable Climate Dynamics of Abnormal East Asian Winter Monsoon: Once-in-a-Century Snowstorms in 2007/2008 Winter. *Climate Dynamics*, 37(7-8).

Yang, S. L., Z. L. Ding, Y. Y. Li, *et al.*, 2015. Warming-Induced Northwestward Migration of the East Asian Monsoon Rain Belt from the Last Glacial Maximum to the Mid-Holocene. *Proceedings of the National Academy of Sciences of the United States of America*, 112(43).

Yokoi, S., Y. N. Takayabu, 2009. Multi-Model Projection of Global Warming Impact on Tropical Cyclone Genesis Frequency over the Western North Pacific. *Journal of the Meteorological Society of Japan*, 87(3).

Zhang, Q., W. Zhang, X. Q. Lu, *et al.*, 2011. Landfalling Tropical Cyclones Activities in the South China: Intensifying or Weakening? *International Journal of Climatology*, 32(12).

Zhou, W., J. C. L. Chan, W. Chen, *et al.*, 2009. Synoptic-Scale Controls of Persistent Low Temperature and Icy Weather over Southern China in January 2008. *Monthly Weather Review*, 137(11).

Zhou, W., X. Wang, T. J. Zhou, *et al.*, 2007. Interdecadal Variability of the Relationship Between the East Asian Winter Monsoon and ENSO. *Meteorology and Atmospheric Physics*, 98.

Zhu, C. W., T. Nakazawa, J. P. Li, *et al.*, 2003. The 30-60 Day Intraseasonal Oscillation over the Western North Pacific Ocean and Its Impacts on Summer Flooding in China During 1998. *Geophysical Research Letters*, 30(18).

澳门环境保护局：“2017 年澳门环境质量状况报告”，2018 年，http://www.dspa.gov.mo/richtext_report2017.aspx。

澳门地球物理暨气象局：“2018 年澳门空气质量简报”，2018 年，https://www.smg.gov.mo/zh/subpage/189/report/iqa-report。

陈隆勋、周秀骥、李维亮等：“中国近 80 年来气候变化特征及其形成机制”，《气象学报》，2004 年第 5 期。

程乘、朱益民、丁黄兴等："中国东部地区夏季降水和环流的年代际转型及其与 PDO 的联系"，《气象科学》，2017 年第 4 期。
邓耀民："澳门气候的变化趋势预测"，澳门地球物理暨气象局，2008 年。
邓耀民、冯瑞权、梁嘉静等："澳门的气候变化情况与 21 世纪变化趋势"，珠三角气候变化和气候预报工作坊，2008 年 12 月 15～16 日。
丁一汇、孙颖、刘芸芸等："亚洲夏季风的年际和年代际变化及其未来预测"，《大气科学》，2013 年第 2 期。
冯瑞权、吴池胜、王婷等："澳门近百年气候变化的多时间尺度特征"，《热带气象学报》，2010 年第 4 期。
冯颖怡、陈兆伟、谭广雄等："香港户外二氧化碳浓度测量分析"，第 25 届粤港澳气象科技研讨会，2011 年 1 月 26～28 日。
谷德军、纪忠萍、林爱兰："影响南海夏季风爆发年际变化的关键海区及机制初探"，《热带气象学报》，2018 年第 1 期。
简茂球、彭敏、罗欣："南海及周边地区晚春初夏降水变异关联主模态及其机理"，《中山大学学报（自然科学版）》，2018 年第 5 期。
李春晖、潘蔚娟、李霞等："南海-西太平洋春季对流 10～30 天振荡强度对南海夏季风爆发早晚的影响"，《热带气象学报》，2017 年第 1 期。
梁延刚、胡文志、杨敬基："香港能见度、大气悬浮粒子浓度与气象条件的关系"，《气象学报》，2008 年第 3 期。
林爱兰、谷德军、郑彬等："南海夏季风爆发与南大洋海温变化之间的联系"，《地球物理学报》，2013 年第 2 期。
卢楚翰、管兆勇、李永华等："太平洋年代际振荡与南北半球际大气质量振荡及东亚季风的联系"，《地球物理学报》，2013 年第 4 期。
米金套："澳门城市景观格局变化与热岛效应研究"（博士论文），北京林业大学，2010 年。
邵勰、黄平、黄荣辉："南海夏季风爆发的研究进展"，《地球科学进展》，2014 年第 10 期。
邵勰、黄平、黄荣辉："基于海温异常的南海夏季风爆发的可预报性分析"，《气象科学》，2015 年第 6 期。
王磊、陈光华、黄荣辉："近 30a 登陆我国的西北太平洋热带气旋活动的时空变化特征"，《南京气象学院学报》，2009 年第 2 期。
伍红雨、杜尧东、潘蔚娟："近 48 年华南日照时数的变化特征"，《中山大学学报（自然科学版）》，2011 年第 6 期。
香港环境保护署："污染物排放趋势（1997 年～2016 年）"，2018 年，https://www.epd.gov.hk/epd/tc_chi/environmentinhk/air/data/emission_inve.html#1。
谢冰："短寿命气候污染物（SLCPs）的有效辐射强迫及对全球气候的影响研究"（博士论文），兰州大学，2016 年。

徐忆菲、彭丽霞、李季等：“1951～2013年我国冬季气温年代际变化与PDO的关系”，《气象科技》，2017年第4期。
杨玉华、应明、陈葆德：“近58年来登陆中国热带气旋气候变化特征”，《气象学报》，2009年第5期。
张莉萍、李栋梁、李潇：“气候变暖背景下南海夏季风建立和结束日期及与其强度的关系”，《热带气象学报》，2014年第6期。

第二章　气候变化的影响、风险和适应

香港和澳门是我国人口密集、经济发展蓬勃、城市化程度高的重要城市。港澳地区处于中国南部沿岸，位于珠江三角洲两侧、珠江流域出海口位置，受亚热带季风气候控制，降水丰富并具有显著的季节特征，台风、暴雨、洪水、热浪等自然灾害频发。港澳地区独特的地理位置条件与发达的社会经济状况，使该地区人民财产对这些自然灾害的暴露度高，灾害造成伤亡破坏的影响和风险高，严重威胁地区人民健康及生命财产安全。气候变化使灾害变得更加频繁和极端，对港澳地区灾害的预估、预防和应对造成巨大挑战，对环境生态品质和人群健康风险构成影响，为社会经济的长期可持续发展带来了不确定性。

港澳地区较早开始对当地天气状况进行系统、科学和持续的观测，例如香港天文台从 1884 年已经开始进行观测并积累了长期气候数据，为研究区域内气候变化提供了宝贵的基础。过去一百多年的观测显示，气候变化对港澳地区造成了广泛影响，降水、气温、台风、海平面等多方面的气候水文特征和事件已经发生了显著变化，并对灾害管理和城市安全造成深远影响，例如极端降水频发、降水强度增高、城市内涝风险增加。降水分布发生变化，为水资源管理带来挑战；热带气旋强度增加，风暴潮频发，海平面显著上升；高温和热浪事件上升。相关数据和研究也证明了气候变化对港澳地区环境生态和人群健康带来的影响，例如空气质量恶化、跨区域空气污染传输、近海环境富营养化和缺氧事件增加、传染性疾病和非传染性疾病增加等。由于港澳地区在社会经济方面的重要性，以及相关灾害影响的广泛性，不少研究对未来港澳地区可能的气候变化、影响和风险进行了科学预测及相关性分析。为适应和应对这些风险，港澳地区根据自身的自然地理条件和社会经济情况，制定和发展了具有港澳特色的适应预防策略和措施。因此，本章将关注气候变化在五个主要方面的影响、风险和适应对策：①洪水灾害与城市安全；②水资源与水安全；③台风灾害、风暴潮和海平面上升；④环境生态质量和保护；⑤人群健康、风险认知与适应策略。

第一节　洪水灾害和城市安全

一、洪灾和城市内涝

香港位处中国南部沿岸，紧邻中国南海（以下简称“南海”），全年雨量充沛，是太平洋周边地区降雨量最高的城市之一。香港年平均降雨量约为2400毫米，且降雨量时间分布极不均匀，历史记录单日降雨量最多可高达534毫米，最大小时降雨量可达145毫米。香港历史上发生过多起洪灾，其中以1972年“六一八”雨灾造成的人员死伤和财产损失最为严重，其原因正是香港连日暴雨，最终导致山洪暴发引发灾难。1891年，香港应对洪灾和保护香港免受水浸威胁的工作由当时的水务及渠务分部负责，其后经数次演变，于1989年成立了渠务署。渠务署自成立后，始终贯彻“防洪三招”，即截流、蓄洪、疏浚，在防洪工作中取得了巨大成就。渠务署于1994年提出了确定香港水浸黑点的系统方法，并推出相应的防洪工程，水浸黑点数目显著减少。由1995年至今，渠务署合共消除了126个水浸黑点，现在只有5个水浸黑点且针对性的改善工程也已分阶段启动，以期尽快消除。现阶段香港已经建立起完善的雨水排放系统，全港约有 2 400 千米地下雨水渠、360 千米人工河道、21 千米雨水排放隧道、4 个地下蓄洪池，大大降低了极端降雨可能带来的洪灾威胁。可见由于香港以往饱受洪涝灾害的影响，政府十分重视基础设施的建设，而随着城市排放系统的逐步完善，香港已经具备应对恶劣天气事件的能力。但是，如何应对气候变化，提高城市安全，推动可持续发展，仍然是目前香港面临的新挑战。

香港由于其特殊的地理位置，经常会受到热带气旋登陆的影响，与之相伴随的往往还有强降雨。例如 2016 年 10 月 19 日，受“莎莉嘉”影响，香港单日降雨量达 223.4 毫米；2017 年 8 月 27 日，受“帕卡”影响，香港单日降雨量达 165.3 毫米；2018 年 9 月 16 日，受“山竹”影响，香港单日降雨量达 167.5 毫米。对现有观测资料分析，很多研究发现，自 20 世纪 80 年代中期以来西北太平洋热带气旋的强度呈现出明显的上升趋势，强热带气旋所占比例不断上升（Elsner *et al.*, 2008; Kang and Elsner, 2012; Walsh *et al.*, 2016; Lee *et al.*, 2020）。同时，詹姆斯 • P. 科辛（Kossin, 2018）指出，全球范围内热带气旋的移动速度在过去 70 年里减慢了大约 10%，其中西太平洋地区为 16%，而在陆地上这一数值甚至达到了 21%。这意味着即使不考虑热带气旋的强度因素，热带气旋在陆地上停留的时间会变得更长，产生持续大降雨的可能性也会变得更高（Lai *et al.*, 2020）。这些数据表明，香港或将面临巨大的城市排水泄洪压力，一旦出现持续强降雨，可能会带来巨大的洪灾风险和不确定性。

气候变暖会导致大气的水汽容量增加，从而提高降雨概率（IPCC, 2013），这也是近年来香港极端降雨越趋频繁的因素之一。例如，2016 年香港总降雨量达 3 027 毫米，为近几年降雨量峰值，高出平均值 2 400 毫米约 26%。此外，全球温度升高也会导致热带气旋中心附近的降雨率增加（IPCC, 2013; Wuebbles *et al.*, 2017）。气候变暖势必导致海平面上升，加剧香港受到的风暴潮的威胁，维多利亚港自 1954 年后十大最高水位的记录中就有四次记录出现在近 10 年中（2008 年“黑格比”，3.53 米；2011 年“尼格”，3.25 米；2017

年“天鸽”，3.57 米；2018 年“山竹”，3.88 米）。海平面的上升也进一步加剧了香港的洪灾风险，特别是沿岸低洼地带。

IPCC 第五次评估报告指出，在气候变暖的大背景下，亚洲东部地区强降雨事件出现的风险将持续上升。1885～2018 年香港降水数据（图 1–2）显示平均降水强度同样呈现上升趋势（Ginn *et al.*, 2010）。以上结论预示了未来香港面临严峻的洪灾风险和防洪挑战。

二、雨洪管理与防洪对策

香港暴雨频发、降雨强度大，北部的城郊、乡村低洼地带、平原以及市区部分老城区易发生雨洪与内涝的灾害（Chan F. *et al.*, 2013; Chan *et al.*, 2014；Wong *et al.*, 2011）。又因为香港的地形以山地为主，大暴雨会引发严重的山体滑坡和泥石流等地质灾害（CEDD, 2020），造成基础设施破坏以及生命财产损失。

在 20 世纪 80 年代，由于城市发展迅速，城郊、乡村地区（包括洪泛平原）被高强度开发，大面积的天然土地变成了硬化的铺装地面，洪水汇流系数逐渐增加、洪水间隔缩短、洪峰增大。然而，部分老城区的排水系统建成时间久远，防洪标准已经不能满足设计标准（大部分标准低于 20 年一遇）（Chui *et al.*, 2006）。上述因素综合，导致香港老城区内涝频发。香港于 1989 年 9 月成立了渠务署，该机构自成立以来竭力提升香港的防洪水平，成效显著。近年来，渠务署根据防洪策略（包括“上游截流”“中游蓄洪”“下游疏浚”）实施了多项措施。并参考国际标准，把市区排水支渠及干渠系统的防洪标准分别提升至 50 年和 200 年一遇（DSD, 2019），相对于许多亚洲城市 1 年到 10 年一遇的防洪标准，香港的防洪标准在国内城市乃至亚洲城市中是较高、较安全的（Chan F. *et al.*, 2012）。

香港渠务署把香港分成了不同区域并编制了《雨水排放整体计划》，针对不同地区提出了不同的防洪方案（如修建截流渠、蓄洪设施和抽水泵等）。防洪工程采用的思路大致可归纳为截流、蓄洪和疏浚三个模式。在半山建造雨水排放隧道，截取中上游雨水经截流隧道直接排放入海或其他河道；在中游地区建造蓄洪池，将部分雨量暂存，从而降低下游洪峰流量；对原有河道进行工程治理或兴建排洪河道和渠道，以加强过流能力，提升防洪能力（DSD, 2019）。通过计划的实施，香港的城市洪水与内涝问题已得到显著改善。但是在气候变化与地理因素的综合影响之下，城市雨洪的问题依然存在且造成了严重的人员和财产损失。例如，2008 年 6 月 7 日和 2010 年 7 月 22 日的两次暴雨引发山洪，造成了两人死亡、5.78 亿港元经济损失（Greenpeace China, 2009）。

总的来说，导致香港城市洪水和内涝频发的原因主要有以下几点：①部分地区防洪标准较低引起的洪水与内涝问题。②排洪系统运行受阻引起的城市洪水与内涝问题。例如集水井、排水渠道等设施被废弃物、垃圾、树木、泥土、石块等阻塞，导致其不能发挥功能。③台风发生时风暴潮引发潮水位升高，导致城市洪水与内涝问题（Lee *et al.*, 2010）。例如港岛东区、九龙东、新界西北等沿岸低洼地带十数年前因海水倒灌引发多次内涝。

事实上，港澳地区现有的排水系统已经比二三十年前大为改善，防洪标准可以媲美海外发达城市如东京、伦敦等，在东南亚地区处于领先地位（Chan F. *et al.*, 2018）。政府与各部门也尽力解决市区内的雨洪问题，采纳了许多创新的改善方案，例如在中上游段采用雨水排放隧道以截取雨水，并在中游段建造地下蓄洪池暂时贮存汇入市区的雨水。该方案既缓解了下游排水系统的防洪压力，也可减低市区与低洼地区（例如香港的旺角、跑马地及湾仔等）的内涝威胁，成为国内乃至亚洲城市在管理城市雨洪和防治内涝方面的成功典范。

但是，因为大湾区的快速发展以及区域气候变化的影响，港澳地区现有的雨水排放系统在未来有再次优化升级的必要。香港特区政府采用的“蓝绿建设”与内地的“海绵城市”理念相近，用以规划城市建设，加强城市应对洪水的能力。地下蓄洪计划是香港“蓝绿建设”的重要成果，结合城市自然地理特征，优化防洪排涝通道和蓄滞洪区，修复自然生态系统，因地制宜推进海绵城市建设，增加城镇建设用地中的渗透性表面。此外，香港在绿化、活化河道和雨水回收利用等方面也充分体现了“蓝绿建设”的概念。在适应气候变化方面，香港已经有一定的基础，尤其是在应对城市内涝方面的工作方面，但仍需加强公共部门基建，应对气候变化带来的各种威胁。与香港比邻的澳门也在积极借鉴、采纳、实施多项防洪措施，例如 2017 年经历了“天鸽”重创之后，2018 年澳门政府立刻积极开展了内港雨水泵房及雨水涵箱渠的建造工程，加装活动式挡潮闸，设计标准为应对两百年一遇的洪水，同时要求低洼地区的公共停车场在悬挂八号风球或者黑色风暴潮之后一小时内要关闭，以防止洪水淹浸停车场造成的人身安全和经济损失。对于沿岸地区，会加建全长逾两千米、高约 1.5 米、可抵御 20 年一遇洪水的防洪墙，能抵挡台风 2008 年“黑格比”级别的风暴潮。但面对 2017 年 8 月的“天鸽”，因澳门对其预测失误，在香港发出十级飓风讯号前十分钟才悬挂八号风球预警，错失了最佳防洪救灾时机，造成了巨大损失。

三、山体滑坡和泥石流灾害

香港的陆地面积有超过 60%为斜坡和天然山坡。在众多类型的山体滑坡灾害中，由降雨导致大量饱和湿润土壤在重力作用下冲垮斜坡形成泥石流，可能是最难预测和最具有灾害性的。由于山沟的倾斜，泥石流可以移动相当长的距离（Ng *et al.*, 2013）。和世界上其他山区不同，港澳地区人口密集，一个小规模的山体滑坡或泥石流可以造成严重后果。

在全球变暖的大背景下，导致降雨事件的对流可用位能将会显著增加（Singh *et al.*, 2017）。相应地，山体滑坡风险也会增加，香港天然山体滑坡发生密度随标准化降雨强度的增加而呈指数上升。

下列部分于香港发生的滑坡、泥石流事件，反映出极端天气事件对社会经济的影响。幸运的是，这些极端天气没有在香港人口最密集的地方发生，影响人口密集地区的暴雨强度没有特别大。否则，滑坡和伤亡数目会增加更多（Ho *et al.*, 2019）。

1982 年 5 月和 8 月的暴雨引起了超过 1 400 个天然山体滑坡，造成超过 20 人死亡和寮屋区的重大破坏。1982 年 5 月 28 日至 31 日四天内降雨量达到 600 毫米，而 1982 年 8 月 15 日至 19 日降雨量达到了 520 毫米（Hudson, 1993; Lee, 1983; Wong *et al.*, 2006）。

1993 年 11 月的暴雨袭击了大屿山并引发超过 800 个天然山体滑坡以及超过 250 个人造斜坡坍塌，造成道路堵塞、居民需要疏散，24 小时内降雨量超过 700 毫米（Wong *et al.*, 1997; Wong and Ho, 1995）。

1994 年 7 月的暴雨导致了约 200 场滑坡，5 人死亡，4 人受伤，道路堵塞和、居民需要疏散（Chan, 1996）。从 7 月 22 日到 24 日，香港天文台录得降雨超过 600 毫米，并在大帽山录得破纪录的最大小时降雨量 185.5 毫米和 24 小时降雨量 954 毫米。

1999 年 8 月的暴雨引起超过 200 场人造斜坡滑坡，导致 1 人死亡和 3 座公共营房居民需要疏散。新界南部录得 24 小时降雨量超过 500 毫米（Ho *et al.*, 2002）。

2008 年 6 月的极端暴雨引发超过 2400 场天然山体滑坡和泥石流（GEO, 2016），北大屿山高速公路作为市区通往香港机场的唯一道路，被泥石流堵塞超过 16 小时（GEO, 2012）。

香港土木工程拓展署下辖的土力工程处采用了全面的斜坡安全系统来管理滑坡风险，现有系统已经被证明能有效应对暴雨情况。但极端降雨会为香港滑坡风险带来新的挑战，香港需要从策略、政策和技术角度来应对这些挑战（Ho *et al.*, 2019）。香港土力工程处已经从预防、应变和教育三方面做好准备。

传统滑坡灾害风险管理策略主要基于定量风险评估（Quantitative Risk Assessment, QRA），包括灾害及设施脆弱性、暴露性的预测。QRA 是量化预计损失的有效工具，但是应用 QRA 预测极端事件风险的一个挑战是极端事件的坍塌机制可能和普通事件不同，有造成隐含剩余风险的可能。气候变化使工程师认识到，使用过去的经验预测未来滑坡灾害的规模和频率，可能不再是可靠的方式。香港土力工程处的压力测试表明，极端事件会影响现有斜坡安全系统的完整性和表现。因此，香港土力工程处一直致力优化现行斜坡工程的设计标准和预防措施，强化斜坡抵御极端降雨的能力以降低发生严重山泥倾泻事故的概率，建造泥石屏障阻挡泥石流以降低其对毗邻设施的影响，这些举措有助于优化现有斜坡安全系统，从而提高应对未来极端气候的准备和应变能力。

减少滑坡灾害的最直接措施是建造物理屏障（Ng *et al.*, 2016a; 2016b; 2018a; Choi *et al.*, 2019）。但世界其他地区的传统防御性措施未必适用于港澳这类土地昂贵和人口密集的城市，这给港澳地区的设计提出了独特的挑战。2010 年香港特区政府推出了“长远防治山泥倾泻计划”，现在已有数以百计的屏障安装在香港山坡。这个计划一直稳健且可持续地研究山体滑坡的潜在危险，并在适当地理位置安装屏障，保护下游的重要措施。

由于土地短缺和自然环境保护的原因，人口密集地区在斜坡基部安装大型泥石坝（Lo, 2000）未必是最合适的方法（Ng *et al.*, 2016c; Ni *et al.*, 2018; Ng *et al.*, 2019a; 2019b）。相反，能够和周边自然环境相配的、相对较小的柔性防护网可能是更加可持续的选项（Ng *et al.*, 2018b; Song *et al.*, 2018）。尽管使用多层防护网是一个具有潜力的方案，但如何在山涧内优化对下层屏障的冲击负荷仍然缺乏科学指引。为构建科学指引，阐明防护网间溢流机制的基础研究是十分必要的（Kwan *et al.*, 2015）。香港土力工程处已在该方面开展研究，该项目的研究结果有助工程师设计应对泥石流的减缓措施。但是因为极端滑坡事件的潜在规模较大，单单依赖物理工程对策并不可行。

统计学习和深度学习已经被开发并成功应用于自然灾害预测。这些人工智能技术也可被应用于升级现有的滑坡预警方法。数据驱动下人工智能和机器学习的模型建立需要高质量数据来支持预测，而香港拥有丰富和高质量的降水和滑坡数据。现有的降水和滑坡历史数据对建立和训练深度学习模式，构建滑坡模式和极端降雨事件之间的关系至关重要。但是，正如上文指出，历史数据可能不足以用于预估极端事件。当前只能继续从覆盖范围广的自动雨量站来获取香港的实时数据并继续为滑坡预警模型分析降水与滑坡的关系，但同时也应注意采用历史数据方法的潜在风险。

预测气候变化带来的影响存在许多不确定因素，因此不能完全依靠工程方案应对极端降雨引致的滑坡风险，提升政府的紧急应变能力同样重要。香港土力工程处与香港天文台共同负责管理山泥倾泻警报系统。一旦预测到暴雨可能导致山泥倾泻，香港天文台便会发出警告，以提醒市民注意暴雨期间的山泥倾泻危险，而发出该警告也会启动各政府部门的紧急应变系统，以便迅速调动资源处理事故。

此外，公众关注和掌握相关知识可提高社会应对气候变化导致滑坡威胁的韧性。因此，决策者、科学家和工程师需要采用崭新的方式和公众沟通。例如教学课程中引进增强实景、虚拟实景和可视化技术，让公众有机会置身于滑坡情景中体验其破坏力，从而增强公众对滑坡灾害的认识。香港土力工程处在加强香港斜坡安全系统的公众教育与信息沟通方面做了很多努力（Tam and Lui, 2013）。

综上所述，现有港澳地区气候情景的预测，明确支持了相关结论：降雨量增加因而极端事件发生频率也将增加，随之滑坡风险也会增加且急需管理。香港土力工程处已经检视了气候变化降雨情景下现有的滑坡紧急管理系统。该检视展示了极端滑坡的潜在规模和现有斜坡风险系统需要改善的方面。更重要的是，明确了不能单单依靠物理工程措施来管理极端降雨风险，因为这些物理工程措施往往非常昂贵。加强应急准备、设计应对灾害体系、恢复受影响设施的能力是应对极端天气带来滑坡灾害的重要策略方针。香港已经收集了丰富的降雨和滑坡数据，并增加了公众教育以提高社区对极端滑坡灾害的应对韧性。

第二节 水资源和水安全

香港受亚热带季风气候控制，温暖潮湿、降水充沛，平均年降雨量约 2 400 毫米，但降雨量年内的季节分布不均匀（80%以上集中在 5～9 月的雨季），年际变化较大，大部分的雨水会经排水系统进入大海。由于受地形地质条件和城市人口集聚等因素的影响，香港本地水资源相当缺乏。地形方面，香港特区约 1 100 平方千米的陆地中近 80%为山地丘陵，缺乏低洼平地和大江大湖，充沛的雨水很难被收集和利用。地质条件方面，香港分布最广泛的是不透水花岗岩和火山岩，地下水补给困难，没有可观的地下水存储可供利用。香港的年均总水资源量仅约 10.8 亿立方米，人均约 140 立方米，远远不能满足本地需求，属严重缺水地区。第二次世界大战以后，随着香港人口的增加，缺水问题日益严重，停水成为 20 世纪 50、60 年代困扰香港社会的常态问题，特别是 20 世纪 60 年代初遇上世纪大旱，最严重的 1963 年 6 月，政府实施 4 天供水 4 小时的停水措施（沈灿、季冰，1997）。香港自 1965 年起从东江引水至深圳水库并入境香港，自 1982 年，香港便享有 24 小时无间断供水（香港水务署，2011）。目前东江水占香港食水供应量的 70%～80%，总用水量（包括海水冲厕）的 50%，已成为香港供水的主要来源，也是目前资源、环境和经济效益最优的重要水源（香港水务署，2018）。过去半个多世纪香港依靠三大水源，即东江调水为主，收集本地雨水为次，尽量使用海水冲厕为辅，成功地确保了供水安全和支持可持续发展。

气候变化下海平面上升、降水规律变化，加上人类活动等因素，使得咸潮上溯的风险增加，对澳门的供水安全构成威胁（徐爽、侯贵兵，2017）。2004～2005年枯水期咸潮上溯，影响澳门、珠海等珠江三角洲城市供水，因此我国首次实施了全流域应急调水，通过流域上游的主要水利枢纽联合调度，实现调水压咸、缓解灾情（全球水伙伴中国委员会，2016）。咸潮上溯对澳门等珠江口沿岸城市的供水安全构成重大威胁，相关部门针对这一问题做了大量工作，取得一定成果。

认识气候变化对本地水资源量的影响，是香港、澳门应对未来气候变化及人口经济快速增长带给长期稳定供水压力的关键。因此，本节分为三部分：第一部分主要讨论东江水资源变化以及跨界调水政策；第二部分主要讨论气候变化下香港本地水资源量的响应，以及维持供水安全实施的措施；第三部分主要讨论气候变化下咸潮上溯对澳门供水的影响以及保持供水稳定采取的适应策略。

一、东江水资源与跨界调水

东江流域受亚热带季风气候控制，降水具有显著的季节特征，夏季降水可占全年的72%～88%，流域径流的补给主要来自降水（Li J. *et al.*, 2016b）。过去多年的观测资料和模型模拟揭示了东江流域水循环要素对气候变化的响应：年降水变化不显著，春冬降水增加、夏秋降水减少，降水天数显著下降（Chen *et al.*, 2011; Liu *et al.*, 2015; 周平等, 2016）；气温显著上升，蒸发皿观测显著下降（Liu *et al.*, 2010）；秋季土壤含水量显著下降，其他季节变化不明显（Zhang *et al.*, 2018）。在气候变化及人类活动（如取水、水库调节、土地利用变化等）的共同作用下，博罗等主要水文站多年观测到的年径流量变化不显著，而枯水期流量明显增加（Chen *et al.*, 2012; 周平等，2016; Wu *et al.*, 2019）。涂新军等（Tu *et al.*, 2015）分析认为水库水量调度降低了东江径流量年际波动的33%，新丰江、枫树坝和白盆珠水库调度对年际波动降低的贡献率分别是 21%、10%和 2%。枯水期的降水增加、潜在蒸发下降、水库季节性调度是东江枯水期流量增加的主要原因（Wu *et al.*, 2018）。

CMIP5 全球气候模式输出显示，东江流域年降水量未来将变化不大或轻微上升（相对变化 10%以内），气温和蒸散普遍上升，但不同全球气候模式间未来的降水预估差异较大（Sun *et al.*, 2015; Yan *et al.*, 2015; Li J. *et al.*, 2016a; Wang *et al.*, 2017）。东江水资源量的未来预估一般采用全球气候模式的未来气候情景作为输出驱动，利用陆面水文模型模拟流域水文过程，推求未来径流量。基于五个 CMIP5 全球气候模式输出驱动 VIC（Variable Infiltration Capacity）水文模型，闫丹等（Yan *et al.*, 2015）认为 2079～2099 年流域枯水期流量将比 1979～1999 年下降，RCP8.5 的下降情况将比 RCP4.5 更明显。由于东江径流量以雨水补给为主，全球气候模式降水输出差异将是水资源量未来预测的主要不确定性因素（Li J. *et al.*, 2016a）。目前水资源量相关变化预估主要基于降尺度统计，李剑锋等（Li J. *et al.*, 2016b）指出降尺度后气候因子日际变率如果改变，即使月或年平均值不变，也可能导致日间流域水文过程改变，该效应会不断累积从而导致水资源量等陆面水文要素长期变化，成为水资源量未来预测的不确定性来源之一。因此，预测东江流域水资源量在未来气候变化下的改变，全球气候模式、陆面水文模型、降尺度方法、未来气候情景等因素是重要变量。

1960～2005 年东江调水方案主要基于固定供水协议（香港立法会秘书处，2017）。鉴于降水的变化以及香港实际用水需要的波动，为减少水资源溢流情况，香港从 2006 年开始采用“统包总额”方式购入东江水，按实际需求弹性输入东江水（上限为每年 8.2 亿立方米），该上限考虑了百年一遇的极旱情况，确保了供水的 99%可靠性（香港水务署，2018）。东江流域是香港、深圳、东莞等城市的主要供水水源，总供水人口达到 4 000 多万，人均水资源量只有 1 100 立方米/年，根据国际标准属于缺水地区。为缓解区域内城市水资源的供需矛盾，确保枯水期的供水安全及河道内生态基本需水，2008 年广东省开始实施东江流域水资源分配方案，香港的用水配额被落实在方案中（涂新军等，2016；王雨，2017）。随着流域内快速人口增长和经济发展，东江水资源已经被高度开发利用，深圳等城市 2020 年的预计用水量将超出分配方案的供水额度，导致用水缺口（香港立法会秘书处，2015；深圳市水务局，2016）。因此，2019 年珠江三角洲水资源配置工程（“西水东调”）开始施工，从西江向珠江三角洲东部区域引水，为东部城市提供应急备用水源。为应对未来水资源量变化的不确定性和用水需求增加的挑战，香港水务署 2008 年发布了全面水资源管理策略，并于 2019 年对该策略完成了检讨，肯定了多元化供水管理的进展，预测香港到 2040 年的水资源需求及供应情况，对未来供水可能遇到的挑战做准备（香港水务署，2008；2018；2019）。同时，东江径流量受新丰江、枫树坝和白盆珠水库的高度调节，未来变化预估及相关适应措施需考虑水库调节的作用。

二、本地水资源与供水安全

过去几十年来，稳定充足和清洁可靠的淡水供应对香港社会的繁荣稳定和经济发展至关重要。香港采用“开源与节流并举，本地与外地水源并用”的“两条腿走路”策略和“软硬兼施”手段，非常成功地解决了城市供水需求，维护供水安全（Chen, 2001）。其中不少水资源开发和利用措施因地制宜，极具特色，在国际上甚至独一无二，备受关注（Yue and Tang, 2011）。在开源方面，政府在 20 世纪城市发展初期兴建水库，将大约三分之一的土地（约 300 平方千米）划为集水区，过去 20 年每年收集的雨水量介于 1.03 亿～3.85 亿立方米之间，可为香港提供 20%～30%的淡水资源。香港的集水区大多位于受高度保护的地区，例如郊野公园。集水区的发展也受到限制，以确保水质不受污染。其次，政府自 20 世纪 50 年代起实施利用海水冲厕的创新方案，建造了全世界最大的双管网城市供水系统。近年来海水冲厕网络覆盖了全港约 85%的人口，每年可节约淡水约 3 亿立方米（约占 2019 年总用水量的 24%）和大量能源（单位海水耗电仅为淡水的三分之二），是全世界唯一全面使用海水冲厕的城市。另外，香港也正推广海水淡化及循环再用水（即再造水、中水重用和雨水回收作非饮用用途）作为新的水资源等。在节流方面，软性措施主要包括采用阶梯式水价和加强宣传，从经济调节和教育引导上鼓励节水，以控制用水需求增大。硬件措施主要是在 2000～2015 年完成了约 3 000 千米老化水管的更换和修复计划，使香港的水管渗漏率从超过 25%降低至 15%。

香港本地水资源严重不足且年内年际波幅极大，引入外地水源是确保供水安全的最主要措施。在经历 20 世纪 60 年代初的大旱缺水后，香港从 1965 年起输入东江水，每年输港水量从 20 世纪 60、70 年代的几千万立方米增长至 20 世纪 90 年代以来的 6 亿～8 亿立方米，东江水占淡水总用水量的比例也从早年的不足 30%增长至近三十年来的 70%～80%。香港与广东省自 1960 年签订第一份供水协议起至 2017 年，两地政府按香港的用水需求一共签订了 11 份供水协议。东江水的输入使香港自 1982 年 6 月起再也不用经历间歇性停水。自 2006 年起，东江供水协议采用“统包总额”的方式，约定每年香港向广东支付一定的总金额从而可获得最多 8.2 亿立方米的东江水，以确保香港在百年一遇的极旱情况下仍能维持全日供水（香港立法会秘书处，2015）。

香港拥有全世界最安全可靠的供水系统之一。水务署负责各种水资源的收集、饮用水和海水的处理和分配供应。供水系统为全港人口提供符合世界卫生组织标准的饮用水，覆盖范围包括人口密集的港九市区、卫星市镇和居民人数较少的乡郊地区。另外，冲厕海水供应系统也不断扩展，已经覆盖大部分市区及新市镇。香港水务署已有近 170 年的历史，政府多年来不断建造、扩充和维护庞大的供水基础设施，以确保香港的供水安全。目前香港的水务设施主要有：①存贮东江和本地原水的两座海上水库和 15 个传统的水库，其中船湾淡水湖是全球第一座在海上兴建的水库。全港 17 个水库的总储水量约为 5.86 亿立方米，相当于全港约六个月的用水量；②引水道、隧道等本地雨水收集系统；③每天可处理 468 万立方米水量的 20 间滤水厂；④超过 190 个不同大小的原水、饮用水和海水抽水站，超过 220 个用于短暂存储饮用水或海水并辅助控制供水水压的配水库以及超过 8 000 千米的配水管，以将饮用水和海水输送给用户（香港水务署，2018）。

过去半个多世纪以来，香港依靠三大水源，即东江调水为主，收集本地雨水为次，尽量使用海水冲厕为辅，成功确保了供水安全、支持可持续发展。随着气候变化对水资源的影响加剧，人口增长与经济发展对水量需求的增加以及粤港澳大湾区城市对区域水资源的竞争，香港的水资源管理和供水安全面临着日益增长的挑战。有鉴于此，水务署在 2008 年推行了全面水资源管理战略，旨在整合多年来富有成效的开源与节流并举、本地与外地水源并用的多渠道、多手段水资源开发利用对策，以确保香港供水的长期稳定和安全（香港水务署，2008）。该策略兼顾用水需求管理和供水管理，重点是"先节后增"，强调节约用水，在控制用水需求增长的同时积极开拓新水源。香港水务署于 2019 年完成了对该策略的检讨，认为用水需求和供水管理取得预期进展，并考虑气候变化影响，进一步预测了至 2040 年的用水需求及供应量（香港水务署，2019）。该报告指出，经过相关举措，过去十年香港全年食水用量控制在约 10 亿立方米，人均食水用量由 1999～2008 年平均每年每人 140 立方米下降到了 2009～2018 年的 133 立方米。

香港的主（东江）、次（本地）两大淡水水源均受到气候变化的影响。珠江流域三大支流中，东江最小，流域人均水资源量最低（仅 1 100 立方米/年，按国际标准属缺水地区），但作为广州、深圳、东莞、惠州、河源等地的主要供水水源，东江同时担负着对香港供水的重要任务，总供水人口超过 4 000 万，如今流域水资源开发利用程度已接近国际公认的 40%的上限。如前所述，东江水资源对香港的重要性不言而喻，气候变化对其影响也十分深远。而香港本地水资源及其管理受气候变化的影响主要集中在以下两个方面：一是本地降雨量和时空分布在气候变化下如何改变并影响雨水收集的效率和水量；二是水库水面蒸发随着气温升高而增加对贮水的影响如何。气候变化对香港本地水资源利用管理造成的影响仍需进一步开展调查研究。

香港与东江流域相邻，受到同样气候条件控制，在干旱年份两地往往同时面临缺水的压力。为了适应气候变化、增加本地水源量及其多样性，全面水资源管理策略提出将香港供水的三支水源增加至五支，即在现有水源的基础上增加海水淡化及循环再用水，相关设施已在有序建设及推行（香港水务署，2008；2019）。在海水淡化方面，政府已经在将军澳着手兴建使用逆渗透技术的海水淡化厂，该厂第一阶段产能可满足香港 5%的总淡水需求，必要时可扩建至 10%的供水量，预计 2023 年建成并投产。在再造水方面，政府自 2006 年起已推行再造水试验计划，先后在昂平和石湖墟等地收集污水，经处理和再造用作非饮用用途。水务署准备于 2023 年开始分阶段在新界东北供应再造水用于冲厕及其他非饮用用途，相关工作正在进行中，预计区内最终再造水总用量可达到每年 2 200 万立方米，是一项既节水又环保的措施。在中水重用和雨水回收方面，政府已制定相关的计划，在新建公营工程项目中安装相关设施。这些项目目前已完成了超过 100 个并计划进一步扩展，其中最受关注的是《安达臣道石矿场发展计划》将建造一个中央中水重用系统，项目预计在 2023 年完成。以上两支新水源都是重要的非传统水资源，均不受气候变化的影响，对增加香港供水安全的气候适应性和韧性非常重要。除此之外，政府已经开展另外两个主要项目，以减少水资源浪费、增强香港的供水安全。一是“智管网”计划，在供水管网中安装感应器以持续监测管网的状况，通过智能技术将管网渗漏减低。二是为了增加本地雨水收集和减轻水浸风险，水务署和渠务署正在合作推行一个水库间转运隧道计划，减少雨季水库溢流从而增加本地集水的有效收集（香港环境局，2015；2017）。

三、咸潮上溯和供水稳定

澳门受亚热带海洋性气候影响，降水充沛，年平均降雨量超过 2 000 毫米，但因面积小，缺乏主要河流，本地淡水资源非常短缺，澳门 96%的使用水源来自西江，由珠海供应（于洋等，2014；海事及水务局，2021）。本地蓄水设施的有效蓄水量为 190 万立方米，主要用于储水和应急调水（海事及水务局，2017）。西江下游的磨刀门水道是澳门和珠海的主要水源地，位于西江的出海口位置。冬春枯水期西江上游来水较少，磨刀门水道容易受海水倒灌影响，原水咸度增加，可能影响澳门的供水安全及稳定。

气候变化下，降水规律也会发生变化，影响枯水期上游来水，同时海平面上升。多种因素共同作用下，海水倒灌、咸潮上溯的风险增加，并威胁澳门的供水安全。而随着澳门的经济发展、人口增多、旅客量增加，澳门的用水量也逐渐增加，对水资源供给带来进一步挑战（于洋等，2014；徐爽和侯贵兵，2017）。珠江降水偏少、海平面上升是引发 2009 和 2010 年枯水期磨刀门水道咸潮上溯的主要原因，相比以往的咸潮入侵，该次咸潮发生的时间提早、咸界上移（孔兰等，2011）。模型模拟显示，珠江口咸潮上溯对海平面上升具有一定敏感性，海平面增加 0.1 米即可导致咸潮状况进一步恶化（Yuan *et al.*, 2015）。因此，未来气候变化下，降水规律进一步变化，海平面进一步上升，咸潮风险将会继续增加。

为应对枯水期的咸潮上溯，澳门及内地水利部门采用多种适应方案，确保澳门供水安全（海事及水务局，2017）。枯水期间珠江流域进行枯水期水量调度，流域的主要水库、泵站等水利设施联合调度，调水压咸，控制供澳原水咸度，积累了相当的抗咸经验。珠海市的西水东调系统保障咸潮期间澳门和珠海的供水安全。同时，有关机构加大了针对咸潮上溯的科研力度，在相关规律机理、预警预报、抑咸防咸等方面积累了相当的研究基础和实践经验，为澳门及珠江三角洲城市的供水安全提供了扎实的科研基础（全球水伙伴中国委员会，2016）。

第三节　台风灾害、风暴潮和海平面上升

一、台风灾害

台风指发生在热带海洋面上，中心风力达到 118 千米/时的热带气旋，其主要生成地区为海水温度较高的西北太平洋及南海地区。过去 50 年（1961～2010 年），每年影响港澳地区的热带气旋约为六个，主要集中于 6～10 月。台风形成后，其中心附近风速可达 240 千米/小时，可导致树木倒塌，引起暴雨、风暴潮等直接灾害，并可能引发泥石流、滑坡等多种次生灾害，有时甚至造成直接人员伤亡（见本章第一节）。

此外，台风来临时常引发极端高温天气、空气质量恶化等情况。台风周边常出现下沉气流，易诱发闷热高温天气（Lee *et al.*, 2015）。2017 年 8 月 22 日，受超强台风“天鸽”影响，香港多处气温上升到 37 摄氏度或以上，创历史新高。台风还可能导致空气质量恶化，包括能见度降低、空气污染物浓度上升等。研究发现，台风抵达香港前（香港天文台发出 1 号戒备信号时），香港地区的 $PM_{2.5}$、PM_{10}、SO_2、NO_2 的浓度将分别增加了 26%、28%、46%、17%（Luo *et al.*, 2018）。当台风途经台湾附近时，珠江三角洲地区向香港地区空气污染物的跨境传输会加强，香港当地空气下沉，能见度下降，臭氧等空气污染物浓度显著增加（Wei *et al.*, 2016; Chow *et al.*, 2018; Lam Y. *et al.*, 2018）。

从长时间尺度看，关于台风年频数的变化趋势尚没有统一的结论，不同观测资料的分析结果出入较大（Lee *et al.*, 2012; Walsh *et al.*, 2016; Lee *et al.*,

2020)，登陆亚洲东海岸的台风数量也没有明显的变化趋势（Chan and Xu, 2009）。但是，近年来登陆台风的强度明显增强（Mei and Xie, 2016），造成的经济损失增大（Wang Y. *et al.*, 2016）。例如有研究估算，2017 年的超强台风“天鸽”在受其影响地区导致的直接经济损失超过 35 亿美元，造成至少 22 人死亡，农田受损 5 841 公顷（Benfield, 2018; HKO, 2019）。2018 年 9 月 16 日，超强台风“山竹”袭击了香港、澳门、广东等地区，横澜岛的 10 分钟平均风速达到 180 千米/时，有 450 人受伤，超过 6 万棵树木倒塌，“山竹”还导致 800 多个航班取消，超过四万户供电受到影响（Choy *et al.*, 2020a; HKO, 2020）。

在预警和应对台风方面，港澳地区民众有一定的能力，尤其是香港特区政府相关部门积累了相当丰富的经验。香港天文台会根据本地实时及预测风力情况尽早发出三号强风信号，教育局会通过传媒及教育局网站宣布所有幼稚园、特殊儿童学校停课。如果需要发出八号烈风或暴风信号，香港天文台会在信号发出前约两小时内通过多种渠道发布预警信息告知市民，并呼吁市民在烈风吹袭前完成所有预防措施。有关应急部门会采取应变措施，提醒受影响的居民，有需要时会进行疏散及救援工作。在八号烈风或暴风信号发出后，教育局会宣布所有学校停课，民政事务总署则会开放临时庇护中心等。例如，为应对 2018 年超强台风“山竹”可能造成的威胁，香港特区政府及早启动跨部门应急，并由保安局协调主持了包括特区政府民政事务总署、房屋署、路政署、警务处、新闻处等 30 多个部门代表出席的跨部门会议，制定紧急应对计划，强化协调及信息互通工作，并指示紧急救援部队及时部署应变（Choy *et al.*, 2020b）；澳门方面则由特区行政长官主持民防会议，提前做好防风措施，尽可能降低台风灾害的影响。

未来在全球变暖的背景下，即使升温幅度控制在《巴黎协定》拟定的 1.5 摄氏度或 2.0 摄氏度的目标下（相对于工业化前水平的全球平均温度升幅），强台风事件的频数及强度都可能继续增强（Huang P. *et al.*, 2015; Knutson *et al.*, 2015; Tsuboki*et al.*, 2015; Walsh *et al.*, 2016; Lok and Chan, 2018; Yang *et al.*, 2018; Wehner*et al.*, 2018; Cha *et al.*, 2020），港澳地区将面临更严峻的台风灾害威胁。由于未来台风预测还存在一定的不确定性，也应加强不同社会发展和气候变化情境下台风风险的预测研究。在日益加剧的强台风威胁下，港澳地区需要进一步提高台风的预警能力，并及时告知民众、有效协调相关部门，减少台风灾害带来的各类损失。

二、海平面上升和风暴潮

海洋受热膨胀和陆地冰川融化会导致全球平均海平面上升。在全球平均气温上升的背景下，根据 SROCC 的推算，在 RCP8.5 情景下，2091～2100 年香港及邻近水域的平均海平面高度（包含本地地壳垂直运动的影响）会较 1986～2005 年的平均值高 0.73～1.28 米。平均海平面的上升趋势预计会持续到 2500 年以后。香港及澳门潮位站的观测表明年平均海平面呈显著上升趋势，而且年平均海平面的上升速率也在增加（Wong *et al.*, 2003; He *et al.*, 2016; Wang L. *et al.*, 2016）。

历史上，热带气旋及其引起的风暴潮曾对香港造成严重破坏和伤亡（例如 1874 年、1937 年、1962 年）。随着海平面上升，这些极端风暴潮的重现期进一步缩短（Lee *et al.*, 2010; Yu *et al.*, 2018）。例如，基于过去资料计算的超过海图基准面 3.5 米的极端风暴潮事件的重现期是 50 年（Lee *et al.*, 2010），但是超强台风天鸽和山竹导致超过 3.5 米的海平面事件连续发生在 2017 和 2018 两年，而目前的重现期已缩短一半（少于 25 年）。这表明了气候变化下风暴潮及极端洪水灾害的威胁正在上升。

潮位变化、天气和地理等因素的综合作用下会导致香港、澳门的海岸侵蚀和洪水灾害。首先，最大季节潮位在约 5～6 月和 10～1 月出现，5～6 月同时也是夏季前中尺度系统的频发时期（Luo, 2017）。这些中尺度系统通常导致珠三角地区水位上升和河道径流增加，容易诱发洪水。其次，大约在同一时段，高空东风急流形成和垂直风切变减弱，导致南海和西太平洋的台风开始发展（Ding and Chan, 2005）。部分热带气旋经过港澳地区并显著提高海平面高度从而导致风暴潮（Yin *et al.*, 2017）。最后，由于回水效应和漏洞形出海口，将会共同作用导致水位升高几米（Tracy *et al.*, 2007）。随着平均海平面的平稳增加，极端洪水事件重现期减少，海岸侵蚀和洪水灾害的潜在风险增加（Yu *et al.*, 2018）。

IPCC 第四次报告第二工作组已经指出，珠三角等亚洲大型三角洲灾害风险的增加具有非常高置信度，因为这些地区人口密集，高度暴露于海平面上升、风暴潮和河道洪水等灾害面前。由于人口经济发展集中，相关灾害对香港澳门造成破坏的风险很高（Tracy *et al.*, 2007）。因此，香港特区政府正推进一系列与气候变化相关的研究以应对未来海平面上升与风暴潮所带来的影响。

最近，基于达特茅斯洪水观测数据库（Dartmouth Flood Observatory）开发和维护的全球大型洪水事件活动集（Global Active Archive of Large Flood Events）报告了洪水破坏情况数据和 IPCC 第五次报告的全球海平面上升预估，有学者（Yu *et al.*, 2018）开展了气候变化下海平面上升对人类影响的定量研究。该研究基于 IPCC 数据，预测到 2100 年全球平均海平面将升高 40～80 厘米，目前 100 年重现期事件的重现期将减少到约 1.5 年（大概三年发生两次）。如果没有长周期的应对措施，到 2100 年前，香港可能每年面对 15～20 人的伤亡，2 万～10 万的人口转移，以及 1 亿～8 亿港币的经济损失。这些数字是参考 1989 年的一宗 30 年一遇的水浸事件，并假设没有采取任何长期适应措施所推算出来的。事实上，香港特区政府在过去 30 年来一直致力舒缓沿岸低洼地区，例如大澳、鲤鱼门的水浸风险。在 2018 年超强台风“山竹”吹袭香港期间，香港并没有死亡纪录。

超强台风“天鸽”和“山竹”分别于 2017 年和 2018 年袭击香港和澳门，

造成了重大破坏和海岸侵蚀，预演了未来极端洪水和风暴潮事件带来的严重风险。尽管“山竹”是更强的风暴，澳门的伤亡人数从 2017 年“天鸽”的 12 人显著减少到 2018 年“山竹”的零伤亡，这表明采用智慧的、有针对性的应对措施来减少城市的暴露度、脆弱性和风险的重要性。

沿海洪水和风暴潮事件能从多方面产生影响，包括海岸侵蚀，海水入侵影响供水系统，堵塞排水系统，土地损失导致海岸线改变（Ayyub *et al.*, 2012），交通阻碍（Dawson *et al.*, 2016），影响地下铁路、停车场、备用电力系统等对洪水非常敏感的地下设施（Tollefson, 2013），改变生态敏感地区的系统组成（Tracy *et al.*, 2007）等。需要更多针对本地的研究来定量和定性的探讨海平面上升、沿海洪水和风暴潮如何对不同行业造成影响，从而能够规划和建设更多有针对性的、智慧地观测，设计适应措施来应对可能的威胁，进一步减少城市的风险，保护大众。

香港和澳门并非能够自给自足的大都市，需要依赖周边环境来生存（Lin, 2014; Sharifi and Yamagata, 2014）。这两个城市依靠珠三角的广州、珠海和中山等邻近城市提供食物、水源和能源补给（Huang *et al.*, 2004）。因此，即使洪水没有正面袭击港澳地区，风暴潮事件如果在任何周边城市引发长时间洪水，也会对港澳地区的供应和社会经济链造成严重干扰和影响。需要建立更好的区域协作规划与应急协议使区域间城市能够互相协调、提供支持，从而进一步提高区域内城市群的整体适应能力（Leichenko, 2011）。

第四节　环境生态质量和保护

一、城市高温与热浪

香港和澳门受夏季风与西北太平洋副热带高压系统的影响，高温天气频发。同时，受到厄尔尼诺南方涛动等影响，华南地区的高温热浪呈现一定的年际涛动特征（Qian C. *et al.*, 2018; Luo and Lau, 2019）。但是从长期尺度上看，港澳地区平均气温、高温天数、热浪等高温风险正日渐加剧。

除气候变化外，城市热岛效应也会加剧城市的高温与热浪风险。城市热岛是指城市地区温度往往高于周围农村地区的现象（Oke, 1982; Giridharan *et al.*, 2004）。由于港澳地区人口密度大、能耗高，城市建筑物密集，城市通风不足，热岛效应非常突出（Siu and Hart, 2013; Peng *et al.*, 2018）。一般来说，香港地区冬季的热岛效应强度高于夏季，差异可能超过 2 摄氏度，热岛强度最大值通常出现在夜间，特别在日出之前，而白天则可能出现负值（Shi *et al.*, 2011; Wang W. *et al.*, 2016）。热岛效应导致香港的城市地区在 1970～2015 年平均温度比农村地区高约 0.87 摄氏度（To and Yu, 2016）。随着持续发展的城市化进程，香港地区的热岛强度可能继续增大（Chen and Jeong, 2018）。

香港与澳门作为人口与建筑物密集的发达城市地区，极易受到高温热浪的影响，一旦遭受高温热浪侵袭，造成的经济损失和人员伤害将十分巨大。高温热浪对人体健康影响很大，容易诱发传染病、心血管疾病、呼吸系统疾病等（Gasparrini *et al.*, 2015; Yi and Chan, 2015; Qiu *et al.*, 2016; Sun *et al.*, 2016, 2018; Tian *et al.*, 2016）。2010～2016 年，香港地区由炎热天气引起的住院人数明显增加（Sun *et al.*, 2019）。当气温高于 28 摄氏度时，每升高 1 摄氏度，死亡风险将增加 2%（Chan E. *et al.*, 2012）。特别是连续的高温事件对人群健康的影响更大，如果连续五天以上出现高温夜晚，人群死亡率风险将增加 7.99%（Ho *et al.*, 2017），而女性和老年群体应对极端干旱事件更加脆弱（Wang *et al.*, 2019）。此外，高温热浪对香港、澳门地区的环境、生态、经济等方面也具有显著影响。高温容易加剧香港地区高浓度的臭氧污染（赵伟等，2019）。高温天气也会增加劳动力成本，温度每增加 1 摄氏度，直接工作时长百分比会降低 0.33%（Yi and Chan, 2017）。随着气候继续变暖，日益增多的高温事件可能加剧电力能源消耗，气温每升高 1 摄氏度，香港地区的电力消耗将增加 4%～5%（Jovanović *et al.*, 2015; Ang *et al.*, 2017）。

目前香港特区政府已采取了一系列措施，试图减缓并积极应对城市高温与热浪风险。例如绘制城市气候环境图可以帮助各部门了解城市热环境的状况特征，辅助政府设立通风标准，改善城市热舒适度。2014 年香港特区政府推出的《可持续发展建筑设计指南》，计划通过街道布局的定向将开敞空间连为一体，设立非建筑退让区来减弱屏风效应，减缓热岛效应及温度继续升高，加强建筑物四周空气流通，以改善香港的建筑环境等（Ren *et al.*, 2011; 任超等, 2012）。港澳气象部门通过电视、广播、网络等多种途径及时发布高温预警，使城市人群特别是脆弱人群能够及时获得气象信息并采取措施进行自我健康防护（Lee *et al.*, 2015）。此外，香港和澳门地区都在积极推动利用可再生能源进行发电，以减缓气候变化的影响（Song *et al.*, 2017）。香港特区政府正鼓励采用水冷空调系统及更多的再生能源技术，提高能源效益和降低耗电量（Wang Y. *et al.*, 2018）。除此之外，香港也推进加种城市树木并通过蒸腾作用进行有效降温（Tan *et al.*, 2016; Kong *et al.*, 2017）。另有研究表明，使用相变材料建造屋顶可以有效降低建筑物温度达 6.8 摄氏度（Yang *et al.*, 2017），与密集型绿色屋顶相比，粗放型绿色屋顶更经济，而且降温效果更佳（Peng and Jim, 2015）。

二、空气质量与跨界输送污染

气候变化通过不同的途径影响空气质量。例如，区域气候变化可使地面风速减弱，直接导致污染物在区内聚积，继而影响区域的空气质量。除了直接影响外，气候变化还会间接影响空气污染物的排放。气候和空气质量之间有着很强的相互作用，已有的科学研究已证明空气污染物可以通过反馈过程来影响气候。

空气污染受天气/气候条件的强烈影响，因此对气候变化很敏感。IPCC

预测，未来城市的空气质量将会持续恶化（IPCC, 2014），并可能是因为反气旋天气的增加，例如迈克·休姆等（Hulme *et al.*, 2002）研究发现气候变化会严重影响香港的空气质量。严鸿霖等（Yim *et al.*, 2019a）采用统计方法分析了 2002～2016 年的 4 次 ENSO 事件和 20 次热浪事件中，当地排放源和跨界空气污染输送对香港颗粒物、二氧化硫和二氧化氮的影响。他们发现两次厄尔尼诺事件期间降雨量增加（2015～2016 年和 2009～2010 年），700 百帕以下的偏北风频率降低，风速总体变强，因而跨界空气污染输送对污染物的贡献在持续减少。相比之下，两次拉尼娜事件（2007～2008 年和 2010～2011 年）期间降雨减少，900 百帕以下的偏北风频率较高，地面风速较弱，有利于远距离污染物的传输和在本地积累，使得跨界空气污染输送成为环境颗粒物浓度增加的主要原因。除了以上这些短期事件影响之外，已有研究还评估了气候变化对空气质量的长期影响。一项研究（Tong *et al.*, 2018）估算，6～8 月污染物会减少，但在其他季节空气污染物的水平都有所增加。远期未来（2090～2099 年）在 RCP8.5 情景下，预估的污染物平均浓度变化更为显著。在不同的气象变量中，地面气温与上述三种污染物的预估变化最相关。例如在 RCP8.5 的情景下，不同季节的地面气温对所有污染物的相对贡献为 56.9%～65.2%。此外，值得注意的其他相互关联的气象因子，包括垂直温度梯度以及温度与露点之间的差值也会对污染物浓度有影响。研究还发现 12 月到次年 2 月、3 月到 5 月的高污染频率将会增加。污染物浓度大于近年第 95% 分位数的发生比例预计将会分别增加 6.4%～9.6%。

刘倩等（Liu *et al.*, 2013）运用大气数值模拟的方法预估了在 A1B 情景下，2000～2050 年 10 月的气候变化与排放变化对华南地区地面臭氧的影响。此研究的结果表明，由于气候变化而导致的辐射和地面温度的变化将导致 2000～2050 年的异戊二烯和单萜排放量显著增加；地面温度高于 40 摄氏度还可能会抑制生物性排放的产生。但是由于 10 月的地面温度预估结果并没有显示出特别高温的状态，因此在未来条件下生物源排放很有可能会持续增加。由于排放增加，异戊二烯浓度预计会增长 30～80 ppt。研究预测由于气候变化，下午的平均地面臭氧将会增加 1.5 ppb；而由于人为排放的变化，即使大湾区南部的臭氧会减少，但大湾区整体的臭氧平均浓度仍将会增加 6.1 ppb。在人为排放的综合效应下，大湾区下午的地面臭氧浓度将会增加 11.4 ppb。此研究结果强调，尽管人为排放变化的影响较大，但气候变化对臭氧的影响仍然重要。

城市化导致的区域气候变化会影响空气质量。根据李蒙蒙等（Li M. *et al.*, 2016）的估算结果，在粤港澳大湾区，城市变暖使臭氧浓度在白天减少了 1.3 ppb，在夜间增加了 5.2 ppb。这种臭氧变化主要是由于风速变大和大气边界层高度升高而导致氮氧化物的稀释，通过抑制氮氧化物的滴定作用使得夜间臭氧增加，而通过减弱光化学产生过程使得白天臭氧减少。另一项研究（Wang *et al.*, 2009）评估了 2001 年 3 月粤港澳大湾区城市化对二次有机气溶胶形成的影响，并估算出城市化会导致氮氧化物和挥发性有机化合物浓度分别减少 4.0 ppb 和 1.5 ppb。温度升高叠加风速降低会导致臭氧和硝酸根浓度分别增加 2～4 ppb 和 4～12 ppt。上述结果显示，城市化效应会对区域气候有较大影响，继而影响臭氧浓度，在高度城市化的珠三角地区，跨界空气污染非常严重。汪梦雅等（Wang *et al.*, 2020）估算跨界的空气污染输送约占臭氧引起过早死亡的 46%。严鸿霖等（Yim *et al.*, 2019b）估算城市化增加了臭氧，这意味着臭氧引起的过早死亡（1 100 例死亡）增加了 39.6%。他们的研究还发现，城市化也将改变大气对排放的反应，从而体现了土地使用政策、城市气候适应战略和空气质量政策之间的强有力的相互作用，需要采取有益的战略和政策。严鸿霖等（Yim *et al.*, 2019b）提出了一个“精准环境管理”概念，该概念强调在制定城市环境政策时，应考虑城市的特定大气条件和组成。

三、珠江河口近海水体富营养化和缺氧趋势

河口是河水和海水混合的重要过渡地带。河口水动力系统主要由淡水径流和海水入侵组成（Pritchard, 1967），形成非常独特的河口生态系统生物化学过程。珠江河口是中国南部海岸的亚热带出海口，连接南海北部大陆架，每年平均有大约 10 000 立方米/秒的淡水径流量（Zhai *et al.*, 2005）。珠江河口面积大约 1 900 平方千米，从出海口延展到外围大于 60 千米，其宽度从上游的 10 千米到下游的 60 千米，除了重力环流，该宽度足够导致陆架环流对河口底部产生影响（Zu and Gan, 2015）。由于夏季西南季风、冬季东北季风和显著的河道流量季节性变化，珠江河口的物理和生物化学进程具有明显的季节性特征。大约 80%的珠江径流集中在 4～9 月的雨季（Zhai *et al.*, 2005），约一米的潮汐变动为季节性进程带来了更大的波动。

当氧气补给不足以维持水体中的氧气消耗时，缺氧情况会发生。当溶解氧浓度低于 2～3 毫克/升时，就会出现缺氧情况（Chu *et al.*, 2005; Dai *et al.*, 2006; Rabalais *et al.*, 2002）。由于近三十年来工业和农业的快速发展和城市化，尽管已经建造了大型污水处理装置，水体富营养化和缺氧风险在珠江河口、香港水域和沿海过渡区仍然普遍快速增加。以往研究及实地调研证明了该区域大部分地区均出现底部缺氧的现象（Qian W. *et al.*, 2018; Su *et al.*, 2017）。大部分水体富营养化和缺氧集中在夏季，主要是因为浮力上升和营养负荷都在这个时期达到峰值。加上有利的风力和水动力条件，浮游植物的快速生长频繁在沿海过渡区表层上发生（Lu and Gan, 2015），并引发底层严重和持续的氧气短缺。

在气候变化的影响方面，图 2–1a 显示了香港水域南部和沿海过渡区东面横澜岛站 1990～2015 年风速和风摩擦速度记录。风速和风摩擦速度在此周期总体下降，表明这段时期沿海上升流强度的下降趋势，以及由风力控制的水

柱稳定性持续增加这为表层浮游植物的增长和下层缺氧状态的持续提供有利的物理条件。香港南部水域底层盐度减少和温度增加进一步证明了上升流的下降趋势（图 2–1b）。由沿海上升流供给的营养物质有下降趋势，表层NO_3^{-1}浓度上升和相应的浮游植物生物质增长主要是由于河流输入的增加。

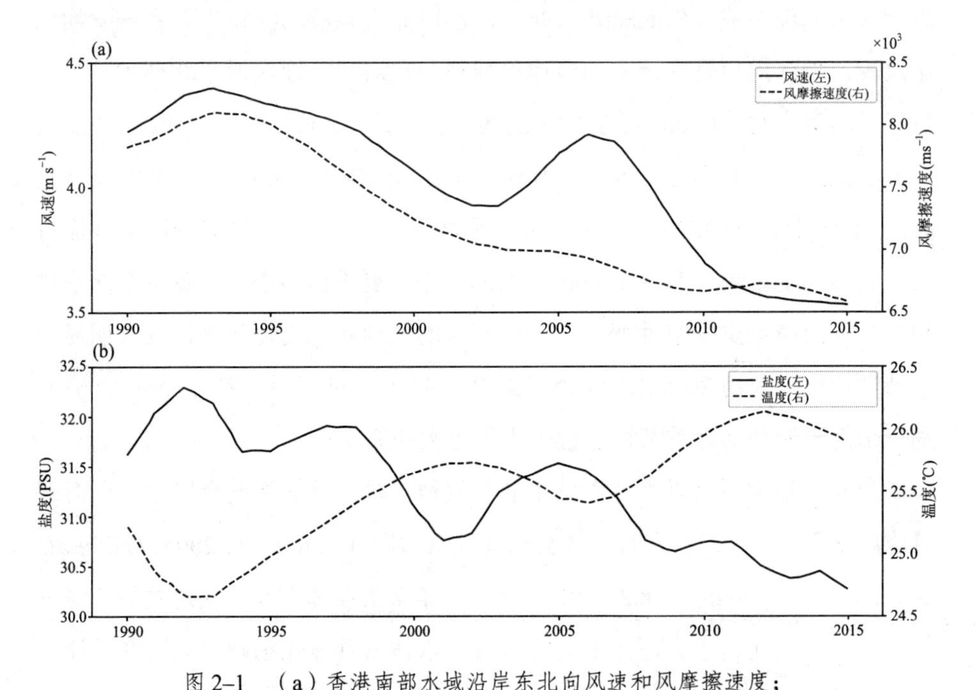

图 2–1 （a）香港南部水域沿岸东北向风速和风摩擦速度；
（b）香港南部水域底层平均海水温度和盐度

资料来源：香港天文台，香港环境署。

水柱通气是水底溶解氧的重要来源。海洋通气变化通常和气候变化有关（例如全球变化），并且受水柱层化和上升流等各种过程影响（Shepherd *et al.*, 2017）。气温增加导致水温增加（图 2–2），而风速减弱（图 2–1a）也加剧了水柱增强，导致了显著的生态影响。

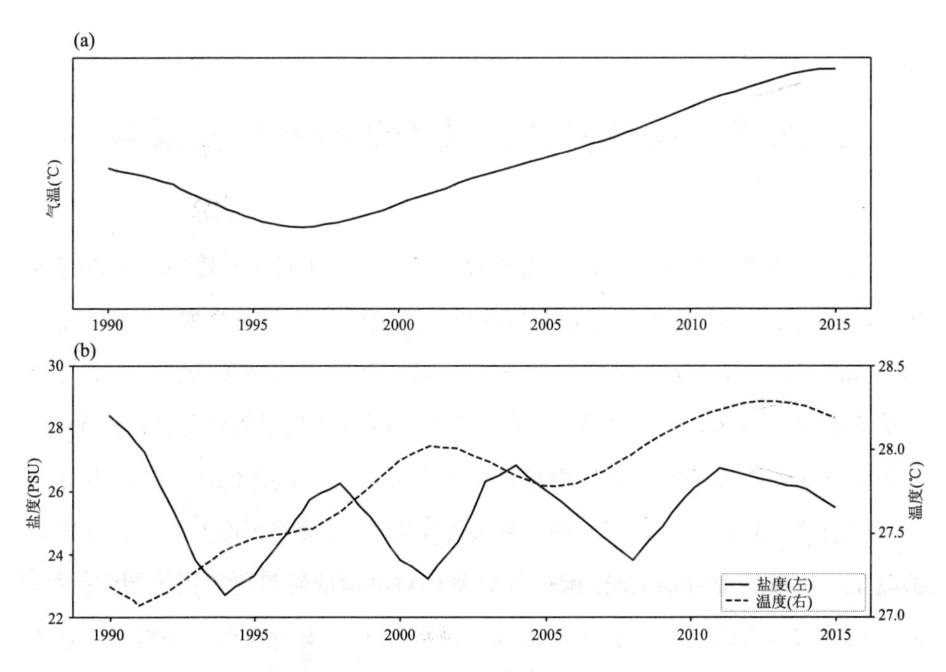

图 2–2　（a）香港南部水域气温；（b）香港南部水域表层平均海水盐度和温度

资料来源：香港天文台，香港环境署。

综上所述，根据海洋、大气和珠江径流的长期观测资料，珠江河口与邻近香港沿海水域的上层富营养化和下层缺氧在雨季都有增加趋势。水体富营养化和缺氧已经达到令人担忧的水平，并导致该区域海水酸化。这种趋势主要是来自珠江的营养物质增加，风速减弱和气温上升等气候变化现象造成的。因此，开展沿海生态系统和相关的物理——生物化学机制研究对于该区域气候变化和快速社会经济发展十分迫切，有助于应对该区域的沿海环境问题。最近 OCEAN-HK 项目（https://ocean.ust.hk/）致力于制定科学的策略来应对珠江河口和香港水域的水体富营养化和缺氧问题。

第五节　人群健康、风险认知和适应策略

气候变化是人类迄今为止面临的规模最大、范围最广、影响最为深远的挑战，也是21世纪全球健康最大的威胁。港澳地区处于亚热带季风气候区，由于人口流动、人口结构改变、城市热岛效应和空气污染等问题，气候变化严重威胁着该地区居民的健康。当前在港澳地区有关气候变化与健康的研究，也多注重研究气象要素与居民健康之间的关系。由于港澳地区的气候变化研究有相当长的时间跨度，其间的气象变量对居民健康的平均影响及极端影响都能在量化模型中得到反映。衡量气候变化对健康的影响可运用不同的指标，由最轻微的行为改变，到急诊、住院，甚至到最严重的死亡。尽管大多数人可能没有出现症状或只出现轻微症状，但部分人口的健康可能会遭受疾病的严重影响，包括传染病和非传染病（Chan, 2017; 2019）。近年由于气象要素如温度、湿度和雨量持续不稳定，极端天气现象和自然灾害事件又日趋频繁，人为因素、社区环境和卫生服务设施所带来的互动影响更是复杂多变，因此，气候变化对人群健康的研究目前仍处于起步阶段。本节将评估气候变化对香港和澳门居民健康风险的影响，并提出应对气候变化健康风险的公共卫生策略。

一、极端气温对人群死亡的影响

气候变化与中国居民的健康息息相关（Huang Z. *et al.*, 2015），并对医疗卫生系统造成了巨大负担。陈英凝等（Chan E. *et al.*, 2012）的研究显示高温与香港地区人群死亡人数呈现出显著相关。在日平均温度超过28.2摄氏度的炎热季节，气温每升高1摄氏度，死亡人数将上升1.8%。高温天气同样会导

致心血管疾病和呼吸系统疾病的死亡人数显著上升，而女性、社会经济状况较差者和已婚人群属于高危人群。与高温不同，寒冷季节的低温对逐日死亡人数的影响存在滞后效应，可延迟约三周（Goggins *et al.*, 2013a）。当日平均温度下降 1 摄氏度，其后三周累计死亡人数上升 3.8%。连续数天的低温天气，较气温波动的天气会发生更多人死亡，而老年人为低温天气的敏感人群。

除极端气温外，伴随快速的城市发展、市区的高密度建设、交通工具和工业活动等因素，热力排放不断增加，同样对城市气候带来重大的影响。高威廉等（Goggins *et al.*, 2012a）的研究发现，城市热岛效应与人群健康也有直接关系。在炎热天气下（平均温度高于 29 摄氏度），较低的风速与死亡人数增加呈显著正相关，而城市热岛指数高的地区两者相关性更强。

二、极端气温对人群患病的影响

极端气温除了会提升居民死亡风险外，对人群患病也会产生影响。随着温度的变化，空气中相对湿度和污染程度也会随之发生改变，进一步加剧极端气温对人体健康的影响，从而导致相关疾病发病率上升。在 1998～2009 年的炎热季节（6～9 月），当气温高于 29.0 摄氏度时，气温每上升 1 摄氏度，香港的入院人数将增加 4.5%；而在寒冷季节（11～3 月），当温度介于 8.2～26.9 摄氏度时，气温每下降 1 摄氏度，入院人数上升 1.4%（Chan E. *et al.*, 2013）。

（一）传染性疾病

气候变化将直接或间接影响经水、食物与病媒传播的传染病的扩散传播，从而影响传染性疾病的风险。气候变化使得气温与海水温度升高、降雨增加，导致经水传播的肠道传染病的发病风险增加；另外，气候变化会改变虫媒的地区分布，增加虫媒的繁殖速度与侵袭力、缩短病原体的外潜伏期，从而导致虫媒传染病的发病率增加。由于城市地区人口密集、国际交流密切以及城市热岛效应等原因，城市人群更容易受到气候变化的影响。当前研究证实温度变化与香港的传染病住院人数呈显著相关，在炎热季节（6～9 月），当 10 天内平均温度超过 28.5 摄氏度时，气温每增加 1 摄氏度，各类传染病相关的住院人数将增加 4.5%；而在寒冷季节（11～3 月），平均温度从 26.9 摄氏度每下降 1 摄氏度，相关住院人数将增加 1.0%（Chan E. *et al.*, 2013）。

1. 肠道传染病

温度和湿度可影响肠道传染病病原体的繁殖及其在外界环境中的生存时间，温度升高将有利于细菌性病原体在外界环境中的存活与繁殖，增加肠道传染病传播的机会。有研究显示，香港地区手足口病与气温呈现非线性关系，当平均温度处于 8～20 摄氏度或高于 25 摄氏度时，因手足口病入院的人数随温度升高而增加；当相对湿度高于 80%时，手足口病入院人数同样急剧上升。另外，手足口病与降雨量、强风和太阳辐射量等也呈现显著正相关（Wang P. *et al.*, 2016）。

2. 沙门氏菌病

沙门氏菌病的发病与相对湿度、降雨和气温有显著相关性。香港的一项研究（Wang P. *et al.*, 2018a）显示，在各种传染病病因中，炎热天气对因沙门氏菌住院的人数影响特别显著，日均温度每升高 1 摄氏度，14 天内沙门氏菌住院率将增加 11.4%。除温度以外，其他气象因素也与沙门氏菌感染住院总人数相关，如平均相对湿度与之呈显著的 S 形关联，当平均相对湿度从 73%上升至 84%，滞后 10 天的住院人数呈显著上升趋势。滞后 10 天的平均风速与沙门氏菌住院人数呈显著的负相关。以 13 摄氏度为基准线，30.5 摄氏度天气下对应的因沙门氏菌病入院的累积（滞后 0～16 天）相对风险为 6.13（95%置信区间：3.52～10.67）。当相对湿度高于 60%且降雨量在 0～0.14 毫米之间时，相对湿度、降雨量和因沙门氏菌病住院人数呈正相关。与相对湿度为 60%和零降雨量相比，极高的相对湿度（96%）和微量降雨（0.02 毫米）对应沙门氏菌病的入院累积风险分别为 2.06（95%置信区间：1.35～3.14）和 1.30

（95%置信区间：1.01～1.67）。总之，高温、高相对湿度和轻微降雨均与因沙门氏菌病住院人数呈正相关。

3. 轮状病毒和诺如病毒感染

降雨、温度和湿度的短期变化与香港幼童因轮状病毒、诺如病毒感染住院之间有显著相关性（Wang P. *et al.*, 2018b）。降雨量与轮状病毒感染入院人数呈负相关，但与诺如病毒感染入院人数呈正相关。相比于微降水量，极端降雨量情况下（99.5 毫米，99%分位数）由轮状病毒和诺如病毒感染引起的住院的相对风险分别为 0.40（95%置信区间：0.20～0.79）和 1.93（95%置信区间：1.21～3.09）。降雨量对轮状病毒感染入院人数影响的持续时间较诺如病毒长。另外，轮状病毒和诺如病毒感染的住院人数会随温度升高而下降；诺如病毒感染的住院人数随相对湿度的上升而增加，而轮状病毒感染的住院人数则随相对湿度的上升而下降。

4. 其他传染性疾病

现有研究证实，气候变化将影响港澳地区虫媒传染病的传播，增加当地新发传染病出现的风险等。另外，气象因素与流感发病的关系十分密切，气候变化可能会加剧流感在该地区的流行。

（二）慢性非传染性疾病

气候变化可以通过直接或间接途径影响慢性非传染疾病。气候变化会导致港澳地区居民出现呼吸系统疾病和过敏、心血管疾病和心理疾病的机会上升，见表 2–1。

1. 呼吸系统疾病和过敏性疾病

（1）慢性阻塞性肺疾病

在寒冷季节，老年人患慢性阻塞性肺疾病的入院率与温度和相对湿度之间均呈现先降后升的趋势（Lam H. *et al.*, 2018a）。当温度为 21～22 摄氏度、相对湿度为 82%时入院人数最少。在低温天气下慢性阻塞性肺病入院的滞后时间为 0～20 天。

表 2-1　气候变化对慢性非传染性疾病的影响

<table>
<tr><th>影响</th><th>气候变化现象</th><th>路径</th><th>慢性非传染性疾病</th></tr>
<tr><td rowspan="5">直接</td><td>热浪</td><td>暑热压力升高</td><td>心血管疾病
呼吸系统疾病</td></tr>
<tr><td rowspan="2">炎热天气
干旱
季节异常</td><td>空气污染浓度升高</td><td>呼吸系统疾病
心血管疾病</td></tr>
<tr><td>花粉过多或分布异常</td><td>呼吸道和过敏性疾病</td></tr>
<tr><td>臭氧空洞</td><td>紫外线过量</td><td>癌症
白内障</td></tr>
<tr><td>洪水泛滥</td><td>溺水、房屋倒塌</td><td>伤害</td></tr>
<tr><td rowspan="2">间接</td><td>干旱
洪水泛滥</td><td>粮食歉收
房屋损毁
人口迁徙</td><td>健康状况欠佳
营养不良
心理健康</td></tr>
<tr><td>极端气候事件
洪水泛滥
台风
山火</td><td>创伤性经历
经济困难
痛失亲人</td><td>心理健康</td></tr>
</table>

资料来源：Chan, 2019。

（2）哮喘

在炎热季节，当气温为 27 摄氏度时因哮喘住院的人数最少，之后随着温度升高，该病的住院人数不断增加，在 30 摄氏度时达到峰值，并在 30～32 摄氏度之间达到平稳状态（Lam *et al.*, 2016）。与 27 摄氏度相比，30 摄氏度时哮喘住院人数的累积（滞后 0～3 天）相对风险为 1.19（95%置信区间：1.06～1.34）。在寒冷季节，温度与哮喘住院人数呈负相关。与 25 摄氏度相比，12 摄氏度时哮喘入院人数的累积（滞后 0～3 天）相对风险为 1.33（95%置信区间：1.13～1.58）。冷热两个季节的温度对成年人因哮喘而入院的影响最大，而对 5 岁以下儿童的影响最小。炎热季节湿度和臭氧浓度的增加，寒冷季节的湿度降低均与哮喘病人入院人数上升呈相关性。

（3）过敏

根据一项 2017 年的研究显示，受访香港居民中，大约 24%有过敏史（Lam and Chan, 2018）。在炎热天气下，有 22.4%的皮肤过敏患者和 15.7%的鼻过敏患者表示症状会加重。与无过敏史的人相比，炎热天气期间，有过敏症状的人更容易出现黏液分泌物增多、口腔溃疡情况加剧、睡眠质量变差、情绪低落等状况。

2. 心血管疾病

低温可导致心血管疾病的急诊住院人数上升。研究发现，2005～2012 年 7.15%因心血管疾病急诊住院的患者可归因于低温，合计有 37 285 例患者（Tian *et al.*, 2016）。

（1）心脏衰竭

低温与心脏衰竭的入院率和死亡率的上升密切相关（Goggins and Chan, 2017）。与 25 摄氏度相比，当温度为 11 摄氏度时心脏衰竭病人的入院人数累积（滞后 0～23 天）相对风险为 2.63（95%置信区间：2.43～2.84）；而因心脏衰竭的死亡人数累积（滞后 0～42 天）相对风险为 3.13（95%置信区间：1.90～5.16）。低温影响在老年人群以及新发病例中更显著。相对湿度过高或过低同样与心脏衰竭入院人数上升相关。

（2）中风

当每日平均温度在 8.2～31.8 摄氏度时，出血性中风入院的人数与每日平均温度呈负相关（Goggins *et al.*, 2012b），日气温越低，因出血性中风入院的人数越多。另外，五天内平均温度每下降 1 摄氏度，出血性中风的入院率将上升 2.7%。该影响在老年人和女性人群中更为显著。

每日平均温度与缺血性中风入院人数关系较弱，以连续 14 日平均温度来分析，当平均温度低于 22 摄氏度时，每下降 1 摄氏度，缺血性中风入院率将会上升 1.6%。

（3）急性心肌梗死

高威廉等（Goggins *et al.*, 2013b）的研究显示了气温变化与香港地区急性心肌梗死病人入院率之间的关系。心血管疾病患者在低温天气下，因急性心肌梗塞入院的风险会上升。

在寒冷季节，糖尿病患者的急性心肌梗死入院率与温度呈线性负相关，12 摄氏度天气与 24 摄氏度相比，低温滞后 0～22 天的累积相对风险为 2.10（95%置信区间：1.62～2.72）（Lam H. *et al.*, 2018b）；而非糖尿病患者的急性心肌梗死入院率与温度的关联较弱，累积相对风险为 1.43（95%置信区间：1.21～1.69），且仅在气温低于 22 摄氏度时开始上升。在炎热季节，糖尿病患者急性心肌梗死住院率随温度开始增加，当温度高于 28.8 摄氏度，滞后 0～4 天时 30.4 摄氏度对应的累积相对风险为 1.14（95%置信区间：1.00～1.31）；而非糖尿病患者的急性心肌梗死入院率与气温不存在显著相关。

3. 心理健康

极端天气事件容易对遭受侵害的社区造成重大的心理影响，这可以直接地因为伤痛，或间接地透过身体上的压力（例如酷热天气的影响）、环境恶化及重要社区资产的破坏而导致心理影响。气温升高与心理障碍入院人数增加呈线性正相关，且影响持续约 2 天（Chan E. *et al.*, 2018a）。当气温高于 19.4 摄氏度（第一四分位数）时健康风险显著升高。与 19.4 摄氏度相比，28 摄氏度（第三四分位数）对心理障碍入院人数的累积（滞后 0～2 天）相对风险为 1.09（95%置信区间：1.03～1.15）。

在不同心理障碍类型中，短暂性心理障碍和偶发性情绪障碍与温度呈正相关，而药物相关的心理障碍在温度高于 20 摄氏度时与温度呈正相关。温度与短暂性心理障碍的关系最为显著，累积（滞后 0～2 天）相对风险为 1.51（95%置信区间：1.00～2.27）；其次是偶发性情绪障碍，累积（滞后 0～2 天）相对风险为 1.34（95%置信区间：1.05～1.71）；再次是药物相关的心理障碍，累积（滞后 0～2 天）相对风险为 1.13（95%置信区间：1.00～1.27）。气温较低时，抑郁症和其他非器质性心理疾病入院的风险较低，而焦虑、解离性障碍、躯体形式障碍、精神分裂症和酒精相关的心理障碍则与气温无显著关联。

三、公共卫生应对策略

本部分将总结气候变化背景下，港澳地区居民对健康风险的认知，以及为保护公共健康而采取的应对策略和防护措施。

（一）高温热浪的健康防护措施

2017 年香港的一项电话调查研究显示，45.3%的受访人认为炎热天气不会影响他们的健康，在 65 岁以上的老年人群中该比例为 28.8%（Chan E. *et al.*, 2018b）。在研究中，约 87%的人知晓香港天文台曾发出的酷热天气警告，37.2%～97.5%受访人群采取过至少一种个人防护措施。最常见的措施为喝更多的水（97.5%），而最少的为使用防晒霜（37.2%）。男性受访者较少采取避免晒太阳的措施（优势比：0.29，95%置信区间：0.13～0.65）和涂防晒霜（优势比：0.41，95%置信区间：0.26～0.64）。受教育程度较低的人群较少使用空调（优势比：0.22，95%置信区间：0.05～0.91）。

另外一项大型电话调查研究发现，90.3%和 90%受访者家中分别拥有电风扇和空调（Gao *et al.*, 2020）。在感到炎热时，91.5%受访者通常使用空调，95.1%受访者通常使用风扇。大约一半的受访者（54.3%）报告了当室内温度到达特定数值时会开启降温设备，其中 58.1%和 95.4%的受访者分别会在 28 摄氏度和 31 摄氏度时开启降温设备。总体来说，老年人、受教育程度较低、家庭收入较低、失业/退休/家庭主妇、公屋租客和慢性病患者的家中拥有空调的比例更低，或即使感觉很热也不使用空调，家庭收入与空调使用之间的关系最为密切。上述结果反映了个人的经济或财务状况是影响空调使用的重要因素，该研究结果与其他研究一致（Hansen *et al.*, 2011），一定程度解释了贫困人口为什么易受炎热天气影响。

尽管超过一半的受访者（57.7%）认为热浪可能对健康产生不利的影响，但这种风险认知水平与其家中是否拥有空调以及是否会使用空调之间并无显著关系，即尽管他们认为热浪是一项健康风险因素，但由于各种原因（例如缺乏应对的方法），他们最终并没有将风险认知转化为实际的行动。当然，这也不排除受访者可能采取了其他防护措施进行应对（例如到其他有空调的公共场所避暑）。

（二）低温寒潮的健康防护措施

2016 年香港地区一项电话调查研究发现，45.0%的受访者具有较高的潜在健康风险，主要包括老年人、慢性病患者、独居者和领取政府低收入援助的人士，而且，该群体中有 63.7%的人低估了低温天气对他们的健康影响（Lam *et al.*, 2020）。对比 2016 年发生的极端的寒流，香港市民在 2017 年寒流期间采取了更多的御寒措施。在 2016 年寒流期间经历过不良健康影响的人在 2017 年的寒流期间会穿上更多的衣服（优势比：5.48，95%置信度：1.26～3.83），其中女性（优势比：1.93，95%置信度：1.20～3.11）、在寒流期间感到寒冷的人（优势比：3.72，95%置信度：1.29～10.72）和 70 岁或以上的长者（优势比：4.00，95%置信度：1.12～13.19）更倾向于使用取暖设备。

（三）台风的应急准备措施

由于与气候变化相关的极端天气事件频率和严重程度都在上升，了解家庭备灾能力对于灾害风险的管理工作非常重要。2018 年香港的一项研究探讨了灾害的风险感知、家庭防备与台风“山竹”的即时影响之间的关联（Chan *et al.*, 2019）。在 521 名受访者中，93.9%和 74.3%的受访者分别报告了有采取常规应急准备措施和针对台风“山竹”而作的特殊应急准备措施。33.4%的受访者表示台风“山竹”对其造成了一定程度的影响。在台风期间，认为家中风险较高并且已经实行常规应急准备的受访者更有可能针对台风进行特殊应急准备。但实施特殊应急准备措施与减少台风所带来的即时影响并无显著关系。此研究显示当前的家庭应急准备措施可能不足以应对超级台风的影响，香港需要推行更加有效的备灾策略以应对气候变化相关的极端事件。

（四）气候减缓策略的健康协同效益

许多气候变化的减缓措施在实施时同时会产生公共健康和环境保护的双赢效应。香港的一项电话调查发现（Chan *et al.*, 2017），最常见的健康环境协同效益行为是减少使用包装袋和一次性购物袋，有 70.1%的受访者表示每天都在实行。大约 50%的受访者同时采取了其他环境与健康协同效益行为，包括步行/骑自行车（54.8%）、家庭垃圾分类（50.2%）、减少用电（48.3%）和减少使用空调（44.1%）。大约 30%的受访者每天都在实行的行为，包括减少食肉（33.3%）和每天淋浴少于五分钟（23.7%）。只有不到 10%的人每周吃一顿素食（5.8%）、购买有机食品（4.3%）和在餐馆用餐时自备个人餐具（4.0%）。一般来说，女性和高龄人口更倾向于实行环境与健康协同效益行为。

香港特区政府此前已实施了一系列措施，以促进香港人口的环境与健康协同效益。然而，研究结果显示这些措施的影响存在差异。例如 2015 年实施的塑料购物袋征费计划，每个塑料购物袋收费 0.50 港元，有 70.1%的受访者提到自己每天都会减少使用包装袋和一次性购物袋，表明塑料购物袋征费计划是较为成功的。与此同时，香港水务署通过“齐来悭水十公升”活动，提出多项建议，包括通过缩短淋浴时间达到节约用水的目的。水务署于 2015～2016 年度进行的“家居用水调查”显示，约 98%的受访者采取了一项或以上的节约用水措施，可见市民意识到节约用水的重要性。但也有其他研究结果显示近一半受访者从未缩短过淋浴时间，也从没有控制每天淋浴时间少于五分钟。因此，政府可以考虑开展更多的调查工作以明确公众对于环境与健康协同效益的意识和实践情况，尤其是确认日常实践中执行频率较低的行为。另外，还需要考虑社区和医疗保健从业者在促进环境与健康协同效益的作用。

参考文献

Ang, B. W., H. Wang, X. Ma, 2017. Climatic Influence on Electricity Consumption: The Case of Singapore and Hong Kong. *Energy*, 127.

Ayyub, B. M., H. G. Braileanu, N. Qureshi, 2012. Prediction and Impact of Sea Level Rise on Properties and Infrastructure of Washington, DC. *Risk Analysis*, 32(11).

Benfield, A., 2018. *Weather, Climate & Catastrophe Insight: 2017 Annual Report*. AON Empower Results.

Campbell, S., 2005. *Typhoons Affecting Hong Kong: Case Studies APEC 21st Century COE Short-Term Fellowship Report*. Tokyo Polytechnic University.

Cha, E. J., T. R. Knutson, T. C. Lee, *et al.*, 2020. Third Assessment on Impacts of Climate Change on Tropical Cyclones in the Typhoon Committee Region–Part Ⅱ: Future Projections. *Tropical Cyclone Research and Review*, 9(2).

Civil Engineering and Development Department (CEDD). 2020. *Management of Natural Terrain Landslide Risk*, Information Note 29/2020 (https://www.cedd.gov.hk/filemanager/eng/content_454/IN_2020_29E.pdf).

Chan, E. Y. Y., 2017. *Public Health Humanitarian Responses to Natural Disasters*. London: Routledge.

Chan, E. Y. Y., 2019. *Climate Change and Urban Health: The Case of Hong Kong as a Subtropical City*. London: Routledge.

Chan, E. Y. Y., W. B. Goggins, J. J. Kim, *et al.*, 2012. A Study of Intracity Variation of Temperature-Related Mortality and Socioeconomic Status Among the Chinese Population in Hong Kong. *Journal of Epidemiology and Community Health*, 66(4).

Chan, E. Y. Y., W. B. Goggins, J. S. Yue, *et al.*, 2013. Hospital Admissions as a Function of Temperature, Other Weather Phenomena and Pollution Levels in an Urban Setting in China. *Bulletin of the World Health Organization*, 91(8).

Chan, E. Y. Y., H. C. Y. Lam, S. H. W. So, *et al.*, 2018a. Association Between Ambient Temperatures and Mental Disorder Hospitalizations in a Subtropical City: A Time-Series Study of Hong Kong Special Administrative Region. *International Journal of Environmental Research and Public Health*, 15(4).

Chan, E. Y. Y., A. Y. T. Man, H. C. Y. Lam, 2018b. *Personal Heat Protective Measures During the 2017 Heatwave in Hong Kong: A Telephone Survey Study*. First Global Forum for Heat and Health, Hong Kong, 2018.

Chan, E. Y. Y., A. Y. T. Man, H. C. Y. Lam, *et al.*, 2019. Is Urban Household Emergency Preparedness Associated with Short-Term Impact Reduction After a Super Typhoon in Subtropical City? *International Journal of Environmental Research and Public Health*, 16(4).

Chan, E. Y. Y., S. S. Wang, J. Y. Ho, *et al.*, 2017. Socio-Demographic Predictors of Health and Environmental Co-Benefit Behaviours for Climate Change Mitigation in Urban China. *PLoS One*, 12(11).

Chan, F. K. S., G. Mitchell, O. Adekola, *et al.*, 2012. Flood Risk in Asia's Urban Mega-Deltas: Drivers, Impacts and Response. *Environment and Urbanization ASIA*, 3(1).

Chan, F. K. S., O. Adekola, G. Mitchell, *et al.*, 2013. Appraising Sustainable Flood Risk Management in the Pearl River Delta's Coastal Megacities: A Case Study of Hong Kong, China. *Journal of Water and Climate Change*, 4(4).

Chan, F. K. S., N. Wright, X. Cheng, *et al.*, 2014. After Sandy: Rethinking Flood Risk Management in Asian Coastal Megacities. *Natural Hazards Review*, 15.

Chan, F. K. S., C. C. Joon, A. Ziegler, *et al.*, 2018. Towards Resilient Flood Risk Management for Asian Coastal Cities: Lessons Learned from Hong Kong and Singapore. *Journal of Cleaner Production*, 187.

Chan, J. C. L., M. Xu, 2009. Inter-Annual and Inter-Decadal Variations of Landfalling Tropical Cyclones in East Asia. Part I: Time Series Analysis. *International Journal of Climatology*, 29(9).

Chan, W. L., 1996. *Hong Kong Rainfall and Landslides in 1994 (GEO Report No. 54)*. Hong Kong, Geotechnical Engineering Office.

Chen, X., S. J. Jeong, 2018. Shifting the Urban Heat Island Clock in a Megacity: A Case Study of Hong Kong. *Environmental Research Letters*, 13(1).

Chen, Y. D., Q. Zhang, X. Chen, *et al.*, 2012. Multiscale Variability of Streamflow Changes in the Pearl River Basin, China. *Stochastic Environmental Research and Risk Assessment*, 26(2).

Chen, Y. D., Q. Zhang, X. Lu, *et al.*, 2011. Precipitation Variability (1956-2002) in the Dongjiang River (Zhujiang River Basin, China) and Associated Large-Scale Circulation. *Quaternary International*, 244.

Chen, Y. D., 2001. Sustainable Development and Management of Water Resources for Urban Water Supply in Hong Kong. *Water International*, 26(1).

Choi, C. E., C. W. W. Ng, H. Liu, *et al.*, 2019. Interaction Between Dry Granular Flow and

Rigid Barrier with Basal Clearance: Analytical and Physical Modelling. *Canadian Geotechnical Journal*, 57(2).

Chow, E. C. H., R. C. Y. Li, W. Zhou, 2018. Influence of Tropical Cyclones on Hong Kong Air Quality. *Advances in Atmospheric Sciences*, 35(9).

Choy, C. W., D. S. Lau, Y. He, 2020a. Super Typhoons Hato (1713) and Mangkhut (1822), Part I: Analysis of Maximum Intensity and Wind Structure. *Weather*.

Choy, C. W., D. S. Lau, Y. He, 2020b. Super Typhoons Hato (1713) and Mangkhut (1822), Part Ⅱ: Challenges in Forecasting and Early Warnings. *Weather*.

Chu, P., Y. C. Chen, A. Kuninaka, 2005. Seasonal Variability of the Yellow Sea/East China Sea Surface Fluxes and Thermohaline Structure. *Advances in Atmospheric Sciences*, 22(1).

Chui, S. K., J. K. Y. Leung, C. K. Chu, 2006. The Development of a Comprehensive Flood Prevention Strategy for Hong Kong. *International Journal of River Basin Management*, 4(1).

Dai, M., X. Guo, W. Zhai, *et al.*, 2006. Oxygen Depletion in the Upper Reach of the Pearl River Estuary During a Winter Drought. *Marine Chemistry*, 102(1-2).

Dawson, D., J. Shaw, W. R. Gehrels, 2016. Sea-level Rise Impacts on Transport Infrastructure: The Notorious Case of the Coastal Railway Line at Dawlish, England. *Journal of Transport Geography*, 51.

Ding, Y., J. C. Chan. 2005. The East Asian Summer Monsoon: An Overview. *Meteorology and Atmospheric Physics*, 89(1-4).

Drainage Services Department (DSD), 2019. *Flood Prevention Strategy*. Drainage Services Department, HKSAR.

Elsner, J. B., J. P. Kossin, T. H. Jagger, 2008. The Increasing Intensity of the Strongest Tropical Cyclones. *Nature*, 455(7209).

Gao, Y., E. Y. Y. Chan, H. C. Y. Lam, *et al.*, 2020. Risk Perception of Climate Change and Utilization of Fans and Air Conditioners in a Representative Population of Hong Kong. *International Journal of Disaster Risk Science*, 11.

Gasparrini, A., Y. Guo, M. Hashizume, *et al.*, 2015. Mortality Risk Attributable to High and Low Ambient Temperature: A Multicountry Observational Study. *The Lancet*, 386(9991).

Geotechnical Engineering Office (GEO), 2012. *Geotechnical Engineering Office Report 272: Detailed Study of the 7 June 2008 Landslide on the Hillside above North Lantau Highway and Cheung Tung Road, North Lantau*. Geotechnical Engineering Office of the Civil Engineering and Development Department, HKSAR.

Geotechnical Engineering Office (GEO), 2016. *Report on Natural Terrain Landslide Hazards in Hong Kong*. Geotechnical Engineering Office of the Civil Engineering and Development Department, HKSAR.

Ginn, W. L., T. C. Lee, K. Y. Chan, 2010. Past and Future Changes in the Climate of Hong Kong. *Acta Meteorologica Sinica*, 24(2).

Giridharan, R., S. Ganesan, S. S. Y. Lau, 2004. Daytime Urban Heat Island Effect in High-Rise and High-Density Residential Developments in Hong Kong. *Energy and Buildings*, 36(6).

Goggins, W. B., E. Y. Y. Chan, 2017. A Study of the Short-Term Associations Between Hospital Admissions and Mortality from Heart Failure and Meteorological Variables in Hong Kong: Weather and Heart Failure in Hong Kong. *International Journal of Cardiology*, 228.

Goggins, W. B., E. Y. Y. Chan, E. Ng, *et al.*, 2012a. Effect Modification of the Association Between Short-term Meteorological Factors and Mortality by Urban Heat Islands in Hong Kong. *PLoS One*, 7(6).

Goggins, W. B., E. Y. Y. Chan, C. Y. Yang, 2013b. Weather, Pollution, and Acute Myocardial Infarction in Hong Kong and Taiwan. *International Journal of Cardiology*, 168(1).

Goggins, W. B., E. Y. Y. Chan, C. Yang, *et al.*, 2013a. Associations Between Mortality and Meteorological and Pollutant Variables During the Cool Season in Two Asian Cities with Sub-tropical Climates: Hong Kong and Taipei. *Environmental Health*, 12(1).

Goggins, W. B., J. Woo, S. Ho, *et al.*, 2012b. Weather, Season, and Daily Stroke Admissions in Hong Kong. *International Journal of Biometeorology*, 56(5).

Greenpeace China, 2009. *The "Climate Change Bill": Economic Costs of Heavy Rainstorm in Hong Kong*. Greenpeace China: Hong Kong.

Hansen, A., P. Bi, M. Nitschke, *et al.*, 2011. Perceptions of Heat-Susceptibility in Older Persons: Barriers to Adaptation. *International Journal of Environmental Research and Public Health*, 8(12).

He, Y. H., H. Y. Mok, E. S. Lai, 2016. Projection of Sea-Level Change in the Vicinity of Hong Kong in the 21st Century. *International Journal of Climatology*, 36(9).

Hong Kong Observatory (HKO), 2019. *Hong Kong Observatory: Tropical Cyclones in 2017* https://www.hko.gov.hk/publica/tc/TC2017.pdf.

Hong Kong Observatory (HKO), 2020. *Hong Kong Observatory: Tropical Cyclones in 2018* https://www.hko.gov.hk/en/publica/tc/files/TC2018.pdf.

Ho, H. C., K. K. L. Lau, C. Ren, *et al.*, 2017. Characterizing Prolonged Heat Effects on Mortality in a Sub-Tropical High-Density City, Hong Kong. *International Journal of Biometeorology*, 61(11).

Ho, K., S. Lacasse, L. Picarelli, 2019. *Slope Safety Preparedness for Impact of Climate Change*. CRC Press.

Ho, K. K. S., H. W. Chan, T. M. Lam, 2002. *Review of 1999 Landslides (GEO Report No. 127)*. Hong Kong, Geotechnical Engineering Office.

Huang, P., I. I. Lin, C. Chou, *et al.*, 2015. Change in Ocean Subsurface Environment to Suppress Tropical Cyclone Intensification Under Global Warming. *Nature Communications*, 6.

Huang, Z., H. Lin, Y. Liu, *et al.*, 2015. Individual-level and Community-level Effect Modifiers of the Temperature–mortality Relationship in 66 Chinese Communities. *BMJ Open*, 5(9).

Huang, Z., Y. Zong, W. Zhang, 2004. Coastal Inundation due to Sea Level Rise in the Pearl River Delta, China. *Natural Hazards*, 33.

Hudson, R. R., 1993. *Report on the Rainstorm of August 1982 (GEO Report No. 26)*. Hong Kong, Geotechnical Engineering Office.

Hulme, M., G. J. Jenkins, X. Lu, *et al.*, 2002. *Climate Change Scenarios for the United Kingdom: The UKCIP02 Scientific Report*, Tyndall Centre for Climate Change Research, School of Environmental Sciences, University of East Anglia. https://artefacts.ceda.ac.uk/badc_datadocs/link/UKCIP02_tech.pdf, 2002.

IPCC, 2013. *Climate Change 2013: The Physical Science Basis. Contribution of Working Group I to the Fifth Assessment Report of the Intergovernmental Panel on Climate Change*. Cambridge University Press, Cambridge, United Kingdom and New York, NY, USA.

IPCC, 2014. *Climate Change 2014: Impacts, Adaptation, and Vulnerability. Part B: Regional of the Intergovernmental Panel on Climate Change*. Cambridge University Press, Cambridge, United Kingdom and New York, NY. USA.

Jovanović, S., S. Savić, M. Bojić, *et al.*, 2015. The Impact of the Mean Daily Air Temperature Change on Electricity Consumption. *Energy*, 88.

Kang, N. Y., J. B. Elsner, 2012. Consensus on Climate Trends in Western North Pacific Tropical Cyclones. *Journal of Climate*, 25(21).

Knutson, T. R., J. J. Sirutis, M. Zhao, *et al.*, 2015. Global Projections of Intense Tropical Cyclone Activity for the Late Twenty-first Century from Dynamical Downscaling of CMIP5/RCP4.5 Scenarios. *Journal of Climate*, 28(18).

Kong, L., K. K. L. Lau, C. Yuan, *et al.*, 2017. Regulation of Outdoor Thermal Comfort by Trees in Hong Kong. *Sustainable Cities and Society*, 31.

Kossin, J. P., 2018. A Global Slowdown of Tropical-Cyclone Translation Speed. *Nature*, 558.

Kwan, J. S. H., R. C. H. Koo, C. W. W. Ng, 2015. Landslide Mobility Analysis for Design of Multiple Debris-Resisting Barriers. *Canadian Geotechnical Journal*, 52(9).

Lai, Y. C., J. F. Li, X. Gu, *et al.*, 2020. Greater Flood Risks in Response to Slowdown of Tropical Cyclones over the Coast of China. *Proceedings of the National Academy of Sciences of the United States of America*, 117(26).

Lam, H. C. Y., E. Y. Y. Chan, 2018. *The Impact of High Temperature on Existing Allergic Symptoms Among an Adult Population (CCOUC Working Paper Series)*. Hong Kong: Collaborating Centre for Oxford University and CUHK for Disaster and Medical

Humanitarian Response.

Lam, H. C. Y., E. Y. Y. Chan, W. B. Goggins, 2018a. Comparison of Short-Term Associations with Meteorological Variables Between COPD and Pneumonia Hospitalization Among the Elderly in Hong Kong—A Time-Series Study. *International Journal of Biometeorology*, 62.

Lam, H. C. Y., J. C. N. Chan, A. O. Y. Luk, *et al.*, 2018b. Short-term Association Between Ambient Temperature and Acute Myocardial Infarction Hospitalizations for Diabetes Mellitus Patients: A Time Series Study. *PLoS Medicine*, 15(7).

Lam, H. C., A. M. Li, E. Y. Y. Chan, *et al.*, 2016. The Short-term Association Between Asthma Hospitalisations, Ambient Temperature, Other Meteorological Factors and Air Pollutants in Hong Kong: A time-series Study. *Thorax*, 71(12).

Lam, H. C., Z. Huang, E. Y. Y. Chan, *et al.*, 2020. Personal Cold Protection Behaviour and Its Associated Factors in 2016/17 Cold Days in Hong Kong: A Two-Year Cohort Telephone Survey Study. *International Journal of Environmental Research and Public Health*, 17(5).

Lam, Y. F., H. M. Cheung, C. C. Ying, 2018. Impact of Tropical Cyclone Track Change on Regional Air Quality. *Science of the Total Environment*, 610.

Lee, B. Y., 1983. *Hydrometeorological Aspects of the Rainstorms in May 1982*. Royal Observatory.

Lee, B. Y., W. T. Wong, W. C. Woo, 2010. *Sea-level Rise and Storm Surge—Impacts of Climate Change on Hong Kong*. HKIE Civil Division Conference.

Lee, K. L., Y. H. Chan, T. C. Lee, *et al.*, 2015. The Development of the Hong Kong Heat Index for Enhancing the Heat Stress Information Service of the Hong Kong Observatory. *International Journal of Biometeorology*, 60(7).

Lee, T. C., T. R. Knutson, H. Kamahori, *et al.*, 2012. Impacts of Climate Change on Tropical Cyclones in the Western North Pacific Basin. Part I: Past Observations. *Tropical Cyclone Research and Review*, 1(2).

Lee, T. C., T. R. Knutson, T. Nakaegawa, *et al.*, 2020. Third Assessment on Impacts of Climate Change on Tropical Cyclones in the Typhoon Committee Region. Part I: Observed Changes, Detection and Attribution. *Tropical Cyclone Research and Review*, 9 (1).

Leichenko, R., 2011. Climate Change and Urban Resilience. *Current Opinion in Environmental Sustainability*, 3(3).

Li, J., Y. D. Chen, L. Zhang, *et al.*, 2016a. Future Changes in Floods and Water Availability across China: Linkage with Changing Climate and Uncertainties. *Journal of Hydrometeorology*, 17.

Li, J., L. Zhang, X. Shi, *et al.*, 2016b. Response of Long-term Water Availability to More Extreme Climate in the Pearl River Basin, China. International Journal of Climatology, 37(7).

Li, M., Y. Song, Z. Mao, *et al.*, 2016. Impacts of Thermal Circulations Induced by Urbanization on Ozone Formation in the Pearl River Delta Region, China. *Atmospheric Environment*, 127(2).

Lin, J., 2014. *Vulnerable and Lagging Behind: The Case of Hong Kong*. Adaptation to Climate Change in Asia.

Liu, B., J. Chen, W. Lu, *et al.*, 2015. Spatiotemporal Characteristics of Precipitation Changes in the Pearl River Basin, China. *Theoretical and Applied Climatology*, 123(3).

Liu, D., X. Chen, Y. Lian, *et al.*, 2010. Impacts of Climate Change and Human Activities on Surface Runoff in the Dongjiang River Basin of China. *Hydrological Processes*, 24(11).

Liu, Q., K. S. Lam, F. Jiang, *et al.*, 2013. A Numerical Study of the Impact of Climate and Emission Changes on Surface Ozone over South China in Autumn Time in 2000-2050. *Atmospheric Environment*, 76.

Lu, Z., J. Gan, 2015. Controls of Seasonal Variability of Phytoplankton Blooms in the Pearl River Estuary. *Deep Sea Research Part II: Topical Studies in Oceanography*, 117.

Lo, D. O. K., 2000. *Review of Natural Terrain Landslide Debris-resisting Barrier Design*. Geotechnical Engineering Office, Civil Engineering Department, HKSAR.

Lok, C. C. F., J. C. L. Chan, 2018. Changes of Tropical Cyclone Landfalls in South China Throughout the Twenty-first Century. *Climate Dynamics*, 51(7).

Luo, M., N. C. Lau, 2019. Amplifying Effect of ENSO on Heat Waves in China. *Climate Dynamics*, 52(5-6).

Luo, M., X. Hou, Y. Gu, *et al.*, 2018. Trans-boundary Air Pollution in a City Under Various Atmospheric Conditions. *Science of the Total Environment*, 618.

Luo, Y., 2017. *Advances in Understanding the Early-summer Heavy Rainfall over South China*. The Global Monsoon System: Research and Forecast.

Mei, W., S. P. Xie, 2016. Intensification of Landfalling Typhoons over the Northwest Pacific Since the Late 1970s. *Nature Geoscience*, 9.

Ng, C. W. W., C. E. Choi, R. P. H. Law, 2013. Longitudinal Spreading of Granular Flow in Trapezoidal Channels. *Geomorphology*, 194.

Ng, C. W. W., C. E. Choi, A. Y. Su, *et al.*, 2016a. Large-scale Successive Impacts on a Rigid Barrier Shielded by Gabions. *Canadian Geotechnical Journal*, 53(10).

Ng, C. W. W., D. Song, C. E. Choi, *et al.*, 2016b. A Novel Flexible Barrier for Landslide Impact in Centrifuge. *Géotechnique Letters*, 6(3).

Ng, C. W. W., A. K. Leung, R. Yu, *et al.*, 2016c. Hydrological Effects of Live Poles on Transient Seepage in an Unsaturated Soil Slope: Centrifuge and Numerical Study. *Journal of Geotechnical and Geoenvironmental Engineering*, 143(3).

Ng, C. W. W., A. Y. Su, C. E. Choi, *et al.*, 2018a. Comparison of Cushion Mechanisms Between Cellular Glass and Gabions Subjected to Successive Boulder Impacts. *Journal of*

Geotechnical and Geoenvironmental Engineering, 144(9).

Ng, C. W. W., C. E. Choi, R. C. H. Koo, *et al.*, 2018b. Dry Granular Flow Interaction with Dual-Barrier Systems. *Géotechnique*, 68(5).

Ng, C. W. W., A. Leung, J. Ni, 2019a. *Plant-soil Slope Interaction*. CRC Press.

Ng, C. W. W., R. Chen, J. L. Coo, *et al.*, 2019b. A Novel Vegetated Three-layer Landfill Cover System Using Recycled Construction Wastes Without Geomembrane. *Canadian Geotechnical Journal*, 56(12).

Ni, J. J., A. K. Leung, C. W. Ng, 2018. Unsaturated Hydraulic Properties of Vegetated Soil Under Single and Mixed Planting Conditions. *Géotechnique*, 69(6).

Oke, T. R., 1982. The Energetic Basis of the Urban Heat Island. *Quarterly Journal of the Royal Meteorological Society*, 108(455).

Peng, L., J. P. Liu, Y. Wang, *et al.*, 2018. Wind Weakening in a Dense High-rise City due to over Nearly Five Decades of Urbanization. *Building and Environment*, 138.

Peng, L. L. H., C. Y. Jim, 2015. Economic Evaluation of Green-roof Environmental Benefits in the Context of Climate Change: The Case of Hong Kong. *Urban Forestry & Urban Greening*, 14(3).

Pritchard, D. W., 1967. *What Is an Estuary: Physical Viewpoint, in Estuaries*. American Association for the Advancement of Science, Washington D.C.

Qian, C., W. Zhou, X. Q. Yang, *et al.*, 2018. Statistical Prediction of Non-Gaussian Climate Extremes in Urban Areas Based on the First-order Difference Method. *International Journal of Climatology*, 38(6).

Qian, W., J. Gan, J. Liu, *et al.*, 2018. Current Status of Emerging Hypoxia in a Eutrophic Estuary: The Lower Reach of the Pearl River Estuary, China. *Estuarine, Coastal and Shelf Science*, 205.

Qiu, H., S. Sun, R. Tang, *et al.*, 2016. Pneumonia Hospitalization Risk in the Elderly Attributable to Cold and Hot Temperatures in Hong Kong, China. *American Journal of Epidemiology*, 184(8).

Rabalais, N. N., R. E. Turner, W. J. Wiseman Jr., 2002. Gulf of Mexico Hypoxia, Aka "The Dead Zone". *Annual Review of Ecology and Systematics*, 33(1).

Ren, C., E. Y. Ng, L. Katzschner, 2011. Urban Climatic Map Studies: A Review. *International Journal of Climatology*, 31(15).

Sharifi, A., Y. Yamagata, 2014. Resilient Urban Planning: Major Principles and Criteria. *Energy Procedia*, 61.

Shepherd, J. G., P. G. Brewer, A. Oschlies, 2017. Ocean Ventilation and Deoxygenation in a Warming World: Introduction and Overview. *Philosophical Transactions of the Royal Society A*, 375.

Shi, X., C. Lu, X. Xu, 2011. Variability and Trends of High Temperature, High Humidity, and Sultry Weather in the Warm Season in China During the Period 1961-2004. *Journal of Applied Meteorology and Climatology*, 50(1).

Singh, M. S., Z. Kuang, E. D. Maloney, *et al.*, 2017. Increasing Potential for Intense Tropical and Subtropical Thunderstorms Under Global Warming. *Proceedings of the National Academy of Sciences of the United States of America*, 114(44).

Siu, L. W., M. A. Hart, 2013. Quantifying Urban Heat Island Intensity in Hong Kong SAR, China. *Environmental Monitoring and Assessment*, 185(5).

Song, D., C. E. Choi, C. W. W. Ng, *et al.*, 2018. Geophysical Flows Impacting a Flexible Barrier: Effects of Solid-fluid Interaction. *Landslides*, 15(1).

Song, Q., J. Li, H. Duan, *et al.*, 2017. Towards to Sustainable Energy-efficient City: A Case Study of Macau. *Renewable and Sustainable Energy Reviews*, 75.

Su, J., M. Dai, B. He, *et al.*, 2017. Tracing the Origin of the Oxygen-Consuming Organic Matter in the Hypoxic Zone in a Large Eutrophic Estuary: The Lower Reach of the Pearl River Estuary, China. *Biogeosciences*, 14(18).

Sun, Q., C. Miao, Q. Duan, 2015. Projected Changes in Temperature and Precipitation in Ten River Basins over China in 21st century. *International Journal of Climatology*, 35.

Sun, S., W. Cao, T. G. Mason, *et al.*, 2019. Increased Susceptibility to Heat for Respiratory Hospitalizations in Hong Kong. *Science of the Total Environment*, 666.

Sun, S., F. Laden, J. E. Hart, *et al.*, 2018. Seasonal Temperature Variability and Emergency Hospital Admissions for Respiratory Diseases: A Population-based Cohort Study. *Thorax*, 73(10).

Sun, S., L. Tian, H. Qiu, *et al.*, 2016. The Influence of Pre-existing Health Conditions on Short-term Mortality Risks of Temperature: Evidence from a Prospective Chinese Elderly Cohort in Hong Kong. *Environmental Research*, 148.

Tam, R. C. K., B. L. S. Lui, 2013. *Public Education and Engagement in Landslip Prevention and Hazard Mitigation*. Paper Submitted to the International Conference on Post-disaster Reconstruction – Sichuan 5.12 Hong Kong, The Hong Kong Institution of Engineers.

Tan, Z., K. K. L. Lau, E. Ng, 2016. Urban Tree Design Approaches for Mitigating Daytime Urban Heat Island Effects in a High-density Urban Environment. *Energy and Buildings*, 114.

Tian, L., H. Qiu, S. Sun, *et al.*, 2016. Emergency Cardiovascular Hospitalization Risk Attributable to Cold Temperatures in Hong Kong. *Circulation: Cardiovascular Quality and Outcomes*, 9(2).

Tollefson, J., 2013. Natural Hazards: New York vs the Sea. *Nature*, 494(7436).

To, W. M., T. W. Yu, 2016. Characterizing the Urban Temperature Trend Using Seasonal Unit

Root Analysis: Hong Kong from 1970 to 2015. *Advances in Atmospheric Sciences*, 33(12).

Tong, C. H. M., S. H. L. Yim, D. Rothenberg, *et al.,* 2018. Projecting the Impacts of Atmospheric Conditions Under Climate Change on Air Quality over the Pearl River Delta Region. *Atmospheric Environment*, 193.

Tracy, A., K. Trumbull, C. Loh, 2007. *The Impact of Climate Change in Hong Kong and the Pearl River Delta*. China Perspectives.

Tsuboki, K., M. K. Yoshioka, T. Shinoda, *et al.*, 2015. Future Increase of Supertyphoon Intensity Associated with Climate Change. *Geophysical Research Letters*, 42(2).

Tu, X., V. P. Singh, X. Chen, *et al.*, 2015. Intra-annual Distribution of Streamflow and Individual Impacts of Climate Change and Human Activities in the Dongijang River Basin, China. *Water Resources Management*, 29.

Walsh, K. J. E., J. L. McBride, P. J. Klotzbach, *et al.*, 2016. Tropical Cyclones and Climate Change. *Wiley Interdisciplinary Reviews: Climate Change*, 7(1).

Wang D., K. K. L. Lau, C. Ren, *et al.*, 2019. The Impact of Extremely Hot Weather Events on All-cause Mortality in a Highly Urbanized and Densely Populated Subtropical City: A 10-year Time-series Study (2006–2015). *Science of The Total Environment*, 690.

Wang, L., G. Huang, W. Zhou, *et al.*, 2016. Historical Change and Future Scenarios of Sea Level Rise in Macau and Adjacent Waters. *Advances in Atmospheric Sciences,* 33(4).

Wang, M. Y., S. H. L. Yim, G. H. Dong, *et al.*, 2020. Mapping Ozone Source-receptor Relationship and Apportioning the Health Impact in the Pearl River Delta Region Using Adjoint Sensitivity Analysis. *Atmospheric Environment*, 222.

Wang, P., W. B. Goggins, E. Y. Y. Chan, 2016. Hand, Foot and Mouth Disease in Hong Kong: A Time-series Analysis on Its Relationship with Weather. *PLoS One*, 11(8).

Wang, P., W. B. Goggins, E. Y. Y. Chan, 2018a. Associations of Salmonella Hospitalizations with Ambient Temperature, Humidity and Rainfall in Hong Kong. *Environment International*, 120.

Wang, P., W. B. Goggins, E. Y. Y. Chan, 2018b. A Time-series Study of the Association of Rainfall, Relative Humidity and Ambient Temperature with Hospitalizations for Rotavirus and Norovirus Infection Among Children in Hong Kong. *Science of the Total Environment*, 643.

Wang, W., W. Zhou, E. Y. Y. Ng, *et al.*, 2016. Urban Heat Islands in Hong Kong: Statistical Modeling and Trend Detection. *Natural Hazards*, 83(2).

Wang, X., T. Yang, X. Li, *et al.*, 2017. Spatio-temporal Changes of Precipitation and Temperature over the Pearl River Basin Based on CMIP5 Multi-model Ensemble. *Stochastic Environmental Research and Risk Assessment*, 31.

Wang, X., Z. Wu, G. Liang, 2009. WRF/CHEM Modeling of Impacts of Weather Conditions

Modified by Urban Expansion on Secondary Organic Aerosol Formation over Pearl River Delta. *Particuology*, 7(5).

Wang, Y., S. Wen, X. Li, *et al.*, 2016. Spatiotemporal Distributions of Influential Tropical Cyclones and Associated Economic Losses in China in 1984–2015. *Natural Hazards*, 84(3).

Wang, Y., Y. Li, S. Di Sabatino, *et al.*, 2018. Effects of Anthropogenic Heat due to Air-conditioning Systems on an Extreme High Temperature Event in Hong Kong. *Environmental Research Letters*, 13(3).

Wehner, M. F., K. A. Reed, B. Loring, *et al.*, 2018. Changes in Tropical Cyclones Under Stabilized 1.5 and 2.0 °C Global Warming Scenarios as Simulated by the Community Atmospheric Model Under the HAPPI Protocols. *Earth System Dynamics*, 9(1).

Wei, X., K. S. Lam, C. Cao, *et al.*, 2016. Dynamics of the Typhoon Haitang Related High Ozone Episode over Hong Kong. *Advances in Meteorology*, 2016.

Wong, C. M., H. Y. Mok, T. C. Lee, 2011. Observed Changes in Extreme Weather Indices in Hong Kong. *International Journal of Climatology*, 31(15).

Wong, H. N., Y. M. Chen, K. C. Lam, 1997. *Factual Report on the November 1993 Natural Terrain Landslides in Three Study Areas on Lantau Island (GEO Report No. 61)*. Hong Kong, Geotechnical Engineering Office.

Wong, H. N., K. K. S. Ho, 1995. *General Report on Landslips on 5 November 1993 at Man Made Features in Lantau (GEO Report No. 44)*. Hong Kong, Geotechnical Engineering Office.

Wong. H. N., F. W. Y. Ko, T. H. H. Hui, 2006. *Assessment of Landslide Risk of Natural Hillsides in Hong Kong (GEO Report No. 191)*. Hong Kong, Geotechnical Engineering Office.

Wong, W. T., K. W. Li, K. H. Yeung, 2003. Long Term Sea Level Change in Hong Kong. *Hong Kong Meteorological Society Bulletin*, 13(1-2).

Wu, C., C. Ji, B. Shi, *et al.*, 2019. The Impact of Climate Change and Human Activities on Streamflow and Sediment Load in the Pearl River Basin. *International Journal of Sediment Research*, 34(4).

Wu, J., Z. Liu, H. Yao, *et al.*, 2018. Impacts of Reservoir Operations on Multi-scale Correlations Between Hydrological Drought and Meteorological Drought. *Journal of Hydrology*, 563.

Wuebbles, D. J., D. W. Fahey, K. A. Hibbard, *et al.*, 2017. *Climate Science Special Report: Fourth National Climate Assessment (NCA4), Volume I*, U.S. Global Change Research Program, Washington, DC, USA.

Yan, D., S. E. Werners, F. Ludwig, *et al.*, 2015. Hydrological Response to Climate Change: The Pearl River, China Under Different RCP Scenarios. *Journal of Hydrology: Regional*

Studies, 4.

Yang, S. H., N. Y. Kang, J. B. Elsner, *et al.*, 2018. Influence of Global Warming on Western North Pacific Tropical Cyclone Intensities During 2015. *Journal of Climate*, 31(2).

Yang, Y. K., I. S. Kang, M. H. Chung, *et al.*, 2017. Effect of PCM Cool Roof System on the Reduction in Urban Heat Island Phenomenon. *Building and Environment*, 122.

Yi, W., A. P. C. Chan, 2015. Effects of Temperature on Mortality in Hong Kong: A Time Series Analysis. *International Journal of Biometeorology*, 59(7).

Yi, W., A .P. C. Chan, 2017. Effects of Heat Stress on Construction Labor Productivity in Hong Kong: A Case Study of Rebar Workers. *International Journal of Environmental Research and Public Health*, 14(9).

Yim, S. H. L., X. Hou, J. Guo, 2019a. Contribution of Local Emissions and Transboundary Air Pollution to Air Quality in Hong Kong During El Niño-Southern Oscillation and Heatwaves. *Atmospheric Research*, 218.

Yim, S. H. L., M. Y. Wang, Y. Gu, *et al.*, 2019b. Effect of Urbanization on Ozone and Resultant Health Effects in the Pearl River Delta Region of China. *Journal of Geophysical Research: Atmospheres*, 124.

Yin, K., S. Xu, W. Huang, *et al.*, 2017. Effects of Sea Level Rise and Typhoon Intensity on Storm Surge and Waves in Pearl River Estuary. *Ocean Engineering*, 136.

Yue, D. P. T., S. L. Tang, 2011. Sustainable Strategies on Water Supply Management in Hong Kong. *Water and Environment Journal*, 25.

Yu, Q., A. K. H. Lau, K. T. Tsang, *et al.*, 2018. Human Damage Assessments of Coastal Flooding for Hong Kong and the Pearl River Delta Due to Climate Change-related Sea Level Rise in the Twenty-first Century. *Natural Hazards*, 92(2).

Yuan, R., J. Zhu, B. Wang, 2015. Impact of Sea-level Rise on Saltwater Intrusion in the Pearl River Estuary. *Journal of Coastal Research*, 31(2).

Zhai, W., M. Dai, W. J. Cai, *et al.*, 2005. High Partial Pressure of CO_2 and Its Maintaining Mechanism in a Subtropical Estuary: The Pearl River Estuary, China. *Marine Chemistry*, 93(1).

Zhang, Q., J. Li, X. Gu, *et al.*, 2018. Is the Pearl River Basin, China, Drying or Wetting? Seasonal Variations, Causes and Implications. *Global and Planetary Change*, 166.

Zu, T., J. Gan, 2015. A Numerical Study of Coupled Estuary-shelf Circulation Around the Pearl River Estuary During Summer: Responses to Variable Winds, Tides and River Discharge. *Deep Sea Research Part II: Tropical Studies in Oceanography*, 117.

海事及水务局：“澳门水资源状况”，2021 年。

海事及水务局：“2014—2016 澳门水资源与供水”，2017 年。

孔兰、陈晓宏、闻平等：“2009/2010 年枯水期珠江口磨刀门水道强咸潮分析”，《自然资源学报》，2011 年第 11 期。
全球水伙伴中国委员会（Global Water Partnership China）："气候变化下珠江三角洲水问题及其应对与治理措施"，2016 年。https://www.gwp.org/globalassets/global/gwp-china_files/wacdep/_3.pdf。
任超、吴恩融、L. Katzschner 等："城市环境气候图的发展及其应用现状"，《应用气象学报》，2012 年第 5 期。
深圳市水务局："深圳市水务发展'十三五'规划"，2016 年。
沈灿、季冰："香港水资源特征和供需水量平衡研究"，《地理研究》，1997 年第 2 期。
涂新军、陈晓宏、赵勇等："变化环境下东江流域水文干旱特征及缺水响应"，《水科学进展》，2016 年第 6 期。
王雨："一国两制下的跨境水资源治理"，《热带地理》，2017 年第 2 期。
香港环境局："香港气候变化报告"，2015 年。
香港环境局："香港气候行动蓝图 2030+"，2017 年。
香港立法会秘书处："香港的水资源"，2015 年。
香港立法会秘书处："东江水输港概况"，2017 年。
香港水务署："香港的全面水资源管理-持续共享珍贵水资源"，2008 年。
香港水务署："香港供水里程碑"，2011 年。
香港水务署："2016～2017 年报"，2018 年。
香港水务署："全面水资源管理策略 2019"，2019 年。
徐爽、侯贵兵："2015～2016 年珠江枯水期水量调度工作实践"，《中国防汛抗旱》，2017 年第 2 期。
于洋、卢然超、李迎霞："澳门水资源现状与展望"，《给水排水》，2014 年第 2 期。
赵伟、高博、刘明等："气象因素对香港地区臭氧污染的影响"，《环境科学》，2019 年第 1 期。
周平、陈刚、刘智勇等："东江流域降水与径流演变趋势及周期特征分析"，《生态科学》，2016 年第 2 期。

第三章　减缓气候变化

基于气候变化的科学事实（第一章）及其影响（第二章），香港和澳门两个特别行政区作为中国最富裕的地区有着减排温室气体的责任。

不同机构及团体已经制定了各种描述或分类温室气体排放的术语。例如，“直接”排放与“地域”“基于生产”“范畴一”排放等术语相对应，而“间接”排放对应的术语包括“隐含”“基于消费”“范畴二”“范畴三”和“上游”“下游”排放等。

表 3–1　《CSG 协议》中规定的范畴一、范畴二、范畴三温室气体排放定义

核算方法	范畴	定义
基于生产（Production-based，PBA）	范畴一	位于城市边界内排放源的温室气体排放
基于消费（Consumption-based，CBA）	范畴二	在城市边界内使用电网供电、热能、蒸汽和/或冷却产生的温室气体排放
	范畴三	城市边界内活动导致在城市边界以外发生的所有其他温室气体排放

资料来源：Fong *et al.*, 2014。

范畴一排放最广泛应用于约束一国的气候承诺，如《京都议定书》和《巴黎协定》，而中国在评估每个省份的二氧化碳排放时采用范畴二排放（国家统计局, 2016）。对于全球来说，这三个范畴没有区别，因为所有的生产和消费都发生在地球上。然而，当能源服务（范畴二）和商品/服务（范畴三）来源于特定地理区域以外时，它们之间的差距往往会扩大。从人均GDP来看，香港和澳门特别行政区是中国最富裕的地区。由于地域小，与其他省份及海外国家的贸易密集，因此在温室气体排放的三个范畴及相应的缓解方案方面均有显著差异。

本章分别对香港和澳门进行考察。第一节回顾了两个区域的范畴一温室气体排放情况和排放源。第二节侧重于范畴二、范畴三排放。第三节将针对范畴一排放探讨缓解方案。

第一节　温室气体排放及来源

一、香港

自20世纪90年代以来，香港特区政府环境保护署（HKEPD）一直持续编写温室气体排放清单。1998年，香港环保署委托进行了一项“控制温室气体排放研究”，为未来的决策奠定基础（HKEPD, 2000）。该研究根据IPCC于1996年修订的《国家温室气体清单指南》，提供了第一份香港温室气体排放清单。2008年，香港环保署委托进行了“香港气候变化：可行性研究”，回顾并更新了本地温室气体排放及清除量清单（HKEPD, 2010），尽可能替换1996年修订的IPCC指南中的方法，与2006年IPCC指南中的方法保持一致（IPCC, 2006b）。有关这项研究的更详细信息以技术附录的形式刊登于香港环保署网站（HKEPD, 2010）。

香港温室气体排放清单数据主要源自官方统计机构和香港环保署，涵盖《京都议定书》规定的六类温室气体包括二氧化碳（CO_2）、甲烷（CH_4）、一氧化二氮（N_2O）、氢氟碳化合物（HFCs）、全氟化碳（PFCs）和六氟化硫（SF_6）。

IPCC 第三次评估报告中查明的其他温室气体（IPCC, 2001）如三氟化氮（NF_3）、三氟甲基五氟化硫（SF_5CF_3）和卤化醚，已纳入香港环保署 2010 年的研究范围。然而，这三种新的温室气体未被要求报告作为国家排放清单的一部分，因此并未纳入香港公布的年度温室气体排放清单。即使计算这些温室气体，对总排放的影响也不大，因为香港少有与这些温室气体相关的工业活动。普遍认为，香港环保署自 2012 年后，在其年度温室气体排放中采用了 2006 年 IPCC 指南所界定的新的全球变暖潜能值（HKEPD, 2010）。

在香港，二氧化碳排放在六种温室气体中占绝对主导地位，在 2006 年占比 90%以上，甲烷和一氧化二氮分别占 5%和 1%，其余气体排放量太少，可忽略不计（图 3–1）。二氧化碳排放占主导地位的主要原因是发电、运输和其他用途的燃料消耗是香港境内主要的温室气体排放活动。香港境内没有主要导致甲烷排放的大规模农业活动，也没有很多可产生其他温室气体的工业活动。

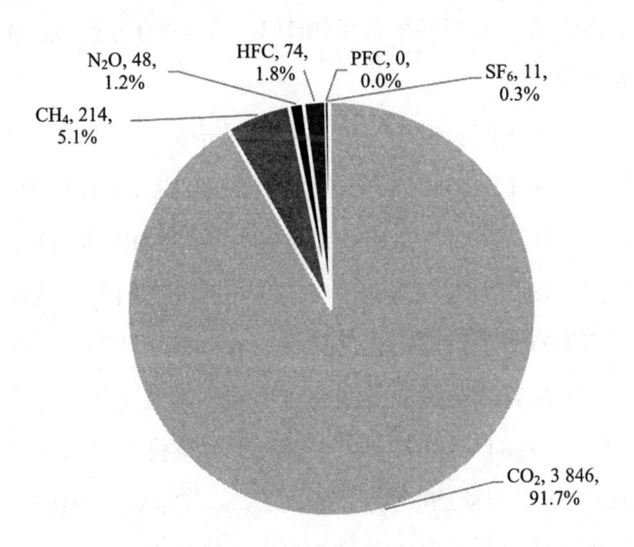

图 3–1　香港 2006 年主要温室气体排放量（万吨 CO_{2eq}）

资料来源：HKEPD，2010。

（一）温室气体排放总量

香港温室气体排放总量从 1990 年的 3 520 万吨 CO_{2eq}，总体增长到 2014 年的 4 490 万吨 CO_{2eq} 达到最高点，然后缓慢下降至 2017 年的 4 070 万吨 CO_{2eq}（图 3–2）。虽然 2017 年温室气体排放总量与 20 世纪 90 年代初的排放总量处于同一区间内，但香港温室气体排放总量曾在 1994～2004 年出现下降，这主要由于 20 世纪 90 年代初的燃料组合转变，在以燃煤为主的发电中结合了核能和天然气。然而，随着时间的推移，燃料组合转变带来的减排效益被香港不断增长的电力消耗抵消，从 1990 年的 85 801 太焦增长至 2017 年的 157 604 太焦，增幅超过 83%（HK Census and Statistics Department, 2019a）。

1990～2017 年，尽管发电的碳消耗降幅大于 30%，但是香港的人均排放量仅下降了约 11%，从 6.2 吨 CO_{2eq} 降至 5.5 吨 CO_{2eq}（图 3–2）。由于整体电力需求增加，香港人口增长了近 30%，从 1990 年约 5 704 500 人增至 2017 年 7 413 100 人；而人均能源消耗增长了约 18%，从 1990 年约 0.039 太焦增长到 2017 年 0.046 太焦。

能源消耗与本地生产总值（GDP）、经济结构密切相关。香港的碳强度已降低了 58%以上，从 1990 年 3.6 吨 CO_{2eq}/100 000 港元 GDP（即 36 克 CO_{2eq}/港元 GDP）降至 2017 年 1.5 吨 CO_{2eq}/100 000 港元 GDP（即 15 克 CO_{2eq}/港元 GDP）（图 3–2）。这在很大程度上反映了这一时期燃料组合从煤炭为主向使用天然气和从其他省份进口的核电的转变。香港从工业型经济向服务型经济的转变，即 GDP 增长由更低的能源强度和碳强度实现的转变也主要发生在 20 世纪 80 年代。1984～1997 年，工业占总体碳排放比重从 33.2%下降到 14.6%，而商业的比重则从 20.0%上升到 29.9%（Chow, 2001b）。这一趋势与工业向珠江三角洲转移及香港转向服务型经济相对应（Chow, 2001b）。

在 2006 年 IPCC 指南所总结的五类主要温室气体排放行业中，香港有四类行业的报告，即能源，废弃物处理，工业过程与产品使用（Industrial Processes and Product Use，IPPU），农业、林业与其他土地使用（Agriculture, Forestry and Other Land Use，AFOLU）。过去25年来，香港温室气体排放清单内不同行业的相对比重并未显著改变，反映出这一时期服务型经济的主导地位相应导致了电力行业在影响香港温室气体排放总量方面起着主导作用。

在香港，能源行业是温室气体排放的最大来源，约占2017年香港温室气体排放总量的89%。其中发电占比最高，约为65%；其次是交通运输的燃料消耗，约占 18%；燃料的其他终端消耗约占 6%。废弃物处理行业的温室气体排放约占7%，IPPU的温室气体排放约占4%，而AFOLU占香港温室气体排放总量的比例不到0.1%。

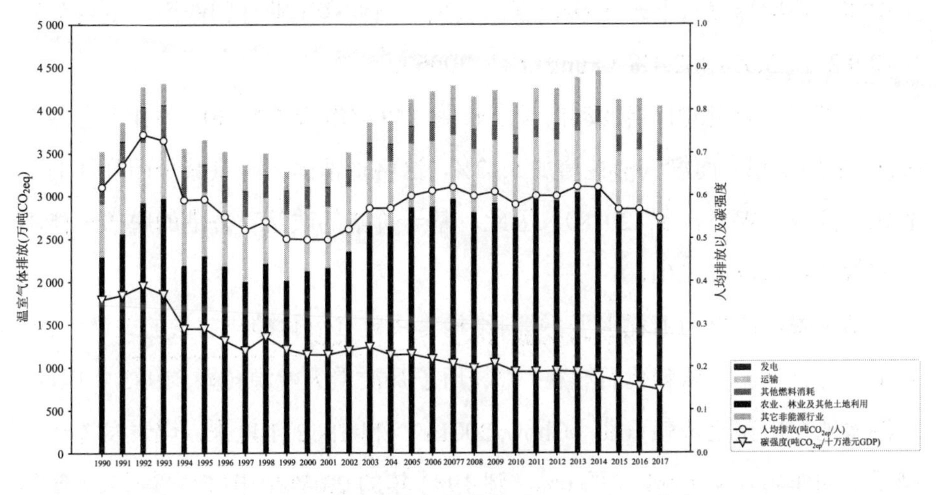

图3–2　香港温室气体排放（1990～2017年）

资料来源：HKEPD，2019a。

（二）电力行业

香港2017年总计4 070万吨CO_{2eq}的碳排放中，能源消费，包括发电、运输及其他终端消耗达到了3 610万吨CO_{2eq}。其中最大的碳排放来源是发电，约占65.4%（2 660万吨CO_{2eq}），而燃气生产仅占2017年香港温室气体排放总量的 0.75%左右（30 万吨 CO_{2eq}）（HKEPD, 2019a）。

本地生产总值和电力消耗紧密相关，在香港尤其如此（Ho and Siu, 2007）。随着经济发展和产业结构转型，过去数十年间，香港各行业在电力消耗中的相对地位发生了显著的变化。1975 年工业和商业用电量占比相当，都约为 38%；但到了 2017 年，工业用电量占比下降到 7%，而商业用电量占比则上升至 66%（Chow, 2001b）。这一趋势与工业向珠江三角洲转移及香港转向服务型经济相对应（Chow, 2001b）。有着炎热潮湿的漫长夏季和温和冬季的香港，社会越富裕，空调的用电量越大。商业和居民用电的季节性波动非常明显（Fung *et al.*, 2006）。因此，气候变暖结合强烈的热岛效应，使香港的耗电量增加。环境温度每上升 1 摄氏度，居民、商业和工业部门的用电量将分别增加 9.2%、3.0%和 2.4%（Fung *et al.*, 2006）。

香港的用电量对价格变动的反应较弱。单位电价剧增 40%预计仅能减少 0.81%的用电量，而燃气消耗增加 5.12%，这将抵消因用电量减少而降低的二氧化碳排放（Woo *et al.*, 2018）。因此，基于价格的政策对香港的脱碳可能不会有效（Woo *et al.*, 2018）。

基于香港的电力消耗水平，燃料结构是决定二氧化碳排放量的关键因素。香港电力行业严重依赖化石燃料，曾经历了几次重大能源转型。1982 年以前，香港本地发电完全依赖石油（Chow, 2001a）。自 1982 年以来，煤炭发挥越来越重要的作用，从 1981 年的 0%，到 1982 年的 23.5%，再增长至 1987 年的 95.2%，而石油的份额锐减（Chow, 2001a）。香港没有本地生产的化石燃料，所有的化石燃料必须从其他省份或海外采购。随着电力消费的增长和能源向煤炭转移，香港自 1981 年开始大量进口蒸汽煤，当年进口量为 5.2 万吨；1982 年跃升至 145.1 万吨，并在 1993 年达到最高点 1182.8 万吨（Chow, 2001a）。后来，因从大亚湾核电站购买核电，使得香港本地发电量大幅减少，从而大幅减少了对煤炭的需求。煤炭进口量在 1993 年达到峰值，1994 年下降到 845 万吨（Chow, 2001a）。相应地，香港的二氧化碳排放量在 1994 年也显著下降。

20 世纪 80 年代大部分煤炭进口来自南非和澳大利亚，南非的份额在 20 世纪 90 年代急速缩减，到 1998 年，大部分煤炭进口来自印度尼西亚和澳大利亚（Chow, 2001a）。

1996 年，天然气占香港本地发电的燃料的比例为 24.0%，1998 年增至 33.1%。天然气一般通过管道或液化天然气罐输送（Chow, 2001a）。由于这一特点，香港完全依赖其他省份供应天然气（HK Census and Statistics Department, 2018）。1995 年开发南海崖城 13-1 气田后，香港依靠该气田供应天然气（Wilburn and Roberts, 1995）。2008 年，中国国家能源局与香港特区政府签署备忘录，支持香港的天然气供应（NEA and HKSAR Government, 2008），除现有供应源外，将通过自西向东的天然气管道和其他省份的液化天然气码头给香港供气。

然而，2002～2015 年煤炭仍然是香港发电的主要燃料来源，占电力部门温室气体排放的比重平均为74.3%，天然气和石油的比重分别为25.1%和0.6%（To and Lee, 2017c）。因此，为了实现政府 2020 年的燃料组合目标，即从 2012 年 53%的煤炭、22%的天然气、23%的核能和 2%燃油及可再生能源组合，到 2020 年转变为 50%的天然气、25%的核能和 25%的煤炭加可再生能源，需要减少燃煤发电，替换为更多的天然气发电，有专家提议用浮式液化天然气码头直接从国际市场进口液化天然气到香港（Environmental Resources Management, 2018）。2019 年，青山发电有限公司及香港电灯有限公司已获香港特区政府批准，在索罟群岛以东海域兴建和运营海上液化天然气码头。该码头能容纳一艘浮式储存再气化装置船和一艘液化天然气运输船停泊在双泊位码头上。储气船上的再气化液化天然气将通过两条独立的海底天然气管道传输至龙鼓滩发电厂和南丫发电厂的天然气接收站（HK Lands Department, 2019）。

（三）运输行业

运输行业是香港温室气体排放的第二大来源，在2017年占排放总量的比重约为18%（720万吨CO_{2eq}）（图3–2）。这个数据只包括香港特别行政区内公路、铁路、海运及民航业内部运输导致的燃料消耗。虽然当前公布的香港特区温室气体排放清单并没有提供更详细的数据源，但香港环保署2010年的报告确实提供了更详细的信息（HKEPD, 2010）。2006年，公路运输是温室气体最大的排放源，占运输行业排放总量的81.9%（图3–3），其次是航运业，占11.2%，而航空业则微不足道，仅占0.3%。虽然香港有重要的海港和国际机场，但由于香港特区面积小，只有一小部分与这两个交通枢纽相关的跨境运输会产生的二氧化碳排放，并记入香港的范畴一排放。

2016年，运输行业消耗了89 819太焦的能源，货运和客运这两个子行业分别贡献了31%和69%（HK Electrical and Mechanical Services Department, 2019）。2006～2016年运输业的能源消耗下降了6.8%，主要由于货运的能源消耗减少了29.3%，而客运能源使用增加了8.5%。在香港的客运方面，汽车与摩托车这一类别消耗的能源最多，为23 344太焦（26%）；其他部分依次为公交车耗能18 918太焦（21%），出租车耗能12 288太焦（13.7%），以及铁路耗能2 951太焦（3.3%）。

在香港的陆路客运方面，2016年电气化列车（地铁和轻轨）平均每天载客470万人次（Lang, 2018）。以汽油或柴油为燃料的13 000辆公交车每天运送数百万乘客。约11.3万辆使用化石燃料的轻型卡车在全港运销食品和消费品往返城市的大街小巷。持牌机动车总数由2010年的607 796辆增至2017年的766 200辆，增幅26.1%，包括摩托车、私家车、公交车、出租车、货车、小巴及政府用车，而2017年日均车辆总行驶千米数为3 740万千米（HK Census and Statistics Department, 2019b）。尽管平均通勤距离较长，但香港的客运任务主要由高能效的轨道交通承担（Leung *et al.*, 2018）。

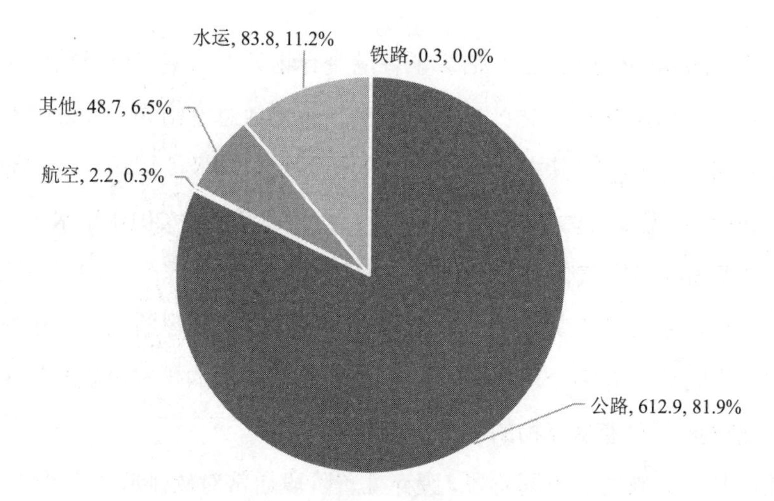

图 3–3　2006 年香港运输业温室气体排放量（万吨 CO_{2eq}）

资料来源：HKEPD，2010。

（四）其他行业

1. 其他燃料消耗

2017 年占排放总量 5.6%的“其他燃料消耗”碳排放（约 230 万吨 CO_{2eq}）包括工业（如制造业和建筑业）、商业和住宅场所的燃料燃烧（HKEPD, 2019a）。1990～2006 年，制造业和建筑业的温室气体排放逐渐减少，而商业和住宅的温室气体排放却在缓慢增加（HKEPD, 2019a）。如果将 2006 年的排放比例应用到 2017 年温室气体排放总量数据，大约 5%的碳排放（200 万吨 CO_{2eq}）来自商业和住宅，而只有 0.7%（27.4 万吨 CO_{2eq}）来自制造业和建筑业。

2. 废弃物处理行业

为香港服务的废弃物处理设施包括三个堆填区、七个废弃物转运站、一个化学废弃物处理中心、一个动物废弃物堆肥场、一个焚化炉及八个污水处理厂（Dong *et al.*, 2017）。废弃物处理的碳排放主要来自：①固体废弃物处理，②废水处理和排放，③固体废弃物的生物处理，④废弃物的焚烧和露天焚烧。

董雅红等（Dong *et al.*, 2017）估算了香港废弃物处理温室气体排放的主要来源。结果显示，2010 年废弃物处理行业温室气体排放总量约为 250 万吨 CO_{2eq}，人均 3.57 千克 CO_{2eq}。其中，超过 67%的排放来自垃圾填埋场。这项研究还显示了 2020 年温室气体排放的预测结果：排放总量将在 2010 年水平上增加 13%，填埋场的温室气体排放量减少，而整个废弃物处理行业的温室气体排放总量增加，主要源于焚化炉中石油产品燃烧和生物处理的运作。敏感度分析表明，当固体废弃物处理量减少 40%时，总体增加的排放可以被抵消，因此强烈建议提高香港废弃物的回收及再利用率。

在产品生命周期的不同阶段，废弃处置阶段通常对碳排放具有更大意义，有人研究了香港使用不同类型的购物袋在不同处置方式下的碳排放，发现与回收或填埋相比，购物袋的再利用降低了碳排放量，购物袋的再利用率增加 5%可带来约 20%的减排效应（Muthu *et al.*, 2011）。

2017 年，废弃物处理行业是香港第三大温室气体排放源，占香港温室气体排放总量 4070 万吨 CO_{2eq} 的 6.9%（280 万吨 CO_{2eq}）（HKEPD, 2019a）。甲烷是其中最主要的温室气体，占废弃物处理行业产生的所有温室气体排放的 90%左右，而一氧化二氮和少量二氧化碳则构成剩余的 10%。固体废弃物处理在废弃物有关的温室气体排放中占比最多，约占废弃物处理行业在 2006 年产生的温室气体排放总量的九成（HKEPD, 2010）。甲烷是固体废弃物处理排放所产生的唯一温室气体。如果将 90%的排放比例应用于 2017 年的排放数据，那么 2017 年垃圾填埋产生的温室气体排放约为 250 万吨 CO_{2eq}。值得注意的是，自 2006 年以来，越来越多的堆填区气体被用于发电（HKEPD, 2017）。

废水处理与排放产生的温室气体排放是废弃物处理行业的第二大排放源，在 2006 年约占废弃物处理行业产生温室气体总量的 9%（HKEPD, 2010）。将这一比例应用于 2017 年的排放总量 280 万吨 CO_{2eq}，计算得出废水处理的二氧化碳排放约为 25 万吨 CO_{2eq}。这部分的温室气体排放仅包括来自生活污水和工业废水处理的甲烷，及来自生活污水处理的一氧化二氮。香港环保署

2010 年报告中的历史数据显示，一氧化二氮排放量约占废水处理及排放产生的温室气体总量的 80%（HKEPD, 2010）。直接二氧化碳排放只产生于废弃物焚烧。香港自 1997 年起不再进行生活固废焚烧，只设有 1 个化学废弃物处理中心。1997～1998 年废弃物焚烧的温室气体排放量出现减少（HKEPD, 2019a）。

3. 工业流程和产品使用

IPPU 行业是香港温室气体排放总量的第二小来源（仅大于 AFOLU 行业），2017 年排放量为 170 万吨 CO_{2eq}，占排放总量 4.3%（HKEPD, 2019a）。IPPU 部门的温室气体排放为以水泥生产排放的二氧化碳为主，还有部分破坏臭氧层物质（Ozone Depletion Substances，ODSs）的替代品使用产生的氢氟碳化合物与全氟化碳，以及用于变压器等电气设备的六氟化硫。

水泥生产起初是香港 IPPU 行业温室气体排放的主要排放源，但到 2006 年最终成为第二大排放源。2002～2005 年，虽然水泥生产继续，但水泥熟料生产（水泥制造过程的一部分，产生二氧化碳）的暂停，呼应了在此期间温室气体排放量的显著减少。2006 年排放量的增加反映了水泥熟料生产的重启。如果将 2006 年 39%的近似排放比例应用于 2017 年 IPPU 行业排放数据 170 万吨 CO_{2eq}，那么 2017 年水泥生产排放的温室气体约为 66.3 万吨 CO_{2eq}。

截至 2006 年，ODSs 替代品在排放源中的占比从较小到最大，已占 IPPU 行业温室气体排放的 53%。ODSs 替代品是香港温室气体排放清单中氢氟碳化合物和全氟化碳的唯一排放源。如果将 2006 年 53%的排放比例应用于 2017 年 IPPU 行业排放数据，那么 2017 年 ODSs 替代品产生的温室气体排放约为 90.1 万吨 CO_{2eq}。2006 年，六氟化硫占 IPPU 行业温室气体排放总量的 8%。ODSs 替代品也是香港温室气体排放清单中唯一排放六氟化硫的类别。如果将 2006 年 8%的排放比例应用于 2017 年 IPPU 行业排放数据，那么 2017 年 ODSs 替代品产生的温室气体排放约为 13.6 万吨 CO_{2eq}。

4. 农业、林业和其他土地使用（AFOLU）

AFOLU 行业是香港温室气体排放总量的最小来源，2017 年仅排放了 3

万吨 CO_{2eq}，占排放总量的 0.1%（HKEPD, 2019a）。该行业还是唯一有能力为碳清除作出贡献的行业，主要来自林地碳储存的潜在收益。鉴于用地类别可能导致温室气体负排放，不适宜采用比例估测法来判断这一类别内各分部门的不同贡献。

AFOLU 行业的温室气体排放主要有：粪肥管理和土壤管理活动产生的一氧化二氮，肠道发酵和生物质燃烧产生的甲烷，使用石灰和生物质燃烧产生的二氧化碳。

作为子行业的畜牧业进一步分为肠道发酵和粪肥管理。甲烷是动物肠道发酵中排放的唯一温室气体，而粪肥管理既排放甲烷也排放一氧化二氮。2006 年，畜牧业约占 AFOLU 行业产生温室气体总量的 56%，其中粪肥管理是主要组成部分，占 AFOLU 行业排放量的 47%。

陆地上的集合排放源和非二氧化碳排放源包括：生物质燃烧、施用石灰、土壤管理以及粪肥管理所间接排放的一氧化二氮。2006 年，这一类别约占 AFOLU 行业产生的温室气体总量的 44%，其中土壤和粪肥管理所间接排放的一氧化二氮是主要组成部分，占 AFOLU 行业排放量的 37%。

温室气体排放是根据林地、耕地和湿地面积计算的，这些是香港现有的用地类型。香港土地有潜力每年储存超过 46 万吨 CO_{2eq}，约占排放总量的 1%。

（五）范畴一排放达到峰值

香港范畴一排放量在 2014 年应已达到峰值，2017 年减少了 9.4%（图 3–2）。2017 年香港范畴一的排放量比 1990 年增长了 15.6%（图 3–2），人口、本地生产总值分别增长了 29.6%、169.3%。这表明人均温室气体排放量和单位生产总值排放量分别大幅下降了 11%和 58%（图 3–2），这主要是由于去工业化（图 3–4）和生产能效的提高。发电约占范畴一排放量的三分之二。20 世纪 70 年代，工业部门是香港最大的电力消耗部门。然而，香港经济逐步的去工业化使工业电力消耗在 2018 年降至 3.08 太瓦时，甚至远远低于 1980 年的水平。同期，商业及住宅领域成为香港最大的用电领域，耗电量分别为 29.46 太瓦时和 11.66 太瓦时。自 2014 年以来，香港总用电量一直保持稳定，在 43.78 太瓦时和 44.80 太瓦时间小幅波动，尽管人口和人均 GDP 持续上升（图 3–5）。

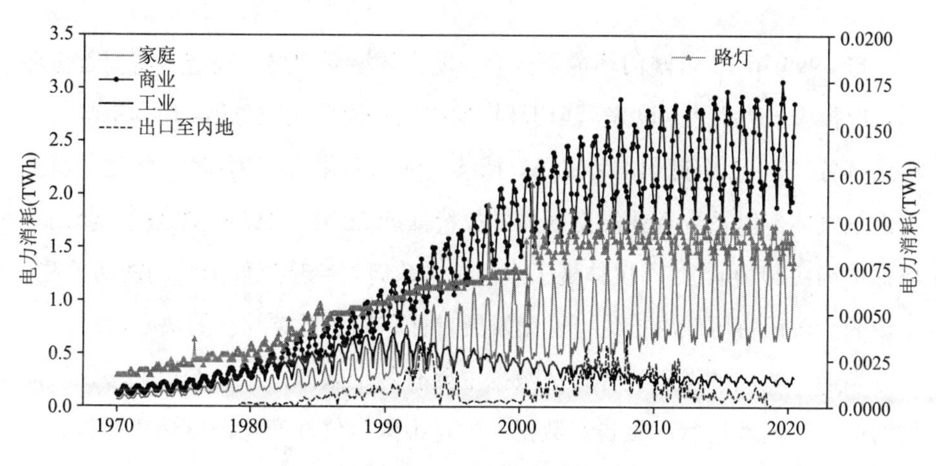

图 3–4　按行业统计的香港每月用电量

资料来源：HK Census and Statistics Department，2019c。

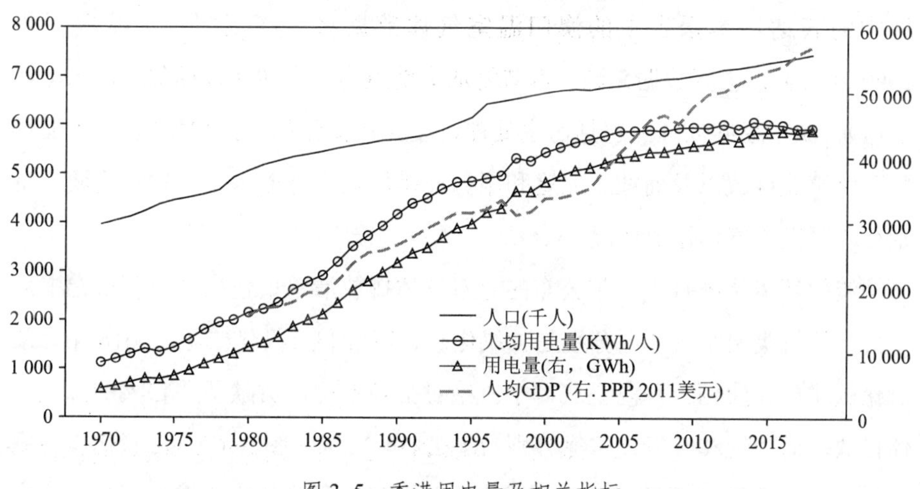

图 3–5　香港用电量及相关指标

资料来源：HK Census and Statistics Department，2019a；IMF，2019。

二、澳门

（一）温室气体排放总量

自 1999 年 12 月澳门环境委员会（现为环境保护局）发布《澳门首份环境状况报告一九九九》以来，澳门环境保护局在澳门民政总署（现为市政署）、交通事务局等 22 个部门的协助下，持续关注二氧化碳、甲烷和一氧化二氮三种温室气体，以及分行业温室气体排放特征近 20 年（1999～2018）（澳门环境保护局，2019）。这不仅体现了澳门特区政府对全球气候形势的密切关注，更表明了澳门特区政府推进可持续发展目标和构建低碳社会的决心。

当前，澳门的温室气体排放数据主要有两个来源：一是澳门环境保护局发布的《澳门环境状况报告》数据；二是中国《气候变化国家信息通报》及《气候变化两年更新报告》数据。两个数据来源由于涉及的具体行业范围有所不同，数据存在一定差异。综合考虑可参考数据的时间跨度以及涉及行业种类的表达，本报告中的澳门温室气体数据将以《澳门环境状况报告1999-2017》为主要数据来源。本节将从本地电力、交通（包括陆运、水运和空运）、城市废弃处理（包括污水处理、废弃物焚化和废弃物堆填）、工业排放、建筑业排放以及商业、家庭和服务业排放六个行业十个类别阐明澳门特别行政区温室气体排放情况。

澳门环境状况报告历年公布的温室气体排放清单仅包括《京都议定书》中规定的三类温室气体，分别是二氧化碳、甲烷和一氧化二氮。2016 年，二氧化碳排放在澳门三种温室气体中占绝对主导地位，贡献了 94.8%的温室气体排放，而甲烷和一氧化二氮分别占 1.9%和 3.2%（图 3–6）。二氧化碳占据主导地位的主要原因是电力行业、运输业和其他用途的化石燃料消耗。澳门的甲烷排放主要来自澳门废弃物堆填及陆上交通，而一氧化二氮则来自澳门污水处理。

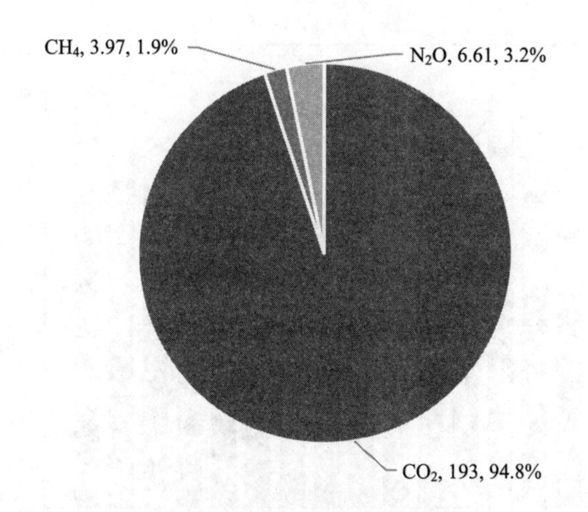

图 3–6 2016 年澳门主要温室气体排放量（万吨 CO_{2eq}）

资料来源：Macao DSPA，2018。

澳门温室气体排放总量（图 3–7），从 2000 年的 173.3 万吨 CO_{2eq}，缓慢增长到 2016 年的 203.6 万吨 CO_{2eq}，并已在 2005 年达到峰值 214.6 万吨 CO_{2eq}，出现这一现象的原因是澳门本地发电产生的温室气体陡然增加。随后温室气体排放量开始回落，并在 2012 年达到历史低谷 146.1 万吨 CO_{2eq}，这主要得益于天然气发电在本地电力中的使用比例增加。2000～2016 年，澳门人均温室气体排放量总体呈下降趋势，从 2000 年的年人均 4.0 吨 CO_{2eq} 下降至 2014 年的 2.5 吨 CO_{2eq}，之后又回升到 2016 年的人均 3.2 吨 CO_{2eq}。

从历年（2000～2017 年）行业平均排放水平来看，本地发电仍是澳门最大的温室气体排放源，排放占比为 38.6%；其次是交通运输业，占比约为 29.0%，其中陆运交通、海运交通和空运交通排放占比分别为 15.1%、10.2% 和 3.8%；废弃物处理行业排放占比约为 14.6%。商业、家庭及服务业温室气体排放占比约为 12.0%。建筑业和工业温室气体排放占比最低，分别为 3.1% 和 2.7%。

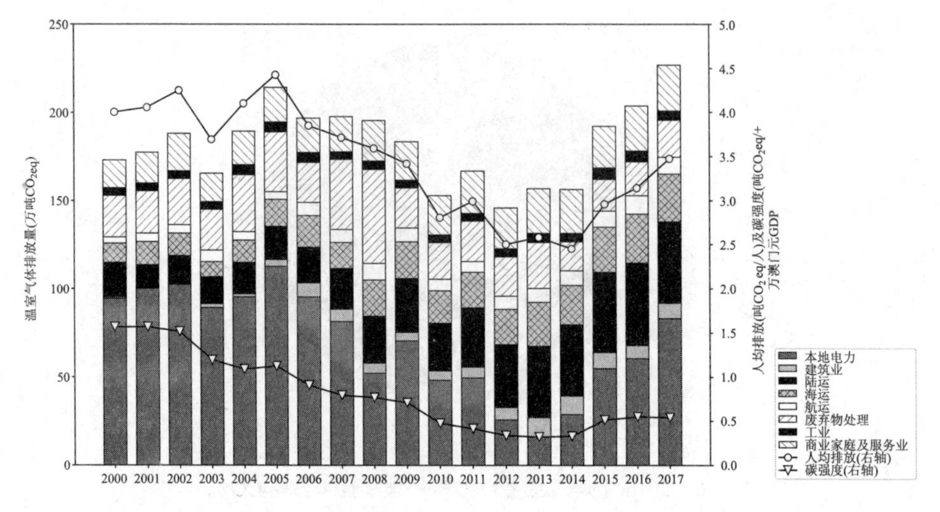

图 3–7　澳门不同行业的温室气体排放

资料来源：Macao DSPA，2018。

（二）电力行业

2016 年，澳门特区的 203.6 万吨 CO_{2eq} 排放当中，约 60.5 万吨 CO_{2eq} 来自澳门本地发电，占全年碳排放总量的 29.7%，是第二大排放源。澳门用电主要由南方电网输入，并以重油和天然气发电作为补充。目前，澳门境内的自有电力设施包括两座发电厂，一个垃圾焚化中心和若干小型光伏发电项目，以上三者共同组成澳门的本地电力供应系统。2016 年，澳门本地发电占总电力结构的 15.0%，垃圾焚化中心发电占总发电量的比重为 3.1%，其余均由南方电网输入（澳门电力股份有限公司，2016）。其中 67.5%的电量用于商业，住宅用电占比约为 22.1%，工业、政府公共照明用电占比分别为 2.5%和 7.9%（澳门电力股份有限公司，2016）。

澳门现存的两座发电厂分别为路环发电厂 A 和路环发电厂 B。其中路环发电厂 A 主要以重油和柴油为燃料，路环发电厂 B 以柴油和天然气为燃料，二者总装机容量为 407.8 兆瓦，均为澳电（CEM）所属设施（澳门电力股份有限公司，2017）。澳门垃圾焚化中心以生活垃圾作为原料，利用其燃烧产生的热能实现发电。满负荷条件下，通过焚烧生活垃圾每小时可产生约 28.7 兆瓦的电力，其中 7 兆瓦（约 24%）用于自身运行，其余 21.7 兆瓦（约 76%）输送至澳电公共电网。另有部分光伏发电系统在路环发电厂等场所已投入使用，2017 年，澳门的光伏系统发电量为 1 448 千瓦时（澳门电力股份有限公司，2017）。

宋庆彬等（Song *et al.*, 2018a）的研究指出 2010～2014 年澳电燃油发电和天然气发电的平均温室气体排放因子分别为 0.71 千克 CO_{2eq}/千瓦时和 0.42 千克 CO_{2eq}/千瓦时，而垃圾焚烧发电平均排放因子为 0.95 千克 CO_{2eq}/千瓦时。天然气发电是澳门最清洁的电力来源。另外，生活垃圾焚烧发电虽然有着相对较高的碳排放，但是由于同时能够实现生活垃圾的有效处理，因此也应作为清洁能源进行鼓励。

澳门环境保护局在其环境报告中公布了澳门本地发电（不包括垃圾焚烧发电）带来的温室气体排放总量（图 3–7）。总体上，澳门本地发电的温室气体排放呈现下降的趋势，从 2000 年的 94.7 万吨 CO_{2eq} 下降到 2016 年的 60.5 万吨 CO_{2eq}，2013 年更是达到了最低值 17.3 万吨 CO_{2eq}。主要原因是澳门本地发电量快速下降，转向从南方电网购入，导致范畴一排放量显著降低。

（三）运输业

澳门有着完整的公交系统，数量庞大的私家车，以及包括外港码头、氹仔客运码头在内的数个港口和澳门国际机场等交通枢纽，他们共同构成了澳门便捷的城市交通系统。2016 年，澳门的交通运输业已经成为第一大温室气体排放源，全年排放量达 83.9 万吨 CO_{2eq}，占总排放量的 41.2%。澳门的交通行业温室气体排放包括陆运、海运以及空运三个类别。其中陆运的排放量最高，占 2016 年总排放量的比重约为 22.4%（45.7 万吨 CO_{2eq}），海运及空运相应比重分别为 14.2%（29.0 万吨 CO_{2eq}）和 4.6%（9.3 万吨 CO_{2eq}）。澳门面积狭小，跨境交通往来密切，但《澳门环境状况报告》结果并未区分本地排放与跨境排放，因此交通业的本地排放和跨境排放均被纳入范畴一。

陆运交通排放特指机动汽车运输的燃料消耗，不包括车辆使用以及巴士站运营等方面的排放。澳门 2016 年陆上交通运输燃料消耗量约为 5 745 太焦，约为整个交通运输业燃料消耗量的 34.0%（Macao DSEC, 2016）。2018 年澳门注册机动车 24 万，其中轻型汽车 10.8 万辆，重型汽车 0.72 万辆，电单车（燃油摩托车）12.4 万辆。陆运交通燃料以汽油为主、轻柴油为辅，并以天然气和电力作为补充（Macao DSEC, 2019）。2000～2016 年，澳门陆运交通的温室气体排放总量由 18.9 万吨 CO_{2eq} 增至 45.7 万吨 CO_{2eq}（图 3–7）。陆路交通温室气体排放总体呈上升趋势，这与澳门私家车及电单车数量的稳步增长趋势基本吻合。

宋庆彬等（Song *et al.*, 2018b）采用生命周期评价方法（Life Cycle Assessment，LCA）测试了基于车辆尾气排放以及能源消耗情况，并计算了澳门公交车和轻型机动车的温室气体排放情况，结果包括能源的间接排放和尾气的直接排放。结果显示，轻型机动车是陆运排放的主要来源，在 2014 年的贡献率约为 48%。2001～2014 年澳门轻型机动车温室气体排放量从 12.5 万吨 CO_{2eq} 增长至 24.8 万吨 CO_{2eq}，增长将近 1 倍（图 3–8）。澳门的公交车温室气体排放也从 2005 年的 3.7 万吨 CO_{2eq} 增长到 2016 年的 5.0 万吨 CO_{2eq}。

海运的碳排放包括轮船使用（包括货轮及客轮）和码头运营等方面所产生的排放。2016 年，澳门的海运交通运输燃料消耗量为 3 092 太焦，其中轻柴油 3 085 太焦，电能 7 太焦，海运交通能源消耗量占运输业总燃料消耗的比重约为 21.4%，占澳门不同领域终端能源消耗量的 9.9%（Macao DSEC, 2016; Macao DSPA, 2018）。2000～2016 年，海运温室气体排放量由 11.1 万吨 CO_{2eq} 增至 29.0 万吨 CO_{2eq}（图 3–7），海运温室气体的排放量逐步上升，但在部分年份有所波动。

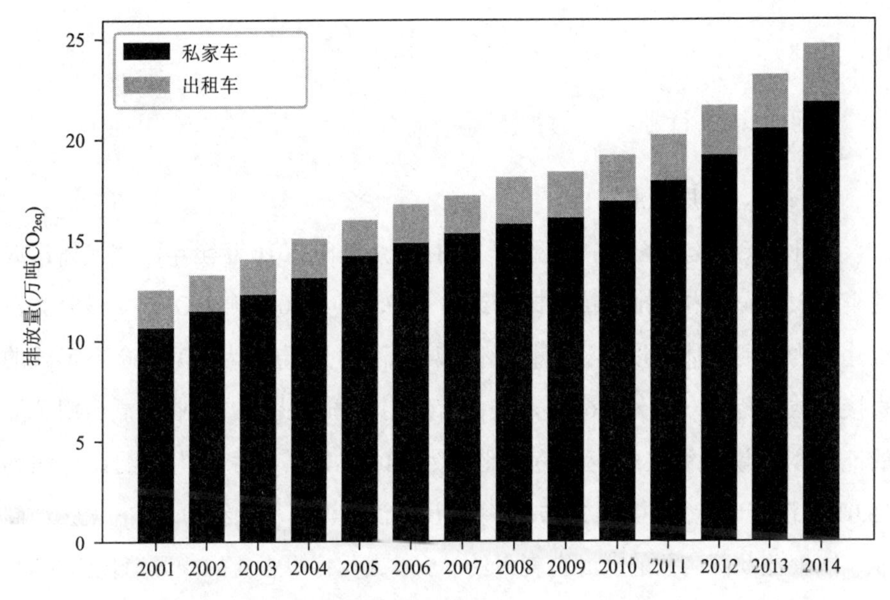

图 3–8 澳门轻型机动车的温室气体排放趋势

资料来源：Song *et al.*，2018c。

空运的碳排放包括普通商业班机和直升机起降过程中的排放，而飞机航行燃油以及机场运营等方面所产生的排放则并不包括在内。2016 年，澳门空路运输总燃料消耗量为 7 970 太焦，其中航空煤油 7 853 太焦，电力 117 太焦，约占运输业总燃料消耗量的 47.2%，几乎全是航空煤油（Macao DSEC, 2017）。2000～2016 年，空运温室气体排放由 3.2 万吨 CO_{2eq} 增长至 9.3 万吨 CO_{2eq}（图 3–7），主要源于澳门国际机场（Macao International Airport, MIA）每年的起降架次不断增长（2000 年 2.9 万架次增长至 2017 年 5.9 万架次）。而澳门的航班单次起降温室气体排放有了显著改善，2016 年已完成每架次碳排放量在 2012 年 0.49 吨 CO_{2eq} 的基础上减少 20%的预定目标，2017 年完成每起降架次碳排放量较 2012 年水平减少 28.7%的目标。对澳门国际机场 2017 年排放来源分布进行统计可知因电能消耗产生的温室气体排放贡献占比最大，达 94.9%（1.94 万吨 CO_{2eq}），其次是冷冻剂 3.7%（762 吨 CO_{2eq}）（澳门国际机场，2018）。

（四）其他行业

1. 商业、家庭及服务业

商业、家庭及服务业中的温室气体排放是澳门 2016 年全年温室气体排放的第三大来源，占总排放量的比重约为 12.75%（26.0 万吨 CO_{2eq}）。该分类指以建筑为主要载体，以能源消耗（以液化石油气和电力为主）为主要形式的直接排放（范畴一）。2016 年，澳门家庭用户和商业、饮食业、酒店服务业的能源消耗量占社会总能源消耗比重为 30.8%，其中家庭用户的比重约为 15.0%（澳门环境保护局，2017）。家庭用户的能源消耗量约为 4 686 太焦；商业、饮食业及酒店业的能源消耗量约 4 966 太焦；博彩业的能源消耗约 7 338 太焦（Macao DSEC, 2016）。由于液化石油气和电力在使用过程碳排放量极低，直接温室气体排放量较小。因此尽管澳门商业、家庭及服务业消耗能源比重较大，但其碳排放比重却不高。澳门商业、家庭及服务业 2000～2016 年温室气体排放量由 15.8 万吨 CO_{2eq} 增至 26.0 万吨 CO_{2eq}（图 3–5）。

2. 废弃物处理行业

目前澳门废弃物处理设施包括四个污水处理厂，一个垃圾焚化中心，一个建筑废料堆填区。其中四个污水处理厂的日平均处理量为 21.1 万立方米；澳门垃圾焚化中心每日能处理约 1 728 吨生活垃圾；澳门建筑废料堆填区自 2006 年启用，已连续使用 13 年。2017 年建筑废料量为 293.3 万立方米，其中海泥量为 140.8 万立方米，飞灰量为 1 924.2 万吨（澳门环境保护局，2018）。废弃物处理行业产生温室气体排放的途径包括污水处理、废弃物焚化及废弃物堆填三种，但此处统计涵盖澳门垃圾焚化中心燃烧城市固体废弃物的排放、废弃物焚烧发电所产生的排放、建筑废料、炉渣以及飞灰堆填所产生的排放情况。2016 年，澳门废弃物处理行业产生的温室气体排放量为 19.4 万吨 CO_{2eq}，占全年总排放量的比重约为 9.5%。废弃物处理行业已成为 2016 年澳门第四大温室气体排放源。

2000～2016 年，澳门因城市废弃物处置产生的温室气体排放水平呈缓慢下降趋势，排放总量由 23.7 万吨 CO_{2eq} 减少至 19.4 万吨 CO_{2eq}（图 3–7）。2008～2010 年出现较大反弹，主要原因是 2008 年焚烧炉进行了扩建，积累废弃物大量焚烧导致排放量快速增加。以上三种废弃物处理方式中，垃圾焚烧是当前澳门废弃物处理过程中最大的温室气体贡献者。近年来，为了有效降低澳门废弃物焚化带来的温室气体排放，2017 年澳门特区政府发布了《澳门固体废弃物资源管理计划（2017～2026）》，制定了澳门未来十年固体废弃物的处理政策、减废目标和行动计划，努力实现在 2026 年将澳门人均城市固体废弃物弃置量减少 30%（相对于 2016 年）的目标。

3. 建筑业

随着经济的快速发展，澳门开展了大量的基础设施建设，建造过程中产生的温室气体排放逐渐成为不可忽视的一部分。2016 年，澳门建筑业的碳排放占全年碳排放的比重约为 4.11%（8.4 万吨 CO_{2eq}）。建筑业产值也从 2000 年的 1.1 亿美元增加到 2015 年的 29.4 亿美元，占澳门 2015 年 GDP 的 6.5% 左右。需说明的是，澳门建筑楼宇的温室气体排放特指建筑业在建设过程中消耗电力和燃料所产生的排放，不包括建筑业所使用的钢筋、水泥、瓷砖等生产过程所产生的排放。

2000～2016 年，建筑业温室气体排放量由 1.3 万吨 CO_{2eq} 增长至 8.4 万吨 CO_{2eq}，增长趋势明显（图 3–7）。其中，2014 年排放量最大，为 10.9 万吨 CO_{2eq}。温室气体的排放量呈现逐年上升的态势，这与澳门基础设施建设增长及楼宇翻新工程增加的社会现状相符。

通过使用 LCA，一项研究（Zhao *et al.*, 2019）估算了 1999～2016 年澳门城市建筑物的温室气体排放情况，包括建设阶段、使用阶段和废弃处置阶段。研究指出澳门建筑物全生命周期的温室气体总排放量已从 1999 年的 152 万吨 CO_{2eq} 迅速增加到 2016 年的 591 万吨 CO_{2eq}。就建筑物的生命周期而言，使用阶段是温室气体排放的主要阶段，占排放总量的 65.8%，其次是建设阶段（33.72%）。此外，研究还比较了不同 LCA 阶段的单位面积建筑温室气体排放情况。结果表明，澳门城市建筑建造阶段的单位温室气体平均排放量为 1.47 吨 CO_{2eq}/平方米，而使用阶段居民建筑和非居民建筑也呈现出明显的差异：以 2016 年为例，居民建筑的单位温室气体排放量为 65.3 CO_{2eq}/平方米/年，而非居民建筑则高达 3 316.4 CO_{2eq}/平方米/年，远高于内地其他省份商业建筑单位排放量。值得注意的是，建筑物废弃处置阶段具有“负价值”，这表明由于材料的回收利用，它通常会带来一些环境效益。

4. 工业

澳门的工业生产碳排放包括以出口再加工为主的纺织制造、建筑水泥生产、水电及气体生产与供应等方面，不包括澳门的污水及废弃物处理工业。工业是澳门温室气体总量排放最小的贡献源，2016 年工业生产的碳排放占全年排放的比重约为 2.7%（5.5 万吨 CO_{2eq}），与之前相比稍有增加，这同澳门工业出现一定萎缩的趋势也相吻合（图 3–7）。

澳门水泥厂是澳门为数不多的大型私营工业企业之一，参与了澳门经济发展过程中一系列标志性项目，如澳门国际机场工程、新大桥等。此外，澳门水泥厂还通过附属公司积极拓展了与内地其他省份及香港特区的业务。陈彬等（Chen *et al.*, 2017a）研究了不同年份澳门水泥生产过程的温室气体排放情况，但其结果与澳门环境状况报告结果出入较大。同一时段甚至出现截然相反的趋势结论，且在结果数值上也表现出巨大的偏差，仅水泥生产就超过了澳门所有工业的温室气体排放值。这是由于澳门环境状况报告仅统计澳门水泥业的温室气体排放，而陈彬等则将澳门消费的进口水泥使用量均包括在内。如 2013 年澳门工业温室气体排放为 5.36 万吨 CO_{2eq}，而水泥消费的相应排放量为 10.3 万吨 CO_{2eq}。

（五）范畴一排放的达峰

澳门范畴一的排放在2013～2017年增长很快，主要原因是本地发电量的大幅上升，因此还没有达峰。对于澳门而言，范畴一排放能否尽快达峰的关键是其电力发展政策，而其他范畴特别是交通运输业也是澳门实现尽快碳达峰的重点行业。如果新增电力需求可用外购电来满足，而本地发电保持在较低比例，同时推动澳门交通车辆的电气化发展，结合节能减废等其他政策措施，就有可能使得澳门范畴一排放尽快达峰，否则该类排放还将持续增加。

第二节 其他核算规则下的温室气体排放

本节聚焦范畴二和范畴三排放，使用前文表3-1中相应的核算方法，讨论它们与范畴一排放的情况差异，并区分了香港与澳门的区域特征。

一、香港

（一）范畴二排放

香港和国内其他省份互相交易电力。在1994年以前，香港曾向内地出口电力，出口量在1993年达到5 776吉瓦时（Chow, 2001a）。1993年，来自大亚湾核电站的电能开始输入香港电网，全部由中华电力香港有限公司运营。而在1994年大亚湾核电站全面投产后，香港成为电力净进口地区，进口电量达到6 430吉瓦时，出口电量1 714吉瓦时（Chow, 2001a）。根据中电集团2018年年报，香港进口电量在2017年已增加到12 426吉瓦时，2018年达到约12 501吉瓦时。

一般认为，核电在其运行阶段（范畴一）的碳排放几乎为零。如果只考虑电力进口的范畴一排放量纳入全港的范畴二排放，则也可计为零。目前化石燃料生成的，用于出口的电力所排放的温室气体已被纳入香港范畴一排放，在核算香港的范畴二排放时，应扣除这部分电力出口时的温室气体排放，香港的用电造成的范畴二排放量应少于范畴一排放量。而随着电力出口的趋势从2006年4 528吉瓦时的峰值下降至2018年的556吉瓦时，直至2019年的0吉瓦时（HK Census and Statistics Department, 2019c），未来将不再有此类扣除的机会。

发电和用电，以及地理距离导致的输电过程，在核算二氧化碳排放方面至关重要。在评估各省政府在实现二氧化碳减排目标方面的表现时，中央人民政府用以下公式来阐明核算规则（National Development and Reform Commission, 2014）：二氧化碳排放量=化石燃料（煤、石油和天然气）产生的二氧化碳排放+进口电力的二氧化碳排放-出口电力的二氧化碳排放。电力的排放系数则是每个省的平均值。

香港特区未使用这套核算方法。但是如果采用这种核算方法，鉴于大亚湾核电站通过专用输电线路连通香港（因此排放系数可只计入核电而不计入广东省的平均水平），香港的二氧化碳排放量应该低于当前报告的范畴一排放量，尤其在大亚湾核电站1994年全面投产后。

从LCA的角度来看，与核电相关的碳排放存在于其建设阶段、使用设备和燃料的生产阶段、退役阶段和废弃物处置阶段。需要参考公开的排放系数来估算大亚湾核电站的LCA碳排放量。许多公开的研究显示，适用于不同价值链组件的假设范畴广泛，导致碳排放估算范畴非常广泛，结果范畴介于3～220克CO_{2eq}/千瓦时（Sovacool, 2008; Warner and Heath, 2012）。

大亚湾核电站不公布其年度碳排放量，但拥有大亚湾核电站75%股份的中电有限公司定期公布该集团温室气体的排放总量，排放主要源于中电集团在其资产的建设、生产和运营过程中的能源使用（2018年排放量约为258 000吨CO_{2eq}）（CGN Power Co., Ltd., 2019），但并非按LCA计算。从LCA的角度来看，2010 年香港本地发电量排放系数为 778.8 克 CO_{2eq}/千瓦时，而大亚湾核电站的核电排放系数为 36.7 克 CO_{2eq}/千瓦时，总体排放系数为 722.1 克 CO_{2eq}/千瓦时（To *et al.*, 2012）。2015 年总体排放系数的估测值为 700.3 克 CO_{2eq}/千瓦时（To and Lee, 2017c）。IPCC 最新的第五次评估报告采用了 Warner 和 Heath 的评估数据（Warner and Heath, 2012），对一些重要的系统参数使用了一致的总系统边界与数值，并得出 4 克 CO_{2eq}/千瓦时～110 克 CO_{2eq}/千瓦时的范畴，中位数为 12 克 CO_{2eq}/千瓦时（IPCC, 2014）。此外，对不同核技术产生的不同碳排放的一些研究给出了压水反应堆的平均温室气体排放系数（大亚湾核电站采用的技术）约为 11.87 克 CO_{2eq}/千瓦时（Kadiyala *et al.*, 2016）。

（二）范畴三排放

1. 基于消费的核算

关于碳泄漏问题的一些近期研究得出结论，从全球技术性核算的角度来看，碳泄漏量似乎并不大，因为“对于大多数碳排放大国而言，核算上的差异似乎相对较小，例如中国（–16%）或美国（+11%）”（Franzen and Mader, 2018），表明 PBA 仍然是一种有效的核算方法，尽管该方法被认为在抑制碳转移行为方面存在不足。

还有一些研究探讨 CBA 的重要性和未来的可能性，得出的结论是“现有的 PBA 模型不太可能很快让位于更具争议性的 CBA 替代方案”（Afionis *et al.*, 2017）。部分原因在于，CBA 方法确实存在挑战，基于相对复杂的输入–输出矩阵，将涉及比 PBA 方法更多的假设，这带来了更多的不确定性和潜在的不准确性。

尽管 PBA 方法更成熟，并被一致作为各国温室气体减排策略的关键支柱，但毫无疑问的是，CBA 能够更好地为气候变化相关决策提供信息，并阻止温室气体（以可能削弱全球减排努力的方式）向没有排放目标的国家转移。毫无疑问，“使各国能够在这样一个系统中运作的制度框架需要进一步发展”

（Fong *et al.*, 2014），但考虑到 CBA 有能力使任何拥有足够消费者或购买力的一方影响低碳产品和制造过程的设计和开发，相关研究指出 CBA 排放日益重要（Hertwich and Wood, 2018），CBA 结果分析（Stockholm Environment Institute, 2017; Liddle, 2018）和提高 CBA 方法准确性的尝试（Chen *et al.*, 2018）越来越多。

建筑物的隐含碳一直是研究的焦点。香港一幢标准高层商业楼宇的平均隐含能源为 94 吉焦/平方米。有学者研究了香港高层商业楼宇的节能潜力，重点关注了建筑材料的回收。他们发现建筑废弃物的适当回收，可以节省下一座新建筑所用材料所需的 53%的能源，而建筑废弃物的再利用可节省 6.2%。基于不同建材的 LCA 结果，他们得出结论，对于含有大量混凝土的建筑部件（如楼板）应实施回收策略，对铝含量高的建筑部件（如窗户）则应采取再利用而不是回收的策略（Ng and Chau, 2015）。此外，在减少隐含能源方面，应优先考虑材料制造和运输过程中的能源消耗，其中运输能耗在香港的案例中需要额外重视（Wang *et al.*, 2018）。甘杰龙等（Gan *et al.*, 2017）通过对香港一座 60 层复合型芯筒-钢臂结构建筑的案例研究，估算了建筑物的隐含碳量。结果表明，传统高炉炼造的结构钢和钢筋占芯筒-钢臂结构占建筑碳含量的 80%，而预拌混凝土仅占 20%。因此他们指出了使用更高效能的建筑材料来减少隐含碳的可能方法。如果钢由电弧炉生产，并以 100%回收的废钢为原料，建筑的碳含量将可减少 60%以上。至于预拌混凝土，利用水泥替代品（35%粉煤灰或 75%炉渣）将可以减少建筑碳含量 10%～20%。更多的研究还指出，建筑的隐含碳主要来自建筑使用周期内不同材料的使用（Chen and Ng, 2016; Dong and Ng, 2015）。香港所谓“零碳大楼”的隐含碳比例可能高达 45%（Ayaz and Yang, 2009）。香港现行的绿色建筑评估计划（BEAM-Plus），因忽视建筑物隐含的温室气体排放而持续受到批评。还有研究比较了新建与翻新住宅建筑中的隐含碳。克雷格·兰斯顿等（Langston *et al.*, 2018）的研究表明，新建项目的隐含碳范畴为 645～1059 千克 CO_{2eq}/平方米，而翻新项目的隐含碳范畴为294～655千克CO_{2eq}/平方米，香港翻新建筑的平均碳含量比新建项目低33%～39%。

香港经济已由制造业转型为服务业，这使碳排放密集型经济活动大大减少。然而，大量温室气体排放隐含在香港消费，却不在香港生产的商品和服务中。因此，对于香港及其他消费和贸易中心，基于消费的核算方法往往得出高得多的温室气体排放水平（Yang *et al.*, 2015）。2004年，香港在基于净进口消费的人均排放方面在全球所有国家与地区中排名第二，每年人均排放量为9.2吨CO_{2eq}，每年总排放量为6 400万吨CO_{2eq}（Davis and Caldeira, 2010）。在这种核算方法下，香港还以每年人均14.8吨CO_{2eq}的排放量位列世界第七大排放经济体（Davis and Caldeira, 2010）。2004年，净进口占香港温室气体排放的62%，而内地净出口排放量为11.47亿吨CO_{2eq}，占总排放量的五分之一以上（Davis and Caldeira, 2010）。另一项研究为香港的消费型排放给出了一个更高的数值，2009年人均排放量为33吨CO_{2eq}，总排放量为2.41亿吨CO_{2eq}（Chen *et al.*, 2016）。内地其他省份是香港消费型排放的最大接受方和出口方（Guo *et al.*, 2015; Liu and Ma, 2011），需要推进更多研究以更好地了解香港基于消费的二氧化碳排放量。

2. 跨境交通

2014年香港货物总吞吐量为3.26亿吨，物流业产生的温室气体排放总量为3 732万吨CO_{2eq}，占香港化石燃料消耗相关温室气体排放总量的41.1%。空运、海运和陆运分别占排放总量的65.5%、33.4%和1.1%（To and Lee, 2017a）。杜伟明（To, 2015）估算了香港物流业的碳强度，显示2012年每百万港元产值对应534吨CO_{2eq}的碳排放。一些重型货物和食品通过燃油动力船只运至香港（Lang, 2018），格雷姆·兰的研究表明，随着这些远洋货船的燃料越来越昂贵或不可得，这些货物的船运物流将减少。香港特区每年空运物流量约500万吨，包括来自世界各地的食品、奢侈品、包裹和邮件（Lang, 2018）。空运是香港物流业最大的温室气体排放源。

在航空旅行方面，香港国际机场每天处理约 1 000 个航班，2017 年运输旅客（“旅客吞吐量”）超过 7 200 万人次（Lang, 2018）。机场雇用了约 73 000 名员工，这些航班和相关收入支持了机场外的许多其他工作和服务。

燃料贸易统计数据可直接揭示香港石油产品重新灌装导致的二氧化碳排放。包括进口化石燃料的所有温室气体排放，香港在 1970 年排放 1 070 万吨 CO_{2eq}，2015 年则增至 9 910 万吨 CO_{2eq}，人均 13.6 吨 CO_{2eq}（To and Lee, 2017b）。

香港大量进口石油产品，包括航空汽油和煤油（2017 年 77.87 亿升），无铅车用汽油（6.41 亿升），汽油、柴油和石脑油（73.98 亿升），燃料油（82.49 亿升）及石油气（378 110 吨）（HK Census and Statistics Department, 2018）。香港没有本地的炼油设施来提炼原油，因此没有原油进口。2017 年，99.9% 的航空汽油和煤油用于飞机舱储存，而船舱则储存 98.4%的燃料油以及 46.4% 的汽油、柴油和石脑油（HK Census and Statistics Department, 2018）。香港国际机场的航班往返国内其他地区与海外，航运也大部分是跨境的。因此，空运和航运的大部分相关排放不计入范畴一，而应计入范畴三。

根据估算，香港净进口的化石燃料（包括 CO_2、CH_4 和 N_2O）在跨境交通类温室气体排放中占主导地位，过去三十年中每年占比达 99.3%～99.4%。其中甲烷和一氧化二氮各占 0.3%。过去三十年来，与能源有关的范畴一排放量与化石燃料净进口隐含排放量的比例一直在稳步下降，从 1990 年的 82.6% 降至 2017 年的 36.9%。石油产品的净进口，尤其是用于跨境航运和航空的石油产品，导致了与能源有关的范畴一排放的大部分差异。2017 年净进口的煤炭、天然气和石油产品隐含碳排放量分别为 2 580 万吨、660 万吨和 6 550 万吨 CO_{2eq}（图 3–9），能源消耗也导致了 3 610 万吨 CO_{2eq} 的范畴一排放量（HKEPD, 2019a）。

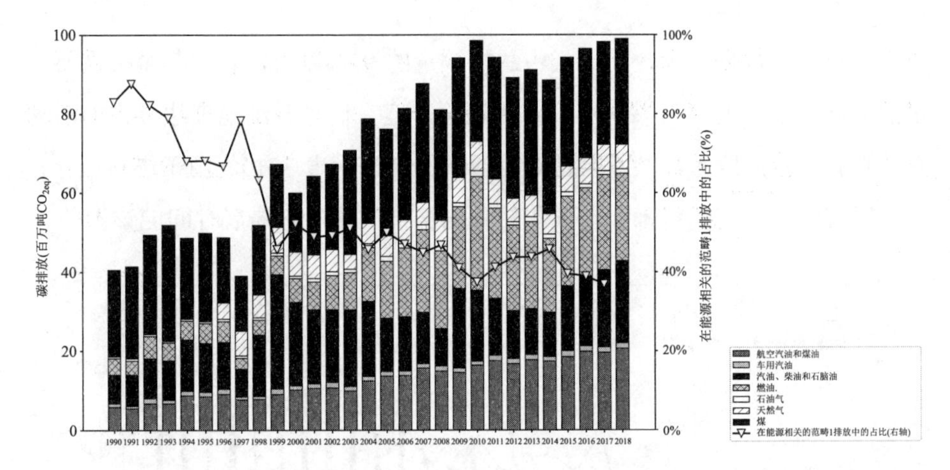

图 3–9　香港净进口化石燃料（包括二氧化碳、甲烷和一氧化二氮）的隐含温室气体排放量

数据来源：HK Census and Statistics Department, 2019a；IPCC, 2006b；IPCC, 2013。

3. 范畴二和范畴三排放峰值

范畴一排放涵盖香港本地生产的电力，无论是供当地消费还是出口，本研究采用 36.7 克 CO_{2eq}/千瓦时（To *et al.*, 2012）作为大亚湾核电站的排放系数，来估算香港的范畴二排放，结果如图 3–10 所示。香港范畴二排放量一直低于范畴一排放量，因为香港长期以来向内地其他地区出口燃煤电力，并大量进口低碳核电，但 2019 年的情况发生逆转，主要因为 2019 年 1 月至 10 月香港没有向其他省份出口电力，而核电进口保持稳定（HK Census and Statistics Department, 2020）。然而由于核电的排放系数较低，香港范畴一和范畴二排放之间的差异至今一直很小，未来应该也会如此。

1990～2017 年，香港的范畴二排放量仅增加 19.0%，从 3 520 万吨 CO_{2eq} 增至 4 070 万吨 CO_{2eq}（图 3–10）。相比香港 2014 年的排放水平，2017 年减少了 9.3%，这表明范畴二排放量也已在 2014 年达到峰值。

香港没有定期研究范畴三的排放，部分原因是其定义较为模糊，没有达成明确的共识。如图 3–10 所示，出于统计数据的可获得性，本报告将香港的范畴三排放量定义为“化石燃料净进口隐含的温室气体排放”与“非能源排放”的总和。数据显示，前一类占总排放量的 95%以上，表明其至关重要。虽然大部分进口的化石燃料用于跨境运输，但它们与香港地理边界内的活动有清晰和直接的联系，故符合范畴三排放的定义（表 3–1）。由于燃料储存，并非所有进口化石燃料都会在同年转化为温室气体，但滞后时间应该不长。

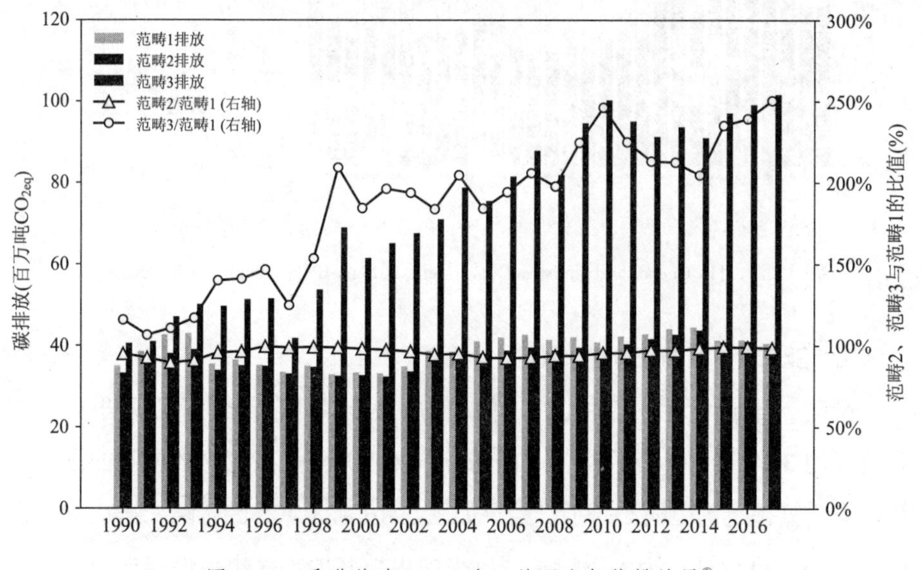

图 3–10　香港范畴一、二和三的温室气体排放量①

没有任何迹象表明香港的范畴三排放量将很快达到峰值。香港大部分石油产品消耗用于跨境运输，特别是香港国际机场及世界级的海港。石油产品的净进口及其隐含的二氧化碳排放量持续快速增长，还没有趋于平缓的迹象

①范畴一排放数据来自官方报告（HKEPD, 2019a）。范畴二排放量的计算公式为：（1）范围 1 排放–（2）向其他省份出售电力隐含的排放+（3）从其他省份购买电力的隐含排放。第（2）类排放=发电产生的范围 1 排放×向其他省份出口电力/本地发电总量。第（3）类排放量=从其他省份购买的核电×核电排放系数。考虑到统计数据的可得性并为界定明确边界，范畴三排放量包括化石燃料净进口（主要用于跨界运输；详情见图 3-9）和其他范畴二排放量（包括电力净贸易和非能源排放）。需要注意的是，范畴三排放量的核算在很大程度上取决于界定的范围。

（图 3–9）。根据最新数据，2017 年，香港范畴一排放量为 4 070 万吨 CO_{2eq}，其中运输业为 723 万吨 CO_{2eq}，在过去 20 年中略有下降（图 3–2）。然而，2017 年净进口的石油产品隐含碳排放量为 7 243 万吨 CO_{2eq}，在 20 年内翻了一番。1990～2017 年，香港的范畴三排放量增加了 142.3%（图 3–10）。香港范畴三排放量在 1990 年比范畴一排放量高出 20.2%，但 2017 年已经扩大至 151.9%（图 3–10）。对比 2014 年数据，虽然范畴一和范畴二排放量在 2017 年分别下降了 9.4%和 9.3%，但范畴三的排放量增加了 10.7%（图 3–10）。

二、澳门

（一）范畴二排放

澳门超过 70%的电力从内地其他省份购买，此部分温室气体排放应归结为澳门范畴二排放（Song *et al.*, 2018a）。澳门与内地的电力交易由来已久，可追溯至 20 世纪 80 年代。1985 年至今，澳电成为澳门公共供电服务的专营机构。自 2005 年以来，澳电加大了从南方电网进口电量，购入电力逐渐成为澳门电力消费主要来源。2013 年时，澳门购入的电力占社会总电力需求的 92%（约 40.6 亿千瓦时）（Chen *et al.*, 2017a）。2017 年澳门总电力供应有所增加，为 54.2 亿千瓦时，本地发电量 14.7 亿千瓦时（其中垃圾焚烧发电 1.7 亿千瓦时），从南方电网进口电力占比约 73.0%（39.5 亿千瓦时）（Macao DSEC, 2019）。2018 年进口电力占比又回升到 88.2%。中国政府从 2010 年开始公布各区域电网平均 CO_2 排放因子，因此本报告以南方电网 2010 年至 2012 年平均 CO_2 排放因子（566 克 CO_2/千瓦时）为澳门购买电力的排放因子进行温室气体排放量进行计算，其结果如图 3–11 所示。2018 年澳门外购电力的温室气体排放量为 277.9 万吨 CO_{2eq}，约为 2000 年排放量的 25 倍（2000 年排放量为 11.0 万吨 CO_{2eq}）。

与香港相同，澳门的范畴二温室气体排放定义为范畴一加上电力贸易所隐含的排放量。与香港非常不同的是，澳门的范畴二排放量远大于范畴一，从 2000 年的 184.3 万吨增加到 2017 年的 450.5 万吨，大幅增长了 144.4%，其中外购电占比为 79.9%（图 3–11）。

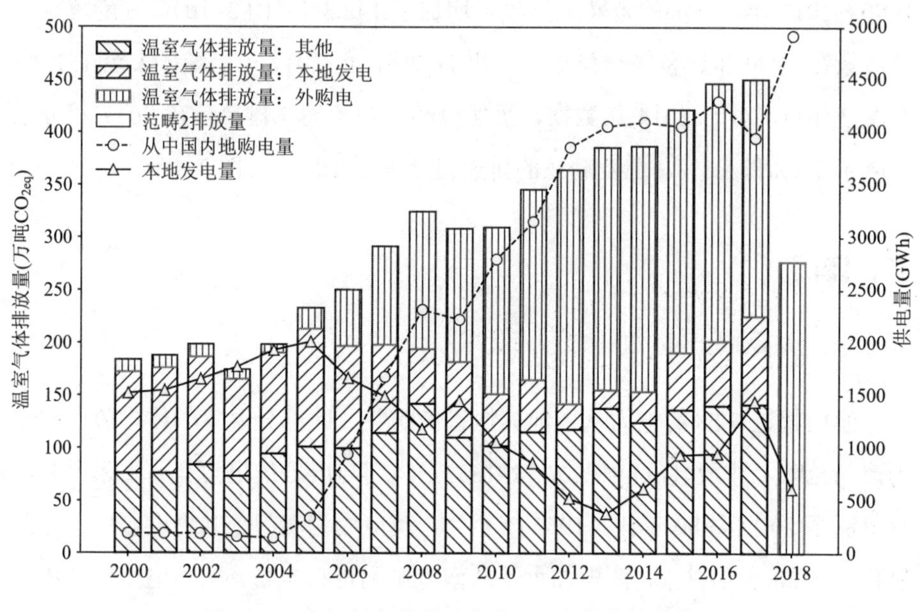

图 3–11 澳门电力供应及范畴二温室气体排放情况

资料来源：外购电力数据来自历年《澳门统计年鉴》；排放因子来自国家气候战略中心；范畴一排放数据（本地发电和其他）来自 Macao DSPA（2018）。

（二）范畴三排放

1. 基于消费的核算

一直以来，澳门严重依赖进出口贸易来弥补自身的资源短缺问题。随着经济水平的飞速增长，澳门对外贸易的依赖程度进一步增加，贸易过程不可避免地带来大量的消费碳流动。陈彬等（Chen *et al.*, 2017b）研究了 2000～2013 年澳门外部贸易产生的碳流入和碳流出情况，考虑的外部贸易包括商品贸易和服务贸易。澳门是典型的异养型经济城市（本地货物几乎全部来自贸易进口），同时也是国际著名的休闲旅游城市，因此碳进口主要指商品贸易，而碳出口主要指旅游博彩为主的服务贸易。图 3–12 显示，澳门贸易的碳进口和碳出口均保持增长趋势，两者平衡差总体上为正值，但差距正逐步缩小。

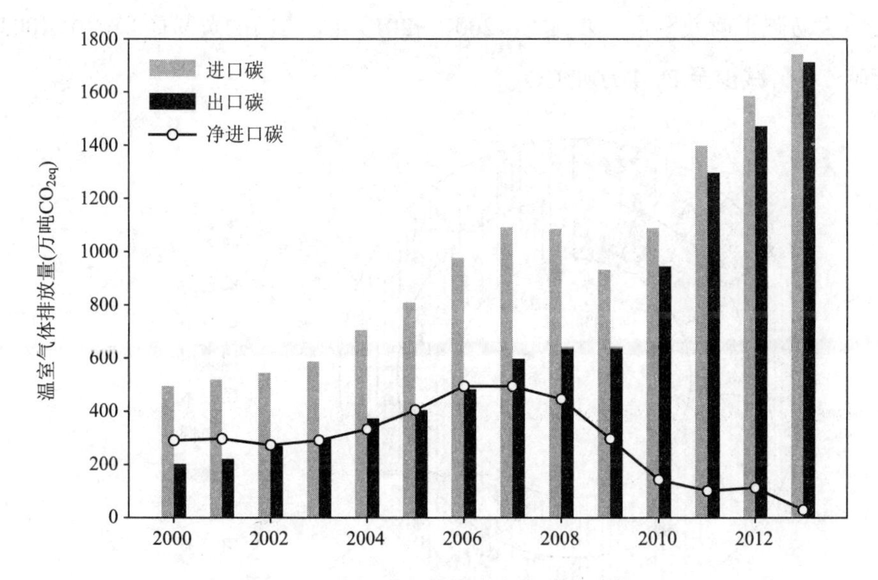

图 3–12　澳门对外贸易隐含的碳流动情况

资料来源：Chen *et al.*, 2017a。

2000～2013 年，澳门贸易碳平衡由 296 万吨 CO_{2eq} 减小至 29.4 万吨 CO_{2eq}。在碳进口方面，其总量由 2000 年的 503 万吨 CO_{2eq} 增加至 2013 年 1 740 万吨 CO_{2eq}。早在 2000 年左右，纺织品和服装业主导着澳门商品贸易碳流入，近年取而代之的是电力、天然气、自来水等能源产品，贸易伙伴以国内其他省份、欧盟和日本为主。通过进一步对澳门的碳进口研究发现，内地作为澳门的主要贸易进口来源，地位进一步巩固，碳进口占比由 2000 年 73.4%增至 2013 年的 83.6%（图 3–13）。

在碳出口方面，澳门的出口服务贸易包括博彩、住宿、购物、邮政和旅游服务等。此外，澳门仍有少量商品出口，主要为纺织品、服装。图 3–12 显示了澳门的服务贸易碳流出保持着自 2000 年以来的增长态势。2013 年澳门贸易碳出口总量达 1 710 万吨 CO_{2eq}，为 2000 年的 2.6 倍。自 2007 年以来，第三产业特别是服务业的蓬勃发展使得澳门的贸易碳出口量显著增加，也使得澳门贸易碳平衡差距进一步缩小。2007～2013 年，澳门的贸易碳平衡由 400.0 万吨 CO_{2eq} 减少至 29.4 万吨 CO_{2eq}。

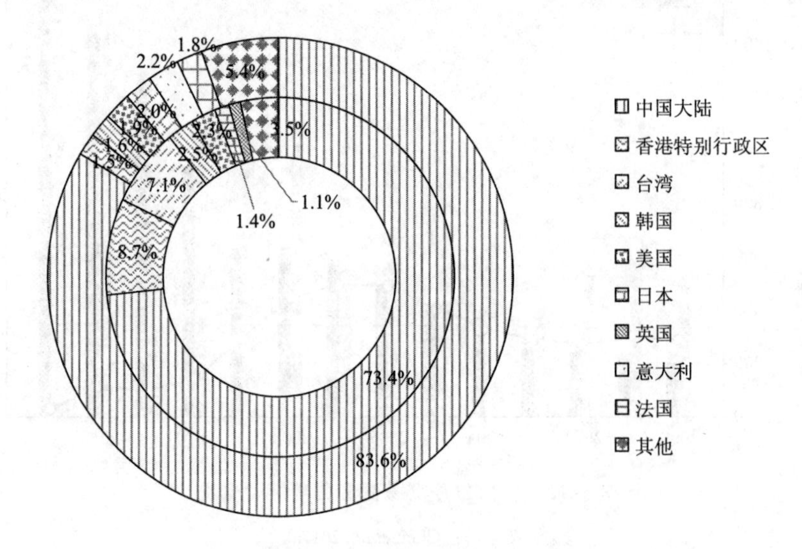

图 3–13　澳门碳进口情况（内圈为 2000 年，外圈为 2013 年）

资料来源：Chen *et al.*，2017b。

尽管 2017 年澳门建筑业碳排放占全年总排放的比重为相对较低的 4%（8.8 万吨 CO_{2eq}），但这一数据并未考虑建筑材料如水泥、钢铁的隐含碳，也未包括建筑使用阶段能源消耗的碳排放。2016 年澳门新建建筑中，水泥的碳排放占总原材料排放比重为 34.6%（约 43.3 万吨 CO_{2eq}），钢铁的碳排放占原材料总排放的 54.4%（约 155.2 万吨 CO_{2eq}）（Zhao *et al.*, 2019）。从建筑生命周期角度来看，澳门建筑的隐含碳比重较高，对建筑废料进行再利用可以有效减少隐含碳排放。以 2016 年为例，通过对建筑废料的回收利用，可减少约

8.6 万吨 CO_{2eq}。

2. 跨境交通

澳门统计暨普查局的统计数据显示，2017 年，澳门共有约 481.9 万车次经由关闸边境站等三个边境站往返内地，是 2000 年（约 205.3 万车次）跨境汽车流量的 2.3 倍。其中关闸边境站作为跨境汽车往来内地的首选，承担了约 72%的跨境汽车流量（Macao DSEC, 2017）。

在航空旅行方面，随着社会富裕水平的提高，越来越多的居民选择更加安全便捷的航空旅行方式，包括往返珠三角的直升机航班和普通商业班机。澳门国际机场统计资料显示，澳门机场航班的起降次数由 2000 年的 2.9 万架次增至 2017 年的 5.9 万架次。机场旅客吞吐量由 2000 年的 323.9 万人次增至 2017 年的 716.6 万人次，但货运吞吐量由 2000 年的 6.81 万吨及 2005 年的 22.72 万吨，减少至 2017 年的 3.75 万吨（MIA, 2020）。

澳门现有 3 个客运码头，分别是外港码头（俗称澳门港澳码头）、内港码头区部分码头和氹仔客运码头。其中内港码头区仅保留部分往返内地的客轮，外港码头和氹仔客运码头主要提供往来香港和内地的客运业务。2017 年，澳门客轮班次达 13.9 万班次，约为 2000 年的 1.8 倍（7.5 万班次）（Macao DSEC, 2017）。

澳门货运码头包括内港码头区和九澳港码头区，内港码头区位于澳门半岛西侧，由 34 个码头组成，以货运、内河运输和渔业为主；九澳港码头区内共有 4 个码头，包括路环电力公司码头、九澳港货柜码头、九澳澳门水泥厂码头和九澳港燃油码头。其中，路环码头仅限于运输发电相关的原料和相关设备；水泥厂码头则仅限于运输水泥生产相关的原材料与相关设备；燃油码头则仅限于运输石油、天然气等危险品。

在温室气体排放方面（图 3–14），澳门跨境交通的温室气体排放总量由 2000 年 35.1 万吨 CO_{2eq} 逐步增至 2018 年 90.3 万吨 CO_{2eq}，陆运跨境交通由于数据无法获取，故不包含在内。总体而言，跨境交通类温室气体的排放情况波动较大，2000～2007 年增长趋势明显，温室气体排放量由 35.1 万吨 CO_{2eq} 增至 83.0 万吨 CO_{2eq}。而从 2008 年开始，跨境交通温室气体排放总量开始下降，后在 2012 开始回升。水运交通的温室气体排放量趋势较为稳定，2000～2018 年平均排放量为 17.1 万吨 CO_{2eq}，占跨境交通温室气体排放总量的平均比例为 26.6%。跨境交通的航空运输变化趋势与跨境交通温室气体排放总量趋势基本一致。2000 年跨境交通的航空运输温室气体排放量为 24.4 万吨 CO_{2eq}，而 2018 年温室气体排放量约为 2000 年的 2.9 倍。

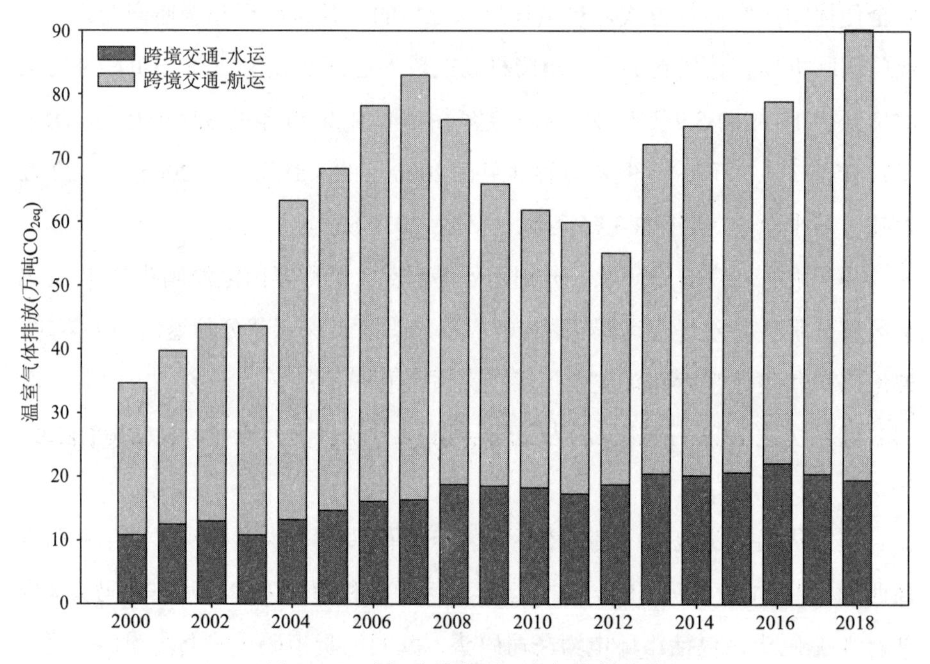

图 3–14　澳门跨境交通温室气体排放量

资料来源：《澳门统计年鉴》；IPCC, 2006a。

根据历年《澳门统计年鉴》及 IPCC 的数据，本研究估算了澳门净进口化石燃料的温室气体排放量（包括 CO_2、CH_4 和 N_2O）。整体而言，澳门进口化石燃料所带来的温室气体排放量有所增加，由 2000 年的 164.7 万吨 CO_{2eq}

增加至 2018 年的 191.7 万吨 CO_{2eq}。其中 2004 年至 2007 年温室气体排放水平较高，年平均为 234.8 万吨 CO_{2eq}，后逐渐下降到 2012 年的 144.6 万吨。在 2018 年，汽油、普通煤油、航空煤油、轻柴油、重油、石油气和天然气的占比分别为 13.3%、0.3%、36.9%、26.4%、6.1%、6.2%和 10.6%。跨境航空和海运能源消耗带来的温室气体排放所占比例也在不断增长，从 2000 年的 21.2%增长到 2018 年的 46.9%。

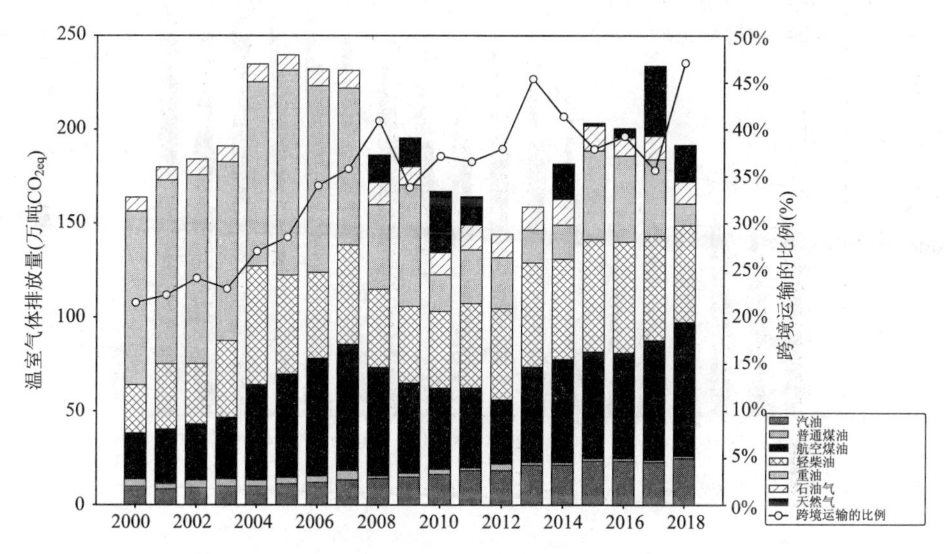

图 3–15　澳门历年化石隐含温室气体排放量（包括二氧化碳、甲烷和一氧化二氮）

资料来源：《澳门统计年鉴》；IPCC, 2006a；IPCC, 2013。

3. 范畴二和范畴三的排放达峰

随着澳门大幅增加从南方电网购电，其范畴二与范畴一排放量的比值也大幅增加，由 2000 年的 106.4%增加至 2014 年的 248.5%，之后随着本地发电和范畴一排放的增加，这一比例又减小到 198.6%（图 3–16）。澳门范畴二温室气体排放还没有展现达峰的迹象，但随着内地的减排努力、特别是南方电网温室气体排放因子的不断下降，外购电所带来的温室气体排放预期将在未来出现显著下降，从而推动范畴二的达峰。范畴三为范畴二加上跨境水运和航空能源使用所隐含的温室气体排放。由于外购电的主导效应以及澳门并非重要的航运中心，范畴三的排放趋势和范畴二类似，还没有出现达峰的迹象（图 3–16）。

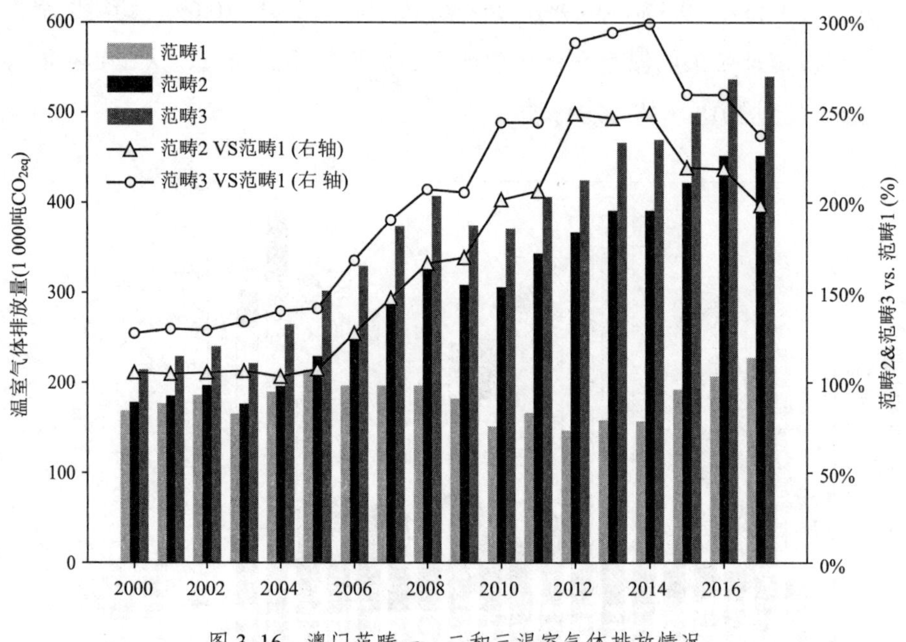

图 3–16　澳门范畴一、二和三温室气体排放情况

第三节　减排情景

本节的分析将只关注范畴一排放。对于港澳地区特别是澳门而言，一个方便的减排策略是将范畴一排放转移到范畴二或者范畴三，因为电力需求可以从外界尤其是内地其他省份购买来替代本地发电。如此，范畴一排放将会下降而基于本地和进口电力温室气体排放强度的不同，范畴二排放可能上升。当进口商品和服务替代本地生产时，范畴三的排放可能会上升而范畴一也会

下降。

如上一节所述，香港和澳门的范畴三排放远大于范畴一排放。由于各自与内地电力贸易的不同情况，香港的范畴二排放略小于范畴一排放，而澳门的范畴二排放远大于范畴一。从此现状出发，本报告不考虑把增加电力进口作为香港范畴一排放的减排措施，但强调其在澳门减排中的重要性。

一、香港

（一）现有情景

据了解，在香港环保署于2010年开展的研究中还包括一项减排评估，以2005年为基准年预测2030年的可能情景（HKEPD, 2010）。在这项减排评估中，存在基准情况与以下三种不同目标的情景：情景1——“空气质量目标情景”，采取措施达到新的空气质量标准；情景2——“加速情景”，在情景1的基础上增加提高能源效率和减少能源需求的额外措施，尤其是在建筑业和运输业；情景3——“激进情景”，在情景2的基础上加快香港电力系统与邻近地区的融合。这些假设的详情载于香港环保署2010年研究报告中。这项评估的主要结论包括：到2030年，三种情景下碳排放总量都将低于2005年4 200万吨CO_{2eq}的水平，降幅6%～36%；三种情景下单位生产总值的碳强度将远低于2005年0.03千克CO_{2eq}/港元，到2020年降幅达37%～57%，到2030年降幅达到56%～70%；三种情景下人均碳排放量也将低于2005年的人均6.16吨CO_{2eq}水平，到2020年降幅达10%～38%，到2030年降幅达23%～48%。值得注意的是，碳强度的这两个指标下降，尤其是在2020～2025年的主要因素，是假定大部分现有的燃煤电厂机组未来停止使用。

这项评估进一步证实，虽然香港正在减少碳排放，但为达到所要求的更大减排量，以下措施必不可少：进一步将进口核电的份额提高到至少35%（情景2），甚至高达50%（情景3）；现有的所有商用建筑至少节省15%的能源；

在所有新建筑中，至少减少 50%的冷气需求，节能 50%；新车的能效比 2005 年提高 20%，所有汽油含 10%乙醇，所有柴油含 10%生物柴油；充分利用所有垃圾填埋气体及废水处理产生的气体；有足够的综合废弃物管理设施来管理香港所有城市固体废弃物，以及至少两个有机废弃处理设施。这项评估假设只使用当时已证实的技术。

继香港环保署 2010 年研究结论后，香港特区政府于 2017 年宣布了碳减排目标：到 2030 年，碳排放与 2005 年水平相比减少 26%～36%（更多详细信息请参阅第四章）。另外值得注意的是，可持续发展委员会已完成应香港特区政府邀请所展开的“长远减碳策略”公众参与活动，并已于 2020 年 11 月向特区政府提交报告，认为香港应在 2050 前逐步迈向净零碳排放。基于这个愿景，可持续发展委员会在八个主要范畴提出了 55 项建议以支持落实有关策略，八个主要范畴分别是目标，生活方式，教育、培训与研究，建筑环境，能源，交通，城市规划与管理，金融。2020 年特区政府的《施政报告》宣布了香港特区将致力争取于在 2050 年前实现碳中和。为此，特区政府将会在 2021 年年中更新《香港气候行动蓝图》，定下更进取的减碳排放策略和措施（Council for Sustainable Development, 2019）。

（二）未来情景 2040

本报告参考了 IPCC 最近关于 1.5 摄氏度气候变暖的报告（IPCC, 2018）所设计的减排方案，该报告要求全球在实现径直减缓斜坡后的 2040 年达到二氧化碳零净排放。由于香港和澳门在提供负排放方面的现实制约，例如难以利用生物燃料和生态系统，本报告将根据最新的温室气体排放官方数据，探索必要的措施，使范畴一排放量比 2017 年的排放水平降低 90%。现有的关于香港和澳门的减缓设想的研究很少，而本章无意进行完全原创研究。因此，在规划实现缓解方案的必要措施时，本报告试图采用“稳定楔子”的概念，为香港和澳门未来的缓解措施确定重大举措（Pacala and Socolow, 2004）。在情景分析中，本报告只分析可提供重大减排机会的必要技术措施，而不分析经济、政治和社会行为。为了达到更详细和更深入研究的目标，今后还需要展开更多关于香港和澳门的建模分析。

2017 年，能源消费和废弃物处理的排放量分别为 3 610 万吨和 280 万吨 CO_{2eq}，其余 180 万吨来自工业加工和其他来源（HKEPD, 2019a）。为开发一条可能的减排途径，即到 2040 年，香港的温室气体排放量从 2017 年的 4 070 万吨减少 90%，至 410 万吨。围绕如何实现 90%的减排目标，本研究设计了六大技术措施，包括暂停化石燃料发电出口，能源消费全面电气化，节能、燃料组合变化带来的低排放系数，捕获和储存（Carbon Capture and Storage, CCS），废弃物焚烧发电。情景分析和减排措施在图 3–17 中进行了总结，并在下文中详细介绍。

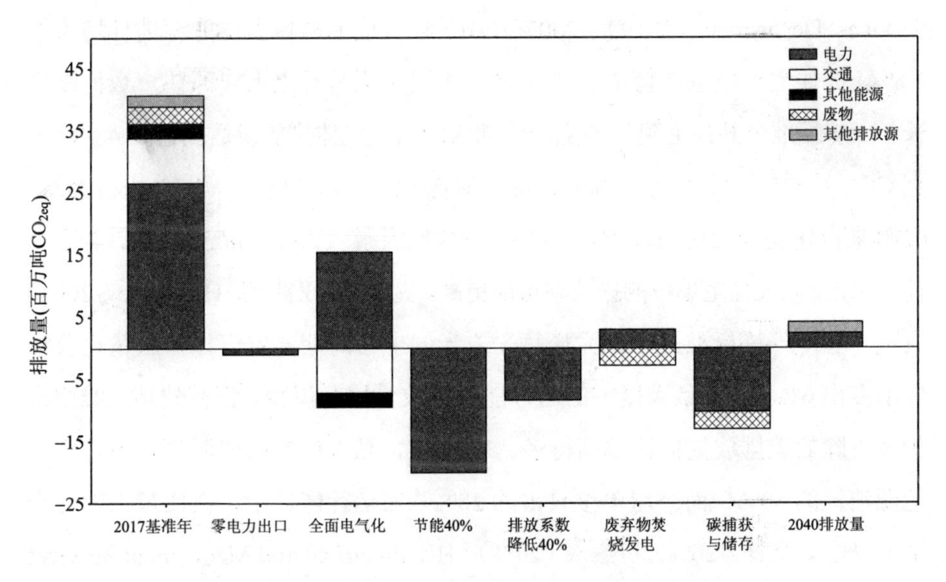

图 3–17　香港范畴一排放的未来情景及减排措施

1. 暂停化石燃料电力出口

到 2040 年，假定香港向内地其他省份出口的化石燃料电力（2017 年为

4 828 太焦或 1 341 吉瓦时）为零，而从中国内地进口的电力（45 274 太焦或 12 576 吉瓦时）维持不变（HK Census and Statistics Department, 2019a）。这将使香港的碳排放减少 100 万吨，减至 3 970 万吨。

2. 能源消耗的电气化

如下分析，香港能源消费全电气化虽然可减少住宅、商业、工业和交通运输业的温室气体排放，但由于香港本地发电基本依靠化石能源，这一措施将导致温室气体排放净增加 590 万吨，达到 4 570 万吨。以下，本文将会分别对两种部门进行论述：①住宅、商业和工业部门；②运输部门。

（1）住宅、商业和工业部门的电气化

住宅及商业建筑，占香港终端能源消耗的绝大部分。住宅约占香港总能源消耗的 21%，2017 年耗能达 59 992 太焦（HK Electrical and Mechanical Services Department, 2019）。2007～2017 年，住宅领域人均能源消耗增长约 2.6%，而住宅领域总能耗增长了 9.6%，主要原因是香港人口和住户数持续增长。一栋标准公共住宅建筑的生命周期内的温室气体排放量估测值为 4 980 千克 CO_{2eq}/平方米，其中 85.8%来自运行能耗，12.7%来自材料，1.1%来自翻新，0.3%来自建筑物的使用末期，0.1%来自其他因素（Yim *et al.*, 2018）。与公共住宅相比，私人住宅单位的排放率可能更高。史蒂文 •汉弗莱（Humphrey, 2004）发现，香港一栋标准的私人大楼每平方米的生命周期温室气体排放量比公共住宅多出 6%。虽然私人楼宇的运行能耗排放量比公屋少，但其建造、维修维护及拆除的碳排放量估计要高得多。2017 年，私人住宅的能源消耗占住宅总能源消耗的 50%，而公屋单位只占约 28%。在运行能耗中，占比最大的为空调（27%）、烹饪（26%）和热水（24%）（HK Electrical and Mechanical Services Department, 2019）。因此应尝试和评估各项碳减排措施，如使用节能设备、可再生能源、可回收材料、应用自然采光和通风的生态设计（Yim *et al.*, 2018）。

商业在总能耗中占更大比例，为 125 158 太焦，约占 2017 年总能耗的 44%。

在商业领域，空调也是最大的能耗终端，占总能耗的25%，其次是烹饪（13%）、照明（12%）、热水和制冷（12%）（HK Electrical and Mechanical Services Department, 2019）。一项对香港十幢高层办公楼能源利用进行的生命周期分析的研究结果表明，在50年的研究时期内，这些办公楼的平均生命周期能耗为51.78吉焦/平方米～73.64吉焦/平方米。建筑运行消耗占78%～89%的能源总消耗，其余部分为隐含能耗（Wang *et al.*, 2018）。由于空调及照明是商业领域的两大主要耗能终端，对香港商场节能决策的研究结果显示，因为人们的耐热度低于耐冷度，香港商场经理们倾向将室内气温设置为略低于适中温度，从而延长顾客在商场的停留时间（Kwok *et al.*, 2017a; 2017b）。商场的照明情况也大致相同。2009～2100年香港的天气状况可能导致办公楼每年的冷气负荷和能源使用量比1979～2008年的水平分别增加10.7%和4.9%（Lam *et al.*, 2010）。

电力是主要的能源类型，2017年住宅和商业部门耗电分别为42 127太焦和104 758太焦，即11 702吉瓦时和29 099吉瓦时（HK Electrical and Mechanical Services Department, 2019）。工业消耗了12 705太焦的能源，包括7 963太焦（2 212吉瓦时）的电力。如果这些行业的所有能源消耗都完全电气化，不考虑能源效率差异，需要55 389吉瓦时的电力。如果采用与2016年本地发电相同的排放系数（731克CO_{2eq}/千瓦时），则意味着排放4 050万吨CO_{2eq}。考虑到使用燃气灶和电烤箱时的能效和行为差异，欧盟委员会的一项研究（BIO Intelligence Service, 2011）表明其净能耗几乎相等。因此，作为粗略估计，可将非电力能耗按等效加热值（1千瓦时=3.6兆焦）转化为电能。这些部门的所有能源消耗全面电气化后，需消耗54 960吉瓦时的电力。

（2）运输部门的电气化

关于低碳交通选择，格雷姆·兰（Lang, 2018）指出，在一些城市，自行车可以承担很大一部分通勤和货品配送的运输量，但与内地许多城市不同，香港自行车通勤及共享单车主要在新界的新市镇出现。目前香港街道上挤满了公共汽车、卡车、汽车和出租车，短期内无法在大多数道路上为安全的自行车交通腾出空间。在道路安全和条件允许的情况下，特区政府将致力缔造“单车友善环境”，推动单车作为休闲及短途代步工具，特别是在设有较全面单车路网的新市镇。就市区而言，由于日常交通非常繁忙，路窄人多，路旁上落客货活动频繁，经常有车辆驶经并需在路旁短暂停留。基于道路安全考虑，政府不鼓励市民在市区以单车作为交通工具。许多住宅建筑都位于山区，除专业自行车手外，其他居民骑车出行相当困难。这样的状况终于随着一项计划得到改变，即扩展自行车道（主要在新界），香港首间无泊位共享单车租赁商在 2017 年 4 月投入服务（Leung *et al.*, 2018）。

车辆拥有量的增加是香港客运能源使用量增加的最大因素（Boey and Su, 2014）。私家车拥有量在 2010～2017 年上升了 33.2%（HK Transport Department, 2018）。一项针对年轻人的调查揭示了香港居民对私家车的态度，“超过七成的受访者表示有意在未来购买汽车”（Cullinane, 2003）。另一项针对车主的调查提出的问题是，汽车一旦被购买会迅速成为必需品，并且这种感觉将随时间的推移而增强（Cullinane and Cullinane, 2003）。平均每位香港车主会在面积约为 20 千米×20 千米的区域内，每年行驶 5 000 千米以上（每天 13 千米以上）（Poudenx, 2008）。王辉等（Wang *et al.*, 2012）采用计量经济学模型，估算了香港私人客运的直接反弹效应，研究发现，1993～2009 年和 2002～2009 年直接反弹效应的幅度分别为 45%和 35%，表明客运的直接反弹效应随着时间的推移呈下降趋势。研究还指出，在评估能源效率措施对减少香港能耗的影响时，需要考虑直接反弹效应。此外，继续发展“以公共交通为本”及“以铁路为骨干的运输系统”的运输策略对减排的影响更大。

设想中，未来香港运输行业的电气化情景下，所有车辆都将以电力为唯一动力。香港运输业 2017 年耗能 88 414 太焦，占总能耗的 31%，其中 3 129 太焦（即 869 吉瓦时）为电能（HK Electrical and Mechanical Services Department, 2019）。根据美国能源部的数据，电动汽车的能源效率约为 59%～

62%，而汽油车的效率为 17%～21%。由于柴油车往往比汽油车效率更高，采用 3.18 的转换率，将运输部门目前的能耗转换为完全依赖电力的未来情景，共需要 8 309 吉瓦时的电力。此外，应采取额外措施以控制日益增长的运输需求，尽量保持能源需求持平。

全面电气化将使这些终端行业的用电量达到 63 269 吉瓦时，比 2017 年水平高出 19 386 吉瓦时。此外，参考 2017 年香港用电量与电力行业自用损耗之比为 10.0%，意味着香港应提供 69 589 千瓦时的电力以满足电力需求。根据前文的假设，香港从内地其他省份购买电力为 45 274 太焦即 12 576 吉瓦时，而出口量为零，因此香港本地发电厂的发电量应达到 57 013 吉瓦时。若采用 2017 年本地发电的排放系数（2 660 万吨/36 917 吉瓦时=721 克 CO_{2eq}/千瓦时）（HK Census and Statistics Department, 2019a; HKEPD, 2019a），能源系统的排放量将为 4 110 万吨 CO_{2eq}；加上废弃物处理产生的 280 万吨和来自其他排放源的 180 万吨排放，未来此类碳排放总量将为 4 570 万吨。

3. 节能

作为全球二氧化碳减排的关键方法，香港应着重提高能源效率和节能技术，加强节能行为。假设到 2040 年香港电力需求减少 40%，达到 41 753 吉瓦时，意味着需要 29 177 吉瓦时的本地发电量和 2 100 万吨 CO_{2eq} 排放，整体而言将使总排放水平降低 2 010 万吨至 2 560 万吨。

4. 排放系数与燃料组合

未来，本地发电量的排放系数也可降低 40%，达到 432 克 CO_{2eq}/千瓦时，与天然气发电量的排放系数相当。这将使本港能源系统的温室气体排放量降低 840 万吨至 1 260 万吨 CO_{2eq}，碳排放总量降至 1 720 万吨。过去 20 年内香港的燃料结构一直保持稳定，其中煤炭约占电力供应的一半，天然气占四分之一，购入核电占四分之一（HK Environment Bureau, 2014）。一项官方建议计划在 2020 年将天然气在香港电力供应中的占比提高至 60%，并将煤炭占比减至 20%以下（HK Environment Bureau, 2014）。一项有关香港消费者支付意愿的调查发现这项计划受到支持，支持率高达 48%～51%，而增加核电进口计划的支持率为 32%～42%（Cheng *et al.*, 2017）。提高天然气占比的计划最终得到采纳，但为降低天然气对电费产生的影响，以及未来燃料组合方案的灵活性，天然气在香港电力供应中的占比在 2020 年提高至 50%，而非原先建议的 60%。未来几十年，随着已退役的燃煤发电厂被天然气发电厂取代，香港的燃料组合将发生重大变化（HK Environment Bureau, 2017）。

能源向可再生能源的过渡对于全球减少温室气体排放至关重要。然而，在人口稠密的香港，能源过渡可能面临三大挑战，包括：①可持续未来的公众教育及公众参与；②协调利益相关者的经济利益；③新兴技术的采纳能力（Ng and Nathwani, 2010）。在官方计划中，可再生能源不会在香港未来的燃料组合中扮演重大角色。香港的两家电力公司曾提出在香港水域建一两个海上风电场的方案，但该计划尚未实现。建筑集成光伏系统已在香港特有的阳光及建筑情况下得到评估，位置和方向是影响光伏系统性能关键因素（Lu and Yang, 2010; Yang *et al.*, 2004）。重新和两家电力公司协商管制计划协议后，香港开始实施上网电价以鼓励生产可再生电力，并为三种不同规模的发电系统设定了每度电的电费率：①10 度及以下为 5 港元；②10 度以上至 200 度为 4 港元；③200 度以上至 1 000 度为 3 港元。上网电价水平会每年讨论（HK Offshore Wind Limited, 2006）。

5. 碳捕获和储存（CCS）

CCS 技术可用于实现生态系统楔和负排放（National Academies of Sciences Engineering and Medicine, 2019）。在此分析一个替代方案，该方案提出并评估了香港继续使用煤炭，但利用 CCS 技术以减少二氧化碳排放量的情形（Xu and Liu, 2015）。目前，一条 800 千米长的天然气管道正用于从南海的油田向香港供应天然气。该方案提出在香港燃煤发电厂捕获二氧化碳，并通过管道反向输送至枯竭的天然气崖城 13-1 气田，进行天然气回收和二氧化碳储存（Xu and Liu, 2015）。该气田的二氧化碳储存能力足以支持这一方案

（Zhang *et al.*, 2014）。

假定 CCS 技术可以减少香港发电厂 85%的 CO_2（IPCC, 2005），即 1 070 万吨。应用 CCS 之后，香港能源系统的碳排放量为 190 万吨 CO_{2eq}，排放总量为 650 万吨。

6. 废弃物燃烧发电

2017 年香港向堆填区倾倒了 390 万吨城市固体废弃物（HKEPD, 2019b），废弃物处理碳排放量为 280 万吨 CO_{2eq}。假设到 2040 年采用 CCS 技术并利用这些废弃物焚烧发电，焚烧后，1 吨城市固体废弃物可产生 0.7～1.2 吨的二氧化碳排放，约 300～600 千瓦时的电力（Johnke, 2001）。然而，根据香港兴建中的综合废弃物管理设施的设计估算，预计焚烧 1 吨城市固体废弃物将会产生 0.2 吨的二氧化碳排放，其发电的排放系数为 0.5 千克 CO_{2eq}/千瓦时。因此，在考虑到该设施可避免将城市固体废弃物倾倒在堆填区，预计在没有应用 CCS 的情况下，处理每吨城市固体废弃物仍可整体减少约 0.4 吨的温室气体排放。采用其平均值估算，废弃物焚烧发电的排放系数为 2.1 千克 CO_{2eq}/千瓦时，远高于化石燃料发电。因此，总量为 390 万吨的城市固体废弃物的焚烧将会产生 370 万吨 CO_{2eq} 的碳排放和 1 763 吉瓦时的电力。由于堆填区几乎占香港废弃物处理的温室气体排放的全部（Dong *et al.*, 2017），预计 2017 年废弃物处理产生的 280 万吨 CO_{2eq} 的碳排放将被完全消除，废弃物产生的电力可取代本地的化石燃料发电。如果不应用 CCS，废弃物焚烧发电可使温室气体排放量净增加 15 万吨。应用 CCS 后，化石燃料及城市固体废弃物焚化发电量将排放 230 万吨 CO_{2eq}，即再减排 240 万吨，并使总排放量达到 410 万吨 CO_{2eq}。

二、澳门

澳门境内的直接排放（范畴一排放）可以利用能源电气化、节能减废、增加外购电来实现减排目标。如前文所述，澳门的电力发展政策是范畴一排放达峰和大幅减排的关键。全面电气化可以使用外购电逐步取代其他本地排放，节能可以减少外购电需求，减废使得本地废弃物处理带来的温室气体排放降低。这一策略等同于将范畴一排放转移到范畴二，而外购电的隐含温室气体排放取决于外购电量和排放因子。与香港的情况不同，澳门的范畴一与范畴二的排放和减排趋势都与内地其他地区特别是南方电网有密切的关系。

（一）全面电气化和节能减废

1. 运输业

运输业是澳门重要温室气体排放源，特别是陆运交通和水运交通。二者主要使用燃料为汽油和轻柴油，它们 2016 年温室气体排放占比分别为 22%和 14%。当前，澳门从进口新车、在用车、车用燃油及推广环保车辆等方面推进了各项改善政策。为进一步实现交通低碳，政府还对车用无铅汽油及轻柴油标准进行了规定（澳门特别行政区政府，2016），并于 2012 年推出了环保车辆税务优惠措施（澳门特别行政区政府，2018）以鼓励居民优先选择环保汽车。在参考外地成功经验基础上，特区政府推出《淘汰重型及轻型二冲程摩托车资助计划》，淘汰了澳门 52%的二冲程摩托车。随着各项政策的实施，传统能源车辆在实现温室气体减排方面的空间已经非常小。新能源汽车将成为交通低碳策略的核心，尤其是纯电动汽车。纯电动汽车理论上将实现使用阶段的零直接排放，将减排责任实际上转移到了电力生产。

2. 废弃物管理

纸张/卡纸、塑料和有机物是澳门城市固体废弃物的主要物理成分，2016 年三者占固体废弃物的比例分别为 24.4%、21.0%和 38.5%（澳门环境保护局，2017）。一方面，与日俱增的废弃物已逐渐逼近澳门垃圾焚化中心最大处理能力；另一方面也正不断地刷新废弃物处理行业的温室气体排放量。此外，焚化之后的飞灰以及建筑废料也在不断压迫着澳门捉襟见肘的可用土地面积。

2017 年澳门特别行政区政府发布了《澳门固体废弃物资源管理计划（2017～2026)》，制定了澳门未来十年的固体废弃物减量目标，计划到 2026 年实现人均固体废弃物弃置量在 2016 年基础上减少 30%（0.63 千克/人）的目标。通过推动源头减废、各类资源废弃物（包括厨余、纸类、塑胶及电子产品等）回收再利用计划及胶袋收费等多项措施工作，有效减少澳门垃圾焚化中心固体废弃物处理量，进而减少温室气体排放。对于固体废弃物中的厨余垃圾等有机物成分，可考虑将其转化为生物能源进行再利用。

3. 工业、建筑业、商业、家庭用户

随着居民生活水平的提高以及技术的进步，电器电子产品已经成为家庭及商业活动中不可或缺的部分。《澳门特别行政区能源效益状况调查研究报告》(澳门能源业发展办公室，2019）指出当前居民在电器电子产品使用方面已经有了相对较高的节能行为，但仍有一定的提升空间。为配合“世界级旅游休闲中心”的发展定位，特区政府一直致力于居民低碳生活倡议的宣传。特区每年举办澳门国际环保合作发展论坛（MIECF）加强国际环保合作，加强居民环保意识宣传（澳门国际环保合作发展论坛，2019)。针对澳门蓬勃发展的酒店业，澳门环境保护局自 2007 年开始设立了“澳门环保酒店奖”，积极鼓励本地酒店业的绿色发展。针对居民、企业和学校，澳门环境保护局每年通过开展“环保 Fun”“低碳节日”“绿色企业伙伴计划”等活动，鼓励居民“绿色出行”，将低碳生活倡议贯彻到居民生活的方方面面。此外，澳门环境保护局通过颁布“澳门环境保护规划（2010-2020)”、公共部门环境管理计划等措施，努力突出公共部门作用，提高社会的环境绩效。

对澳门来说，电器电子产品的能源消费，尤其是空调产品的消费已经成为重要的能源消费来源，其能源效率的提升已经成为城市节能减排的重要举措。为此，澳门特区政府建立了“环保与节能基金”资助使用更加高效的环保和节能产品（澳门能源业发展办公室，2019)。对于空调产品来说，2017 年仅非政府写字楼/办公室和酒店设定温度有所下降，居民家庭、饭店的空调温度反而有所上升。夏天，空调每调高 1 摄氏度，能源效率将提高 7%～10%，而室内外温差越大，能源效率提高的将越明显，因此澳门空调节能空间巨大（参考内地 26 摄氏度的空调温度设定平均值，将有超过 20%的节能空间）（Song *et al.*, 2017）。此外，节能能源标识的推广也有助于居民选择更加节能环保的电器电子产品，以同样 268 升的家用电冰箱为例，能效为 1 级的产品比 5 级产品每天可省电约 0.7 千瓦时。

（二）电力发展

一直以来，本地电力行业都是澳门温室气体的重要排放源。在未来的减排情景下，澳门外购电的比重应该还需要持续上升。澳门可全部改用天然气以取代重油作为发电的主要燃料。宋庆彬等（Song *et al.*, 2018a）的研究表明，澳门天然气和重油发电的排放因子分别为 0.42 千克 CO_{2eq}/千瓦时和 0.71 千克 CO_{2eq}/千瓦时，每千瓦时可以减少 41%的温室气体排放。此外，可再生能源如太阳能的广泛使用也能为澳门带来减排收益，但其比例和潜力小，2017 年澳门光伏系统发电量仅为 1 448 千瓦时（澳门电力股份有限公司，2017）。

参考文献

Afionis, S., M. Sakai, K. Scott, *et al.*, 2017. Consumption-based Carbon Accounting: Does It Have a Future? *Wiley Interdisciplinary Reviews: Climate Change*, 8.

Ayaz, E., F. Yang, 2009. Zero Carbon Isn't Really Zero: Why Embodied Carbon in Materials Can't Be Ignored, https://www.di.net/articles/zero_carbon/.

BIO Intelligence Service, 2011. *Domestic and Commercial Ovens (Electric, Gas, Microwave), Including Incorporated in Cookers: Task 3 Consumer Behaviour and Local Infrastructure*, https://www.eceee.org/static/media/uploads/site-2/ecodesign/products/lot22-23-kitchen/lot22-task3-final.pdf.

Boey, A., B. Su, 2014. Low-carbon Transport Sectoral Development and Policy in Hong Kong and Singapore. *Energy Procedia*, 61.

CGN Power Co., Ltd., 2019. *Environmental, Social & Governance Report 2018*.

Chen, B., Q. Yang, J. Li, *et al.*, 2017a. Decoupling Analysis on Energy Consumption, Embodied GHG Emissions and Economic Growth—the Case Study of Macao. *Renewable and Sustainable Energy Reviews*, 67.

Chen, B., Q. Yang, S. Zhou, *et al.*, 2017b. Urban Economy's Carbon Flow Through External Trade: Spatial-temporal Evolution for Macao. *Energy Policy*, 110.

Chen, G. W., T. Wiedmann, Y. F. Wang, *et al.*, 2016. Transnational City Carbon Footprint Networks-Exploring Carbon Links Between Australian and Chinese Cities. *Applied Energy*, 184.

Chen, Y., S. T. Ng, 2016. Factoring in Embodied GHG Emissions When Assessing the Environmental Performance of Building. *Sustainable Cities and Society*, 27.

Chen, Z. M., S. Ohshita, M. Lenzen, *et al.*, 2018. Consumption-based Greenhouse Gas Emissions Accounting with Capital Stock Change Highlights Dynamics of Fast-developing Countries. *Nature Communications*, 9.

Cheng, Y. S., K. H. Cao, C. K. Woo, *et al.*, 2017. Residential Willingness to Pay for Deep Decarbonization of Electricity Supply: Contingent Valuation Evidence from Hong Kong. *Energy Policy*, 109.

Chow, L. C. H., 2001a. Changes in Fuel Input of Electricity Sector in Hong Kong Since 1982 and Their Implications. *Energy Policy*, 29.

Chow, L. C. H., 2001b. A Study of Sectoral Energy Consumption in Hong Kong (1984-97) with Special Emphasis on the Household Sector. *Energy Policy*, 29.

Council for Sustainable Development, 2019. *Support Group on Long-term Decarbonisation Strategy*.

Cullinane, S., 2003. Hong Kong's Low Car Dependence: Lessons and Prospects. *Journal of Transport Geography*, 11.

Cullinane, S., K. Cullinane, 2003. Car Dependence in a Public Transport Dominated City: Evidence from Hong Kong. *Transportation Research Part D: Transport and Environment*, 8(2).

Davis, S. J., K. Caldeira, 2010. Consumption-based Accounting of CO_2 Emissions. *Proceedings of the National Academy of Sciences of the United States of America*, 107.

Dong, Y. H., A. K. An, Y. S. Yan, *et al.*, 2017. Hong Kong's Greenhouse Gas Emissions from the Waste Sector and Its Projected Changes by Integrated Waste Management Facilities. *Journal of Cleaner Production*, 149.

Dong, Y. H., S. T. Ng, 2015. A Life Cycle Assessment Model for Evaluating the Environmental Impacts of Building Construction in Hong Kong. *Building and Environment,* 89.

Environmental Resources Management, 2018. *Hong Kong Offshore LNG Terminal*.

Franzen, A., S. Mader, 2018. Consumption-based Versus Production-based Accounting of CO_2 Emissions: Is There Evidence for Carbon Leakage? *Environmental Science & Policy*, 84.

Fung, W. Y., K. S. Lam, W. T. Hung, *et al.*, 2006. Impact of Urban Temperature on Energy Consumption of Hong Kong. *Energy*, 31.

Fong, W. K., M. Sotos, M. Doust, *et al.*, 2014. *Greenhouse Gas Protocol: Global Protocol for Community-Scale Greenhouse Gas Emission Inventories: An Accounting and Reporting Standard for Cities*.

Gan, V. J. L., J. C. P. Cheng, I. M. C. Lo, *et al.*, 2017. Developing a CO_2-e Accounting Method for Quantification and Analysis of Embodied Carbon in High-Rise Buildings. *Journal of Cleaner Production*, 141.

Guo, S., G. Shen, J. Yang, *et al.*, 2015. Embodied Energy of Service Trading in Hong Kong. *Smart and Sustainable Built Environment*, 4.

Hertwich, E. G., R. Wood, 2018. The Growing Importance of Scope 3 Greenhouse Gas Emissions from Industry. *Environmental Research Letters*, 13.

HK Census and Statistics Department, 2018. *Hong Kong Energy Statistics Annual Report*.

HK Census and Statistics Department, 2019a. *Hong Kong Energy Statistics Annual Report*.

HK Census and Statistics Department, 2019b. *Hong Kong Annual Digest of Statistics*.

HK Census and Statistics Department, 2019c. *Table 127: Electricity Consumption*.

HK Census and Statistics Department, 2020. *Hong Kong Energy Statistics Annual Report*.

HK Electrical and Mechanical Services Department, 2019. *Hong Kong Energy End-Use Data 2019*, https://www.emsd.gov.hk/en/energy_efficiency/energy_end_use_data_and_consumption_indicators/hong_kong_energy_end_use_data/.

HK Environment Bureau, 2014. *Planning Ahead for a Better Fuel Mix: Future Fuel Mix for Electricity Generation Consultation Document*.

HK Environment Bureau, 2017. *Hong Kong Climate Action Plan 2030+*.

HKEPD, 2000. *Greenhouse Gas Emission Control Study*.

HKEPD, 2010. *A Study of Climate Change in Hong Kong-Feasibility Study*.

HKEPD, 2017. *Landfill Gas Utilization*.

HKEPD, 2019a. *Greenhouse Gas Emissions in Hong Kong*.

HKEPD, 2019b. *Hong Kong Waste Treatment and Disposal Statistics*.

HK Lands Department, 2019. *Proposed Construction of Hong Kong Offshore Liquefied Natural Gas (LNG) Terminal*.

HK Offshore Wind Limited, 2006. *Hong Kong Offshore Wind Farm in Southeastern Waters*.

HK Transport Department, 2018. *The Annual Traffic Census 2017*.

Ho, C. Y., K. W. Siu, 2007. A Dynamic Equilibrium of Electricity Consumption and GDP in

Hong Kong: An empirical Investigation. *Energy Policy*, 35.

Humphrey, S., 2004. *Whole Life Comparison of High-rise Residential Blocks in Hong Kong.*

IMF, 2019. *World Economic Outlook Databases - October 2019 Edition.*

IPCC, 2001. Climate Change 2001: *The Scientific Basis. Contributing of Working Group I to the Third Assessment Report of the Intergovernmental Panel on Climate Change.* Cambridge University Press, Cambridge, United Kingdom and New York, NY. USA.

IPCC, 2005. *IPCC Special Report on Carbon dioxide Capture and Storage.* Prepared by Working Group Ⅲ of the Intergovernmental Panel on Climate Change. Cambridge University Press, Cambridge, United Kingdom and New York, NY. USA.

IPCC, 2006a. *2006 IPCC Guidelines for National Greenhouse Gas Inventories, Volume 2 Energy.* Prepared by the National Greenhouse Gas Inventories Programme, Eggleston, H.S., L. Buendia, K. Miwa, *et al.* (eds). Published: IGES, Japan.

IPCC, 2006b. *Guidelines for National Greenhouse Gas Inventories.* Prepared by the National Greenhouse Gas Inventories Programme, Eggleston, H. S., L. Buendia, K. Miwa, *et al.* (eds). Published: IGES, Japan.

IPCC, 2013. *Climate Change 2013: The Physical Science Basis. Contribution of Working Group I to the Fifth Assessment Report of the Intergovernmental Panel on Climate Change.* Cambridge University Press, Cambridge, United Kingdom and New York, NY, USA.

IPCC, 2014. *Climate Change 2014: Mitigation of Climate Change. Contribution of Working GroupⅢ to the Fifth Assessment Report of the Intergovernmental Panel on Climate Change.* Cambridge University Press, Cambridge, United Kingdom and New York, NY, USA.

IPCC, 2018. *Global Warming of 1.5 ℃. An IPCC Special Report on the Impacts of Global Warming of 1.5 ℃ Above Pre-industrial Levels and Related Global Greenhouse Gas Emission Pathways, in the Context of Strengthening the Global Response to the Threat of Climate Change, Sustainable Development, and Efforts to Eradicate Poverty.* World Meteorological Organization, Geneva, Switzerland.

Johnke, B., 2001. *Chapter 5.3: Emissions from Waste Incineration*, in: IPCC Good Practice Guidance and Uncertainty Management in National Greenhouse Gas Inventories.

Kadiyala, A., R. Kommalapati, Z. Huque, 2016. Quantification of the Lifecycle Greenhouse Gas Emissions from Nuclear Power Generation Systems. *Energies*, 9.

Kwok, T. F., Y. Xu, P. T. Wong, 2017a. Complying with Voluntary Energy Conservation Agreements (I): Air Conditioning in Hong Kong's Shopping Malls. *Resources, Conservation and Recycling*, 117.

Kwok, T. F., Y. Xu, P. T. Wong, 2017b. Complying with Voluntary Energy Conservation Agreements (Ⅱ): Lighting in Hong Kong's Shopping Malls. Resources. *Conservation and*

Recycling, 117.

Lam, J. C., K. K. Wan, T. N. Lam, *et al.*, 2010. An Analysis of Future Building Energy Use in Subtropical Hong Kong. *Energy*, 35(3).

Lang, G., 2018. Urban Energy Futures: A Comparative Analysis. *European Journal of Futures Research*, 6.

Langston, C., E. Chan, E. Yung, 2018. Hybrid Input-output Analysis of Embodied Carbon and Construction Cost Differences Between New-build and Refurbished Projects. *Sustainability*, 10(9).

Leung, A., M. Burke, J. Cui, 2018. The Tale of Two (Very different) Cities–Mapping the Urban Transport Oil Vulnerability of Brisbane and Hong Kong. *Transportation Research Part D: Transport and Environment*, 65.

Liddle, B., 2018. Consumption-based Accounting and the Trade-carbon Emissions Nexus. *Energy Economics*, 69.

Liu, L., X. M. Ma, 2011. CO_2 Embodied in China's Foreign Trade 2007 with Discussion for Global Climate Policy. *Procedia Environmental Sciences*, 5.

Lu, L., H. X. Yang, 2010. Environmental Payback Time Analysis of a Roof-mounted Building-integrated Photovoltaic (BIPV) System in Hong Kong. *Applied Energy*, 87(12).

Macao DSEC, 2016. *Yearbook of Statistics*.

Macao DSEC, 2017. *Yearbook of Statistics*.

Macao DSEC, 2019. *Yearbook of Statistics*.

Macao DSPA, 2018. *Report on the State of Environment of Macao 2000-2017*.

Macao International Airport (MIA), 2020. *Airport Data*.

Muthu, S. S., Y. Li, J. Y. Hu, *et al.*, 2011. Carbon Footprint of Shopping (grocery) Bags in China, Hong Kong and India. *Atmospheric Environment*, 45.

National Academies of Sciences Engineering and Medicine, 2019. *Negative Emissions Technologies and Reliable Sequestration: A Research Agenda*, The National Academies Press, Washington, DC.

National Development and Reform Commission, 2014. *Assessment Method on the Mitigation of CO_2 Emissions per Unit of GDP*, NDRC Climate [2014] No. 1828.

NEA, HKSAR Government, 2008. *Memo on the Supply of Natural Gas and Electricity*.

Ng, A. W., J. Nathwani, 2010. Sustainable Energy Policy for Asia: Mitigating Systemic Hurdles in a Highly Dense City. *Renewable & Sustainable Energy Reviews*, 14.

Ng, W. Y., C. K. Chau, 2015. New Life of the Building Materials – Recycle, Reuse and Recovery. *Energy Procedia*, 75.

Pacala, S. W., R. H. Socolow, 2004. Stabilization Wedges: Solving the Climate Problem for the Next 50 Years with Current Technologies. *Science*, 305(5686).

Poudenx, P., 2008. The Effect of Transportation Policies on Energy Consumption and Greenhouse Gas Emission from Urban Passenger Transportation. *Transportation Research Part A: Policy and Practice*, 42.

Song, Q., J. Li, H. Duan, *et al.*, 2017. Towards to Sustainable Energy-efficient City: A Case Study of Macau. *Renewable and Sustainable Energy Reviews*, 75.

Song, Q., Z. Wang, J. Li, *et al.*, 2018a. Comparative Life Cycle GHG Emissions from Local Electricity Generation Using Heavy Oil, Natural Gas, and MSW Incineration in Macau. *Renewable and Sustainable Energy Reviews*, 81.

Song, Q., Z. Wang, Y. Wu, *et al.*, 2018b. Could Urban Electric Public Bus Really Reduce the GHG Emissions: A Case Study in Macau? *Journal of Cleaner Production*, 172.

Song, Q., Y. Wu, J. Li, *et al.*, 2018c. Well-to-wheel GHG Emissions and Mitigation Potential from Light-duty Vehicles in Macau. *The International Journal of Life Cycle Assessment*, 23.

Sovacool, B. K., 2008. Valuing the Greenhouse Gas Emissions from Nuclear Power: A Critical Survey. *Energy Policy*, 36.

State Council, 2016. *Work Plan for Greenhouse Gas Mitigation in the 13th Five-Year Period*.

Stockholm Environment Institute, 2017. *Consumption-based Accounting Reveals Global Redistribution of Carbon Emissions*.

To, W. M., 2015. Greenhouse Gases Emissions from the Logistics Sector: the Case of Hong Kong, China. *Journal of Cleaner Production*, 103.

To, W. M., T. M. Lai, W. C. Lo, *et al.*, 2012. The Growth Pattern and Fuel Life Cycle Analysis of the Electricity Consumption of Hong Kong. *Environmental Pollution*, 165.

To, W. M., P. K. C. Lee, 2017a. A Triple Bottom Line Analysis of Hong Kong's Logistics Sector. *Sustainability*, 9(3).

To, W. M., P. K. C. Lee, 2017b. Energy Consumption and Economic Development in Hong Kong, China. *Energies*, 10.

To, W. M., P. K. C. Lee, 2017c. GHG Emissions from Electricity Consumption: A Case Study of Hong Kong from 2002 to 2015 and Trends to 2030. *Journal of Cleaner Production*, 165.

Wang, H., D. Zhou, P. Zhou, *et al.* 2012. Direct Rebound Effect for Passenger Transport: Empirical Evidence from Hong Kong. *Applied Energy*, 92.

Wang, J., C. Yu, W. Pan, 2018. Life Cycle Energy of High-rise Office Buildings in Hong Kong. *Energy and Buildings*, 167.

Warner, E. S., G. A. Heath, 2012. Life Cycle Greenhouse Gas Emissions of Nuclear Electricity Generation: Systematic Review and Harmonization. *Journal of Industrial Ecology*, 16.

Wilburn, J. S., P. M. Roberts, 1995. Major Asian Subsea Pipeline to Start up This Year. *Oil and Gas Journal*, 93.

Woo, C.K., A. Shiu, Y. Liu, *et al.*, 2018. *Consumption Effects of an Electricity Decarbonization*

Policy: Hong Kong. *Energy*, 144.

Xu, Y., G. J. Liu, 2015. Carbon Capture and Storage for Hong Kong's Fuel Mix. *Utilities Policy*, 36.

Yang, H., G. Zheng, C. Lou, *et al.*, 2004. Grid-connected Building-integrated Photovoltaics: A Hong Kong Case Study. *Solar Energy*, 76.

Yang, Z. Y., W. J. Dong, T. Wei, *et al.*, 2015. Constructing Long-term (1948-2011) Consumption-based Emissions Inventories. *Journal of Cleaner Production*, 103.

Yim, S., S. Ng, M. Hossain, *et al.*, 2018. Comprehensive Evaluation of Carbon Emissions for the Development of High-rise Residential Building. *Buildings*, 8.

Zhang, C., D. Zhou, P. Li, *et al.*, 2014. CO_2 Storage Potential of the Qiongdongnan Basin, Northwestern South China Sea. *Greenhouse Gases: Science and Technology*, 4(6).

Zhao, S., Q. Song, H. Duan, *et al.*, 2019. Uncovering the Lifecycle GHG Emissions and Its Reduction Opportunities from the Urban Buildings: A Case Study of Macau. *Resources, Conservation and Recycling*, 147.

澳门电力股份有限公司：“2016 澳电年报”，2016 年。

澳门电力股份有限公司：“2017 澳电年报”，2017 年。

澳门国际环保合作发展论坛：“年度环保盛事-把握机会参与其中！”，2019 年。

澳门国际机场：“绿色机场”，2018 年，https://www.macau-airport.com/en/about-us/about-mia/green-airport/。

澳门环境保护局：“澳门环境状况报告”，2017 年。

澳门环境保护局：“澳门环境状况报告”，2018 年。

澳门环境保护局：“澳门环境状况报告”，2019 年。

澳门能源业发展办公室：“澳门特别行政区能源效益状况调查研究报告（2007，2011，2013，2017）”，2019 年。

澳门特别行政区政府：“车用无铅汽油及轻柴油标准”，2016 年。

澳门特别行政区政府：“第 256/2018 号行政长官批示”，2018 年。

第四章　气候变化政策与地区合作

香港和澳门特别行政区以极高的人口密度、城市化程度和经济水平而著称。受气候变化影响，港澳地区的气候条件日益严峻，恶劣天气状况给居民生活带来诸多新风险和新挑战。香港和澳门特别行政区现行的减缓、适应和应对气候变化政策从回归前就开始实施，其中一些政策走在世界前列，为中国及世界其他地区提供了有益的借鉴。回归后，香港和澳门地区的气候政策仍然持续发展，不断增加的区域合作机会也为气候变化的减缓、适应与应对带来诸多帮助。

本章首先介绍香港和澳门应对气候变化的治理结构及两地主要的气候政策，然后描述和评估相关政策。最后，本研究将着重介绍香港和澳门地区政策的独特之处，回顾区域合作框架，并就加强气候变化研究和应对能力、开展地区合作提供建议。

第一节　气候变化政策综述

一、　香港和澳门特别行政区政府结构及气候变化政策治理结构

在治理结构方面，香港、澳门特区与内地其他省份不同。作为特别行政

区，它们在地方政策决策方面享有很大的自主权，在气候变化相关政策制定方面也不例外。

香港和澳门均为依据中华人民共和国宪法设立的特别行政区（特区），分别被称为香港特别行政区（香港特区）和澳门特别行政区（澳门特区）。两个特区拥有各自的宪制性文件，即《基本法》。该文件由全国人民代表大会颁布，体现了 “一国两制” 的重大方针。特区在当地的日常事务中享有“高度自治权”，对气候变化政策和行动计划的相关决策也有相当大的政策自主性。

香港特区和澳门特区各自由其行政长官领导，行政长官既是特区首长，又是特区政府首脑。两地的政治制度被称为“行政主导体制”，意味着行政机关在行政长官的领导下制定政府政策。“主要官员”（司局级长官）主理各自负责的政策范畴。两个特区都有自己的立法机关，负责提出关乎公众利益的议题、审议立法、审批公共资金。香港特区设有审计署，对政府部门及公营机构的绩效和责任落实进行独立审计并发布报告。部分报告与气候变化相关，如水资源[①]。

虽然香港特区与澳门特区的行政体制相似，但在行政架构上却有组织差异（图 4–1，图 4–2），这是香港特区与澳门特区各自殖民历史的遗留所致。

应对气候变化的决策机构

根据《联合国气候变化框架公约》，减缓气候变化指“减少二氧化碳等温室气体排放和增加碳汇”[②]，适应指“生态、社会或者经济系统回应实际的或

①香港特区审计署署长选择要审计的主题和部门。这些报告可以参阅 https://www.aud.gov.hk/ eng/pubpr_arpt/rpt.htm。例如，在 2015 年 4 月 1 日，审计署发表了一份关于水务署管理供水和用水需求的报告。

②United Nations Framework Convention on Climate Change,“Introduction to Mitigation,” United Nations Climate Change, accessed January 21, 2020, https://unfccc.int/topics/mitigation/the-big-picture/introduction-to-mitigation.

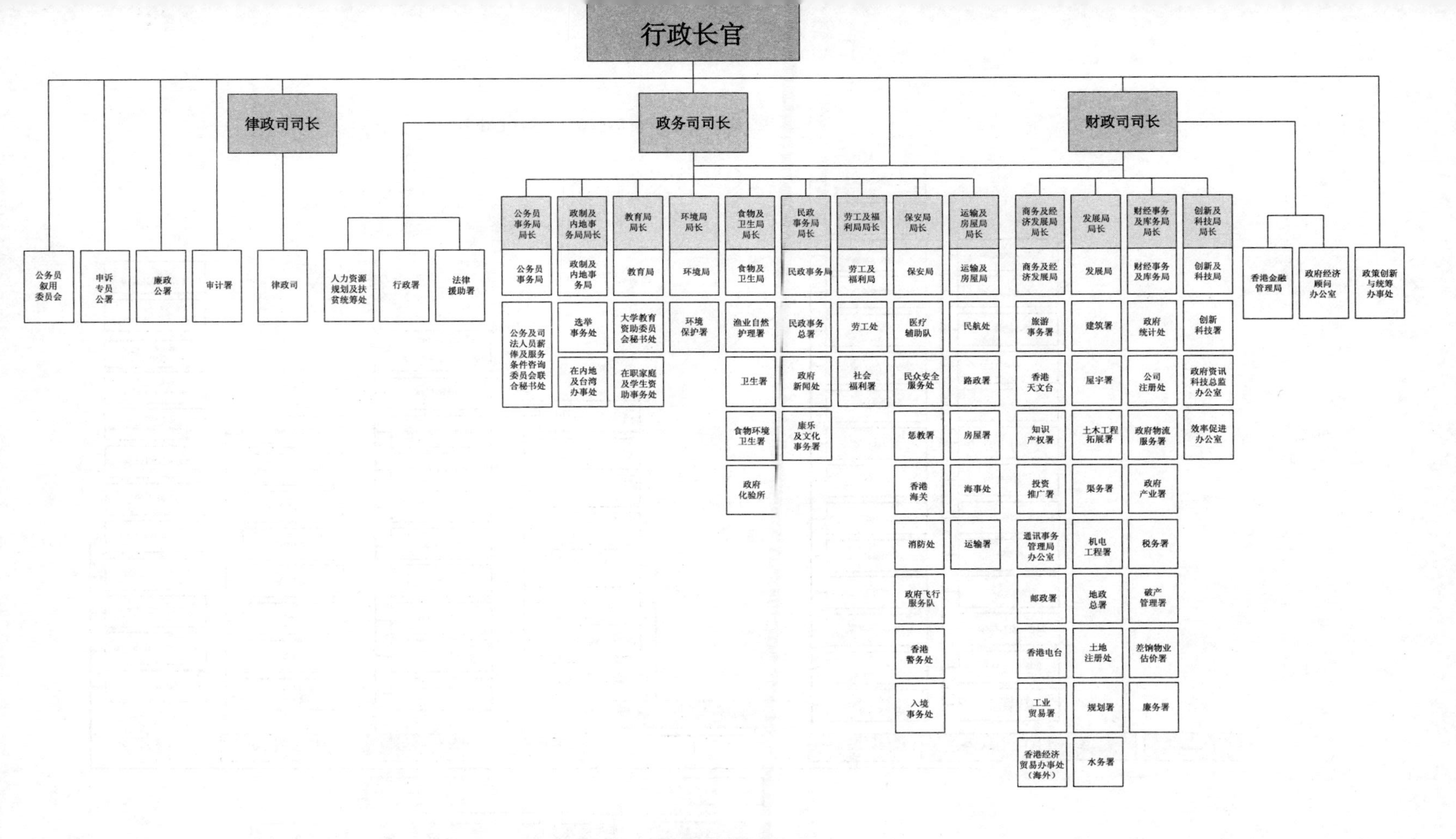

图 4–1　香港特别行政区政府组织图

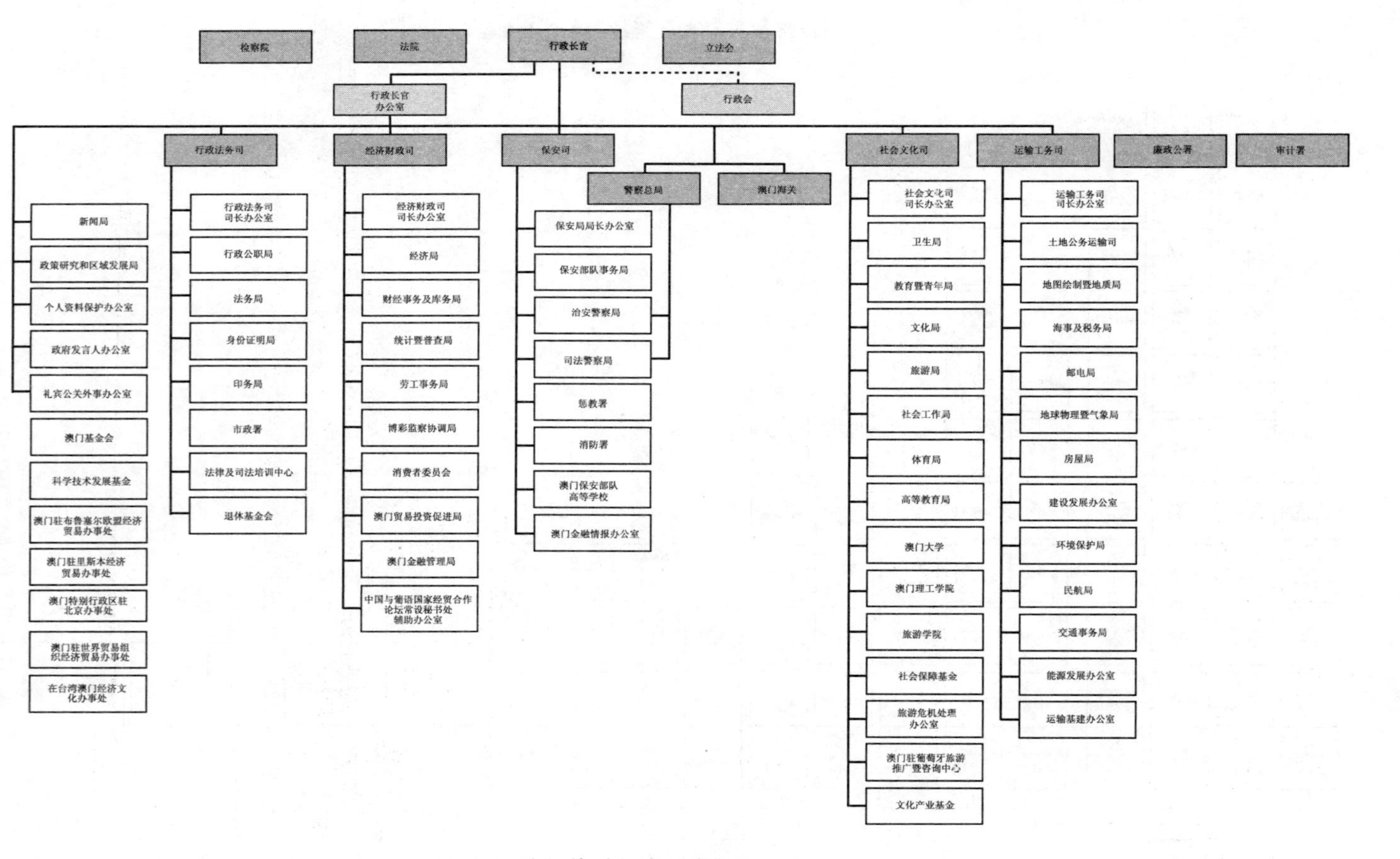

图 4–2　澳门特别行政区政府组织图

者预期的气候变化刺激及其影响而作出的调整”[①]。在本章中，减缓政策是指将气候变化程度降到最低的政策，而适应政策指将气候变化风险降到最低并保护生命和财产的政策[②]。此外IPCC将应对定义为“社会、经济和环境系统应对危险事件、趋势和干扰的能力，以回应或以充足的方式维持其基本功能、特性及结构，同时也保持其适应、学习和改造的能力”（IPCC, 2018）。适应气候变化将增强一个地区应对挑战的弹性，因此，适应和应对经常被放在一起讨论，本章也采取了这种方式。

香港特别行政区环境局局长领导环境局，主要负责香港特区环境、保育、废弃物管理、能源及气候变化在内的整体环境保护政策。环境局局长由环保署协助，但由于香港特区政府应对气候变化的工作涉及减缓、适应、应对等多个范畴，其他政策局和部门也需根据相关职责向环境局局长汇报，比如渠务署进行的污水处理工作、机电工程署[③]开展的节能工作及渔农自然护理署的自然保育工作。此外，建筑物管理由发展局辖下的屋宇署负责，环境局局长担任推动绿色建筑及可再生能源督导委员会主席[④]。

发展局的主要工作包括两个主要范畴：规划、土地及楼宇发展，基础设施建设。其辖下的水务署负责供水、渠务署负责防洪、机电工程署负责机电服务，而土木工程拓展署则负责斜坡安全和建设沿海基础设施等。交通运输

---

①United Nations Framework Convention on Climate Change, “What Do Adaptation to Climate Change and Climate Resilience Mean?”, United Nations Climate Change, https://unfccc.int/topics/adaptation-and-resilience/the-big-picture/what-do-adaptation-to-climate-change-and-climate-resilience-mean.

②参阅 https://www.nature.com/subjects/climate-change-policy。

③机电工程署为环境局的能源政策提供技术支援。

④推动绿色建筑督导委员会成立于2013年1月。2017年，其职责范围扩大至可再生能源并改称为推动绿色建筑及可再生能源督导委员会。委员会主席为环境局局长，他拥有建筑师专业资格，对于推动绿色建筑有突出贡献。屋宇署是发展局辖下的部门，负责监管建筑规范。

及房屋政策由运输及房屋局负责。极端天气事件等紧急事故应对处理则由保卫局负责。民政事务局主管社区关系，食物及卫生局联同卫生署负责公共卫生政策的制定和推行。香港天文台进行有关香港地区气候变化的研究，隶属于商务及经济发展局。

澳门行政区的管辖范围相对较小，治理架构也相对简单。运输工务司司长领导所有与减缓气候变化有关的部门，例如气候科学研究（由地球物理暨气象局负责）①、土地利用、公共工程、交通、房屋和环境保护。能源政策也由运输工务司下的能源业发展办公室负责制定。

粤港两地均设有咨询及智库机构，协助政府制定并推行包括与气候变化有关的各项政策。这些机构是政府施政观念的交流平台，也是收集专家和非政府组织意见的渠道。

二、气候政策的协作与实施：地方、国家和国际的视角

（一）地方气候政策的协作和实施

由于香港特区政府各部门间分工明确，面对气候政策的复杂性问题，协作成为跨领域政策制定和实施的关键。2007 年，香港特区政府成立了由环境局主持的气候变化跨部门工作小组，职责包括监察有关机构和部门制定及推行控制温室气体排放及适应气候变化政策措施的进展。2010 年香港发布了第一个气候变化行动计划，气候变化跨部门工作小组成为政策进程情况的监测平台。随后，香港取消了较低行政级别的气候变化跨部门工作小组，成立气候变化督导委员会，由政务司司长（地位仅次于行政长官的主要官员）担任主席，并于 2016 年 4 月举行了第一次会议，指导、协调、组织气候变化相关应对行动。督导委员会的成员包括 13 个决策局和 3 个部门，其他部门则按需

①参阅 https://www.smg.gov.mo/zh/subpage/429/page/302。

要参加会议。

2015年，澳门特区政府成立了由气象局理事会协调的气候变化跨部门工作小组，环境保护局是其成员之一。第一次气候变化跨部门工作小组会议于2015年3月召开，会上介绍了该小组的职责、结构和运作方法，并回顾了澳门特区政府近年来在应对气候变化方面进行的工作。

(二)关于地方政策的国家计划和国际条约

"一国两制"重大方针赋予了香港和澳门特别行政区充分的行政自主权，可根据各自情况，决定各自气候政策的制定、实施与协调。特区可以选择不受中央政府设置的碳排放限额的约束，设置自己的目标额度。就香港而言，由于其较高的经济发达程度，香港特区政府对减少碳排放设置了较高目标[①]，而澳门特区的碳排放限额与中央政府的规定相同。

至于中央政府签订的国际协议，需要进行磋商才能扩展到两个特区。最初，《联合国气候变化框架公约》和《京都议定书》不适用于香港和澳门[②]，但根据两地特区政府的要求，它们在2003年5月5日开始延伸至香港实施，并在2008年1月14日延伸至澳门[③]。同样的，在同香港和澳门特区政府磋商

① "香港应对气候变化策略及行动纲领"：2010年9月公布的公众咨询文件和2017年发布的《香港气候行动蓝图2030+》指出，香港特区制定了超过中央政府设定的碳排放量减少目标。

②香港特区和澳门特区的《基本法》规定，中华人民共和国缔结的国际协议，中央人民政府可根据香港特区和澳门特区的情况和需要，在征询特区政府的意见后，决定是否适用于香港特区和澳门特区。2002年8月30日，中国通知联合国《京都议定书》将暂时不适用于香港和澳门。

③《联合国气候变化框架公约》于1994年生效，中国是该公约的签署国之一。《京都议定书》于1997年向各国开放，中国于2002年核准了该议定书。当时，国家未将其适用于香港及澳门特区，因为尚未征求两个特区的意见。香港要求遵守《联合国气候变化框架公约》，并于2003年5月5日起生效。《京都议定书》实施后（2005年2月16日），也适用于香港特别行政区。澳门于2007年要求遵守《联合国气候变化框架公约》和《京都议定书》，并于2008年1月14日起生效。

后，《京都议定书》的后续条约——《巴黎协定》生效时，也适用于两个特区。

中央人民政府大力支持特区加入此类国际协议，共同履行国家承诺。香港特区政府定期派官员加入中央政府代表团参加《联合国气候变化框架公约》缔约国会议。2011年，香港特区政府和广东省政府在现有的粤港合作联席会议增设了“粤港应对气候变化联络协调小组”。2019年2月中共中央、国务院颁布的《粤港澳大湾区发展规划纲要》鼓励广东、香港和澳门加强合作，气候变化问题是重要的合作内容（见下文第五节）[①]。同年，双方将粤港持续发展与环保合作小组和粤港应对气候变化联络协调小组合并而成为“粤港环保及应对气候变化合作小组”，推进和深化双方就有关减缓及适应气候变化的科学研究和合作交流。

巴黎协定（2015年）

2016年11月4日，《巴黎协定》继《京都议定书》后生效。其关键条款要求全球采取行动：①尽快达到碳排放全球峰值，并在21世纪下半叶实现碳源和碳汇之间的平衡（即在2051～2100年之间达到“碳中和”）；②把全球平均气温升幅控制在不超过工业化前水平2摄氏度之内，并努力将气温升幅限制在1.5摄氏度之内。全球行动基于“自下而上”的方法，所有缔约方都必须制定适合各国的国家的具体目标和时间表。各缔约方须每五年编制一次国家自主贡献，同时更新的国家自主贡献目标在内容上须较上一份有所进步。

如第三章所述，香港和澳门都有碳强度目标。两个特区应与《巴黎协定》规定的提交时间表保持一致，每五年完成一次气候变化工作进展评估。两个特区至2030年的审查时间表如表4–1所示。中央政府要求两个特区同时提交符合《联合国气候变化框架公约》要求的清单和报告。

---

①见《粤港澳大湾区发展规划纲要》。https://www.bayarea.gov.hk/tc/home/index.html。

表 4–1　《巴黎协议》审查时间表

年份	2017	2019	2020	2024	2025	2029	2030
行动	设定 2030 年碳排放目标	行动工作评估	计划更新（即承诺采取进一步行动）	行动工作评估	下一个气候行动计划（即承诺采取进一步行动）	行动工作评估	下一个气候行动计划（即承诺采取进一步行动）

三、2010 年以来关键气候政策

如第一章中所述，香港天文台于 2004 年为香港提供了 21 世纪气候推算的首份研究，并定期更新气候数据、进行新的研究。如第三章中所述，香港特区政府委托了可行性研究以确定 2008 年减少碳排放的长远措施。此后，香港特区政府公布了多项与气候变化和节能有关的重要政策文件，这些文件共同展示了政策的进程：

香港应对气候变化策略及行动纲领-咨询文件（2010 年 9 月）[①]；

未来发电燃料组合-咨询文件（2014 年 3 月）[②]；

关于 2010 年行动纲领所列措施的进展和成果报告（2014 年 4 月）；

电力市场未来发展公众咨询（2015 年 3 月）；

香港都市节能蓝图 2015-2025+（2015 年 5 月）[③]；

---

①https://www.epd.gov.hk/epd/sites/default/files/epd/english/climate_change/files/Climate_Change_Booklet_E.pdf。

② https://www.enb.gov.hk/sites/default/files/zh-hant/node2606/Consultation%20 Document.pdf。

③《香港都市节能蓝图 2015～2025+》，2015 年 5 月。https://www.enb.gov.hk/sites/default/files/pdf/EnergySavingPlanTc.pdf。

香港气候变化报告 2015（2015 年 11 月）[①]；

香港气候行动蓝图 2030+（2017 年 1 月）[②]；

透过 4T 合作伙伴加强在香港现有建筑物节约能源（2017 年 6 月）[③]。

澳门于 1991 年颁布了《环境纲要法》，确立了澳门应对气候变化的总体纲要和原则。在《京都议定书》适用于澳门后，澳门特区政府于 2008 年发布了《澳门特别行政区政府应对气候变化方案》。澳门有一份重要的政策文件，涵盖了包括气候变化在内的整体环境保护政策：《澳门环境保护规划（2010-2020）》[④]。

第二节　港澳减缓气候变化政策的综述与评估

一、应对气候变化的动力及减排目标

1997～2018 年，香港特区和澳门特区实施减缓气候变化政策的动因主要来自两方面，一是地方政府在改善空气质量方面采取的积极举措，二是受国家层面乃至全球在应对气候变化行动中的积极影响。

第三章介绍了港澳致力于减少空气污染物排放的背景，说明了两地通过

①《香港气候变化报告 2015》，2015 年 11 月。https://www.enb.gov.hk/sites/default/files/pdf/ClimateChangeChi.pdf。

②《香港气候行动蓝图 2030+》，2017 年 1 月。https://www.enb.gov.hk/sites/ default/files/pdf/ ClimateActionPlanChi.pdf。

③《透过 4T 合作伙伴加强在香港现有建筑物节约能源》，2017 年 6 月。https://www.enb.gov.hk/sites/default/files/pdf/EnergySaving_EB_TC.pdf。

④ https://www.dspa.gov.mo/Publications/EnvPlanningBook/ 201209-EnvPlanningBook_PB_TC.pdf。

改用清洁能源，制造业迁移等措施，大幅减少了碳排放。这也说明了减少碳排放是达到空气质量目标的一项重要内容。港澳积极应对气候变化的另一个推动因素，正如上文中提到的，在于国家和全球层面要求制定具体碳减排目标的要求力度越来越大，特区政府对此进行了政策回应。

现今，改善空气质量对于香港和澳门仍十分重要。就香港而言，法律规定政府需要至少每五年检讨一次空气质量指标，并考虑是否在切实可行的情况下收紧指标①。两个特区政府与广东省环境保护厅就改善空气质量开展积极合作。鉴于国家在空气污染防治方面作出的努力，减少碳排放将是粤港澳三地政府努力达成的共同目标。与此同时，国际和全球减缓气候变化的努力也将引起这两个大力推广“低碳生活”社会发展主题的特区政府的持续关注。

二、港澳减缓气候变化的相关政策

2018 年，香港总人口约 745 万，比 1997 年成立特别行政区时增长了约 15%；澳门人口约 67 万人，比 1999 年回归时增长了近 55%。2018 年，香港本地生产总值约为 3 617 亿美元，澳门本地生产总值约为 553 亿美元。两地都是服务业高度发达的经济体，香港经济发展相对更加多元化，而澳门作为全球最大的博彩市场，主要经济部门是博彩业，占其 GDP 的 50%。同时，旅游业也是两地的重要经济支柱。

香港和澳门在缓解气候变化的政策上有两个明确的重点：①改变燃料结构，减少空气污染物和碳排放，这是因为能源工业（发电及煤气生产）是迄今为止空气污染和碳排放的最大贡献者；②节约能源（见表 4–2），正如第二章指出的那样，在城市化高度发展的大都市，建筑物节能尤为重要。例如，

①空气质量标准在香港被称为空气质素指标。

建筑物耗电量约占香港总用电量的 90%。尽管有诸如发展电动汽车等减少汽车排放的计划推出，但这些计划主要针对路边污染物排放，在碳减排方面作出的贡献较少，具体原因将在下一部分阐述。

（一）改变燃料组合

1. 规管电力、燃气行业

正如第二章提到的，香港特区和澳门特区的人口密度都很高，分别为 6 930 人/平方千米（2019 年）和 21 400 人/平方千米（2016 年），其社会经济活动运行依赖持续稳定的电力。香港的主要竞争优势之一就是拥有高度可靠的电力和天然气等能源供应。澳门特区自 2001 年起制定的开放和大力发展博彩业的相关政策，因此提高能源可靠性对于地区发展也至关重要。

香港的电力和煤气业务一直为私营公司运营。香港中华煤气有限公司（煤气公司）于 1862 年成立，是香港历史最悠久的公用事业机构。香港的两家电力公司也同样有着悠久的历史——中华电力有限公司（中电）成立于 1901 年，香港电灯有限公司（港灯）成立于 1889 年。鉴于电力行业巨大的规模经济性，两家公司实际上是其各自服务领域的唯一服务提供者[①]。负责能源供应的这三家公司经营良好，盈利能力强，已在当地的证券交易所上市或是当地上市公司的全资子公司[②]。

---

①煤气公司供应范围广。2017 年，煤气公司在香港拥有 3 632 千米的供气网络，为 188 万楼住宅、商业和工业用户供气。此外，煤气公司也为香港出租车队供应石油气，以及为香港机场供应航空燃油。中电作为两家电力供应商中规模较大的一家，覆盖九龙及新界，而港灯则为港岛及南丫岛供电。目前与中电签订的管制计划协议有效期为 2018 年 10 月 1 日至 2033 年 12 月 31 日，与港灯签订的管制计划协议有效期为 2019 年 1 月 1 日至 2033 年 12 月 31 日。

②这三家公司在香港以外的地区都有广泛的业务。本章只讨论他们在香港特别行政区的运作情况。

表 4–2 香港和澳门特区应对气候变化的主要措施

香港特区减排行动		澳门特区减排行动	
1. 改变燃料组合 a. 发电领域 i）提高本地天然气发电比例 ii）增加零碳能源（包括可再生能源）的比例 b. 交通运输领域 i）支持电动车辆使用 ii）建设充电桩等基础设施 iii）政府车队采用电动车辆 iv）给予电动私家车主税务优惠 v）设立绿色运输试验基金以鼓励交通运输行业尝试使用包括电动汽车在内的绿色创新运输技术 vi）支持电动单层专营巴士的试验 vii）发展铁路作为客运系统的骨干 viii）发展新兴代步工具	2. 促进节能 a. 建筑物节能 i）政府以身作则 ii）推广绿色建筑实践 iii）加强建筑物节能法规建设 iv）推动对现有建筑进行能源审核、改造和重新校验 v）为私人楼宇业主提供税务优惠 vi）通过对话平台，加强与业主的交流 b. 其他领域节能 i）政府以身作则 ii）加强和推广良好实践 iii）提高强制性能源效益标签标准并覆盖更多电子产品 iv）发展水冷式空调系统 v）开展区域供冷系统工程 vi）激励社区行动	1. 改变燃料组合 a. 发电领域 i）提高本地天然气发电比例 ii）开发可再生能源 b. 消费领域 i）发展轻轨 ii）发展电动出租车 c. 交通运输领域 i）提供环保车辆税务优惠 ii）提升机动车辆排放标准 iii）于全澳合适的公共停车场及公共道路泊车位安装充电桩	2. 促进节能 a. 建筑物节能 i）发展建筑物能耗优化技术 ii）设立澳门环保酒店奖 b. 其他领域节能 i）设立澳门节能周 ii）推广校园节能活动

香港特区政府通过与中电及港灯签订的非排他性合约，即“管制计划协议”管理电力业务。管制计划协议为政府监察电力公司财务状况及营运表现提供了行政框架，并制定中电和港灯依据其固定资产平均净值为基数所获取的准许回报率，这一定程度上提高了中电及港灯对于投资数额较大、回报周期较长的发电工厂和设备的投资意愿，保证了供应的可靠性[①]。除了必须遵守管制计划协议中的监管条款外，这两家公司独立于政府，自行管理其业务。

长期以来，煤气公司与香港特区政府之间没有签署类似的合约协议。直到1997年4月，政府才与煤气供应商签订了资料及咨询协议，要求煤气公司向市民披露部分财务、营运、环保和安全资料。更重要的是，煤气公司在调整收费、添置主要的系统项目或更改定额保养月费前，需根据资料及咨询协议的规定征询政府意见。这反映了政府开始关注收费变动带来的影响，并希望有机会向公司提出自己的意见[②]。

香港的这三家能源公司制定了各自的减排目标。中电和港灯的目标是根据它们预期的燃料组合结构的变化设定的，而煤气公司也在2008年自主设定了相应目标[③]。

澳门的电力供应始于1906年，最初由一家私营公司负责。因其经营状况不善，澳门政府于1972年成立了澳门电力公司（澳电），负责供电并兴建了一座新工厂，实际上掌握了专营澳门电力生产和输送的权利。但随着澳电的重组，政府已将其大部分股权转让给国内、海外与本地投资者。1985年，澳

①有关管制规划协议的运作详情，请参阅环境局《电力市场未来发展公众咨询》第一章，https://www.enb.gov.hk/en/resources_publications/policy_consultation/public_consultation_future_development_electricity_market.html。

②请参阅“煤气公司的监察”，https://www.enb.gov.hk/en/about_us/policy_ responsibilities/financial_monitoring.html#monitor2。

③煤气公司承诺，到2020年，香港煤气生产业务的碳排放将比2005年降低30%。到2017年，已实现了23%的减排。碳排放量减少的原因可以归结为天然气的使用和使用堆填气体用于城市煤气生产，以及工厂能源利用效率的提高。

电与澳门政府签订了25年的电力专营合约，该合约对澳门以专营制度在澳门地区进行电力的生产、进出口、输送、分配和出售作了全面规定，特许权还规定了与香港管制计划协议类似的准许回报率和监管条款。2010年，这一专营合约被延长了15年，但由于政府希望保留引入竞争机制的选择权，因此取消了合约的排他性条款。

中电及港灯将于2025年完成其服务范围内新款智能电表的安装工作。香港特区政府一直与他们及煤气公司商讨电力消耗及发电数据的披露问题，以提高能源效益、促进数据开放及发展智慧城市。香港特区政府已与三间公司达成共识，逐步公布相关数据，推动智慧城市发展。

就澳门而言，专营合约能够帮助澳门特区保留一家公共服务提供商，同时也有助于澳电满足快速增长的电力需求并促进创新。除考虑安全、环保、效率和经济等配电原则外，专营合约并不直接涉及碳减排。澳电为减少碳排放做出了自己的努力，采取了系统性的、切实的措施控制其温室气体排放，并在2010年成为了第一家取得了ISO 14064温室气体管理系统认证的公用事业公司，这也是港澳地区第一家获此认证的电力公司。

第三章指出，香港和澳门由于陆地和海洋面积的限制，都不具备大规模商业化可再生能源发电的有利条件（例如发展风能或建设太阳能发电厂）。香港特区政府在2017年估计，香港能够以风力、太阳能及转废为能实现可再生能源的潜力约为本港耗电量的 3%～4%，这些资源可基于先进的商用技术加以开发。电力公司正在研究在香港水域兴建两座风力发电厂的可行性，其发电量预计将提供香港总电力需求的1.5%。在转废为能方面，香港的堆填沼气、污泥焚化和厨余处理会产生一定数量的能源。煤气公司和中电获准使用政府所有的堆填区提取的堆填气体。同时，一座最高处理量达每日3 000公顷的综合废弃物管理设施正在建设中，并将于2025年左右投入使用。届时，转废为能项目预计可提供香港1%～2%的总电力需求。2017年的可再生能源潜力评估未包含尚未开发的项目（如渠务署和环保署正在跟进的污水污泥和食物废弃物的联合处理、波浪能等），也未考虑可能的技术进步（如提高太阳能光伏发电系统的效率）①。

香港特区政府在技术和财政可行的情况下，率先发展了可再生能源，并创造有利条件，鼓励私营机构参与。在公营机构方面，香港特区政府过去几年已预留 20 亿元，为政府建筑物、场地和设施装置小型可再生能源装置②。此外，政府部门正积极考虑发展更大规模的可再生能源项目。例如，特区政府于 2016 年在小蚝湾污水处理厂装设了政府最大的太阳能发电场（达 1.1 兆瓦）。

在私营企业方面，在 2018 年 10 月/2019 年 1 月生效的管制计划协议下，香港特区政府及电力公司正式推行上网电价计划，通过财政激励措施鼓励私营部门投资建造分布式可再生能源系统。上网电价根据发电系统容量的不同，单价分别为 3 港元、4 港元和 5 港元，预计项目投资回收期为 10 年。同时，中电和港灯也向希望支持使用清洁电力的用户出售可再生能源证书③。政府出台了一系列措施，促进和支持公众参与发展可再生能源。例如：放宽村屋天台安装光伏系统的限制；引入“采电学社”为符合资格的学校和非政府福利机构安装小型的可再生能源系统。自 2018 年 4 月起，政府对企业购置可再生能源和建筑物节能装置给予了税务优惠④。这些有力举措都加快了香港可再生能源发展的步伐。

澳门垃圾焚化中心建于 1992 年，解决了澳门特区约 3%的电力需求。为

---

①详见《香港气候行动蓝图 2030+》，第四章。

②香港特区政府预留资金以资助部门在政府建筑物和场地及社区设施等安装可再生能源项目，在 2017～18、2018～19 及 2019～20 年度的财政预算案中，香港特区政府将这笔投资分别由 2 亿港元增加至 8 亿港元及 10 亿港元，总额达 20 亿港元。

③有关上网电价和可再生能源证书的详情，请参阅机电工程署旗下的香港可再生能源网相关网页 https://re.emsd.gov.hk/english/fit/int/fit_int.html。

④《香港特区 2018～2019 财政年度政府财政预算案》，第 168 段。

回收垃圾焚化过程中产生的热量，焚化中心于 2001 年增设了发电设施[①]。由于澳门每天的垃圾产生量已接近焚化中心的处理极限，为此，澳门特区政府于 2006 年又兴建了三台焚化炉。截至 2013 年，垃圾焚化发电占当地发电总量的 23.6%（Song *et al.*, 2017）。

澳门特区一直致力于保障天然气的长期供应，澳电在 2018 年获准建设的新工厂就使用了天然气发电机组。专营合同和政府的部分所有权是政策实施的有效途径。在节能方面，澳门行政公职局与能源业发展办公室自 2010 年以来一直向各政府部门推广节能项目，并通过公众教育活动鼓励私营企业开展节能行动。能源业发展办公室制定的《太阳能光伏并网安全和安装规章》于 2015 年 1 月 26 日正式生效，制定了公共和私人建筑物安装太阳能光伏发电系统及相关设备的规范，设定了太阳能光伏发电系统直接或经配电系统与低压及中压的公共电网连接的安全规格。

2. 推广绿色交通

电动汽车没有尾气排放，用电动汽车取代传统燃料汽车尤其是商用汽车，将有助于改善路边空气质量，减少碳排放。香港特区政府于 2011 年 3 月成立 3 亿港元的“绿色运输试验基金”，以资助相关运输行业（例如出租车、小巴、巴士、货车和轮渡的经营商）试验绿色创新运输技术，这些技术包括发展更清洁的新能源商用车辆和渡轮[②]。为进一步推动运输业界试验及更广泛使用绿色创新运输技术，香港特区政府于 2020 年重新讨论了基金的资助范围，并向基金注入 8 亿港元额外拨款以扩大基金的资助范围至涵盖“试验申请”和“应用申请”两部分。同时，更多新能源商用运输工具被纳入资助类别，包括船只、电单车、非道路车辆等。基金已于 2020 年 9 月改名为“新能源运输基金”。

①澳门特区统计暨普查局：“1990～2015 年统计年鉴”，2016 年。

②有关绿色试验基金的讨论，请参阅香港特别行政区立法会文件《改善路边空气质素的进展》，香港特区环境保护署，2018 年 12 月 19 日。

截至 2020 年 11 月底，基金共批出 196 个试验项目，资助金额共约 1.54 亿港元，其中涉及 163 辆电动商用车（包括单层巴士、小巴、出租车和货车）和 103 辆混合动力车（包括单层巴士、小巴和货车）。

澳门特区政府也制定了改善交通能源利用的政策，包括发布“公共部门绿色出行指引”、推出环保车税务优惠政策及于全澳合适的公共停车场及公共道路泊车位安装充电桩等，以推动环保车辆的使用。同时，环境保护局为配合环保车税务优惠的实施，制定了“新轻型汽车的环保排放标准”，并于 2018 年收紧有关标准，以达到更佳的环保效益。

（1）电动车首次登记税

香港为购买电动车辆提供首次登记税减免，大幅降低了购车者的成本。首次登记税适用于所有类型的新登记车辆。电动商用车、电动单车及电动机动三轮车的首次登记税在 1994 年 4 月 1 日至 2024 年 3 月 31 日期间可获全额减免。对于电动私家车，在 2017 年 3 月 31 日前可以完全减免首次登记税。为了确保在推动电动私家车使用和不增加整体私家车数目之间取得平衡，2017 年 4 月 1 日开始，特区政府设立了一般电动私家车的首次登记税减免额的上限，并于 2018 年 2 月 28 日开始推出新的“一换一”计划，报废其拥有的旧私家车后首次登记一辆新电动私家车的车主，可获得较高的首次登记税减免额。上述电动私家车的首次登记税减免安排将持续到 2024 年 3 月 31 日[①]。政府也致力与商界合作安装电动车公共充电设施，截至 2020 年 9 月底，已有 3 219 个公共充电桩分布全港各区供公众使用[②]。

（2）推广电动及混合动力巴士

由私营公司经营的专营巴士提供约 33%的公共交通服务，是香港公共交

①香港特别行政区政府新闻公报—电动车首次登记税现有减免安排期限延长三年（2020 年 8 月 20 日）。

②有关推动使用电动车辆措施的详情及有关资料，请参阅 https://www.epd.gov.hk/epd/english/environmentinhk/air/prob_solutions/promotion_ev.html。

通系统的重要组成部分。现在，香港特区有约 6 200 辆专营巴士，其中约 95%为双层巴士。混合动力及单层电动巴士的技术已经在香港以外的地方得到利用和推广，比如深圳和其他城市。

由于香港多丘陵地形，夏季天气炎热，巴士需安装空调，在本地环境下的运作测试相当必要。为此，香港特区政府为专营巴士经营商提供资助购置双层混合动力巴士及单层电动巴士作试验。上述试验详情可参阅本章第三部分。至于双层电动巴士，新能源运输基金于 2020 年 11 月批准资助试验两款共四辆双层电动巴士，预计在 2022 年中期可以陆续开展试验。

（3）柴油出租车及公共小巴转用液化石油气

香港出租车及公共小巴分别运输约 7.4%及 15%的公共交通乘客人次[①]。2000～2003 年，香港通过换车资助计划，将柴油出租车转为使用液化石油气。2002～2005 年香港的公共小巴也实施了类似的计划[②]。现在几乎所有出租车（超过 99%）都使用石油气，而在公共小巴中，约 80%已转用液化石油气[③]。相对柴油，石油气是较清洁的燃料。尽管有研究表示液化石油气和柴油汽车的碳排放量大致相同，转用石油气仍能显著减少路边的空气污染物[④]。

尽管出租车车主可以通过绿色运输试验基金申请试验电动出租车，且电动出租车也可获全额减免首次登记税，但在香港试验并不成功。香港多斜坡及在夏季行驶时要提供空调，以及出租车实行的 24 小时轮班制经营模式对电

---

①https://www.td.gov.hk/filemanager/en/publication/ptss_final_report_eng.pdf（参阅此网页的 3.6 和 3.8）。

②https://www.info.gov.hk/gia/general/200311/05/1105158.htm。

③有关石油气车辆计划的具体内容，请参阅 https://www.emsd.gov.hk/en/gas_safety/lpg_vehicle_scheme/。

④Atlantic Consulting 2009. LPG's Carbon Footprint Relative to Other Fuels: A Scientific Review, http://www.shvenergy.com/wp-content/uploads/2015/05/atlantic_consulting_scientific_review_carbon_footprint_june_2009.pdf.

池续航力需求较高，当前车型下能有效配合出租车正常营运的需要[①]。随着日本制造商正在淘汰现有的液化石油气车型，香港的出租车车队将迎来重大变化。一款石油气及电能混合动力出租车已引入香港，并获得了出租车业界的正面评价[②]。

2018 年澳门约有 1 600 辆出租车。同年澳门开始试用电动出租车，目前试验仍在进行中。

（4）柴油商业车辆

香港是重要的集装箱港口和物流枢纽，拥有高水平的公路运输能力可将货物运往香港及内地其他地区。为减少路边空气污染，香港特区政府自 2014 年 3 月至 2020 年 6 月实施了特惠资助计划，共淘汰了约 8 万辆欧盟四期以前的柴油商业车，并于 2020 年 10 月推出一项新计划，目标于 2027 年底前分阶段淘汰约四万辆欧盟四期柴油商业车。

（5）车辆的环境排放标准

2012 年，澳门特区实施了进口新汽车的尾气排放标准，为进口新汽车设定了尾气排放限值。2018 年 11 月，澳门环境保护局和交通事务局收紧了进口新车辆的排放标准，汽油车标准由欧盟四期提升至欧盟六期水平，柴油车由欧盟四期提升至欧盟五期水平，并增加重型天然气车排放标准（须符合欧盟五期水平），进口新重型及轻型摩托车则提升至国家第III阶

---

①LegCo Question, Promoting Use of Electric Vehicles and Hybrid Vehicles, 7 February 2018, https://www.info.gov.hk/gia/general/201802/07/P2018020700429.htm; and The Standard, “End of the Road for Our Only EV Taxi”, 30 May 2018, http://www.thestandard.com.hk/section-news.php?id=196234&sid=11.

②South China Morning Post, “First hybrid Toyota taxis hit Hong Kong roads, but wider use will need incentives, industry says”, 25 January 2019, https://www.scmp.com/news/hong-kong/transport/article/2183725/first-hybrid-toyota-taxis-hit-hong-kong-roads-wider-use.

段水平[①]。预计这一措施将有助于控制澳门的过度交通能源消耗，并减少碳排放。

（二）节能

2007年，香港作为亚太经济合作组织的成员采纳了亚太经合组织的能源强度目标，即以2005年为基年，于2030年前将能源强度降低至少25%。2011年，亚太经合组织提高了此目标—以2005年作基年，于2030年或以前将区域总能源强度降低至少45%。香港特别行政区政府于2015年5月公布了《香港都市节能蓝图 2015～2025+》，并为香港设定了更宏伟的目标，即以2005年作为基年，于2025年以前将能源强度降低40%。为达成目标，香港需在2025年前减少6%的能源使用量。配合于2017年6月发布的《透过4T合作伙伴加强在香港现有建筑物节约能源》，蓝图列出了香港至2025年的节能工作最重要的领域，重点实现建筑物节能。

截至2018年，香港的能源强度已经降低了30%以上。自2012年以来，香港特区政府与社会各界、广大市民一起，开展了大量全民节能活动。其中最突出的是联合公司、组织签署的涉及更多设施的节能协定。随着香港2025年实现降低能源强度目标时间的逼近，这些活动愈发重要，并需要整个社会采取行动。

澳门也有类似的节能活动。2006年6月，一系列鼓励节能的活动随着“澳门节能周”的启动而展开。自此，节能周于每年6月在澳门各地开展，这也是澳门最大的能源活动。澳门节能周2019年的系列活动包括：“齐熄灯，一

①*Macau Daily Times*, New Vehicles' Emissions Requirements Tightened, November 21, 2018. https://macaudailytimes.com.mo/environment-new-vehicles-emissions-requirements-tightened.html.

小时；便服夏，齐节能”；还组织了“5%节能行动”抽奖活动等节能活动①。此外，特区政府指定中小学每年举办“校园节能文化活动”②，并于2011年起推进“能源教育：初中辅助教材”，将节能知识纳入课程③。澳门在节能政策的体系构建方面也作出了努力，例如《澳门特别行政区五年发展规划（2016-2020）》及《澳门环境保护规划（2010-2020）》提出了一系列关于节能减排的行动计划，并有序落实及推进。

1. 新建及已有建筑的节能

正如第三章指出的，在两个特别行政区中，建筑物的能源使用模式及缓解方案展示了高能效的重要性。其中一个重要的举措是让香港的主要业主签署例如《透过4T合作伙伴加强在香港现有建筑物节约能源》的节能约定，共同协助香港实现2025年目标。2016年，环境局及机电工程署将已有建筑节能纳入了长期利益相关者参与的目标区域。普遍认为，建筑物中能源使用的高度透明化将有利于激发人们在节能上的努力。因此，如表4–3所示，从机电工程署对香港各部门电力使用情况的统计可以看出，商用建筑（包括政府和机构建筑）中耗电量居于首位。商用建筑仅占主要建筑物的相对较小部分，但其用电量占全港用电量的20%，因此商业建筑的用电节能工作具有重要意义。由于政府承诺在2015～2020年将降低楼宇能耗的5%，因此要求有关楼宇业主自行设定节能目标。此外，从2018年4月起政府允许企业进一步加速采购节能建筑设备，并提供资本支出税收抵扣，以促进建筑节能④。

---

①Activities During the Macao Energy Conservation Week 2019，资料来源：https://www.gdse.gov.mo/eco2019/index.html。

②澳门能源发展办公室：“校园节能文化活动 2016”，2016年（http://www.gdse.gov.mo/public/chn/ecampus2016/2016/2016.html）。

③澳门能源发展办公室：“能源教育——初中辅助教材”，2011年。

④见《财政司司长2018-19年度预算》，第168段。

表 4-3　香港建筑用电百分比

部门	用途	使用占比
现有商业建筑	400 强商业建筑（公共区域）	11%
	酒店	4%
	地铁办事办公室，车库和车站	1%
现有公共建筑	政府建筑	5%
	学校和大学	4%
	医院管理局	3%
	房屋委员会	1%
	机场管理局	1%
总计	香港电力使用总量的 20%	

由于经济增长及对新商业及住宅空间的需求，香港和澳门将继续增加其建筑物存量。香港于 1995 年颁布的《建筑物（能源效率）规例》，首次推广通过设计实现建筑节能。该规例通过规定最大的总热传递值来规范特定类型建筑物外墙和屋顶的设计和施工。1998 年，香港推行《自愿性香港建筑物能源效益注册计划》，并推动了《建筑物能源效益守则》。该计划规定了照明、空调、电气设施、升降梯及自动扶梯装置的最低能效标准。2012 年《建筑物能源效益条例》与《建筑物能源效益守则》同时颁布。《建筑物能源效益条例》要求在设计新建筑和改造现有规定建筑时遵守《建筑物能源效益守则》[①]。《建筑物能源效益守则》每三年进行一次修订，以反映国际标准的发展和最新技

①规定建筑包括商业建筑、酒店和宾馆、住宅建筑（仅公共区域）、工业建筑（仅公共区域）、复合建筑（非住宅和非工业部分）、复合建筑（住宅或工业部分的公共区域）、教育区域建筑、社区建筑、市政建筑、医院和诊所、政府建筑、机场客运大楼和火车站。

术的进步。迄今《建筑物能源效益守则》已完成两次修订，并实施新标准。预计至 2028 年，《建筑物能源效益条例》可为香港所有的新建建筑物和现有建筑物节省约 270 亿度电，相当于减少排放 1 900 万吨 CO_2。此外，《建筑物能源效益条例》要求商业建筑物业主按《建筑物能源审核实务守则》每十年对“中央屋宇装置”进行能源审核。同时，《建筑物能源审核实务守则》也会定期修订和更新。

与此同时，人们对如第三章范畴二排放标准所述的方式进行设计和建造的绿色建筑的兴趣日益浓厚。如今，香港绿色建筑主要以“绿建环评+”为标准，综合考虑当地环境和特点考虑进行评估①。此外还参考了海外标准，如“能源与环境设计先锋”。自 2011 年，政府规定私人开发项目在“绿建环评+”下注册是开发的必要条件之一。通过使私人开发项目获得具有绿色设施和绿色空间，鼓励建筑的绿色、可持续设计。2010～2019 年 5 月，超过 1 160 个项目已在“绿建环评+”下注册并建造超过 3 300 万平方米占地面积的高标准新建筑物。若同时考虑其他评估标准，则有约 1 300 个项目注册并建造了超过 4 000 万平方米占地面积的新建筑②。

香港特区政府拥有约 8 000 幢建筑，并于 2003 年开始推进为政府建筑物设定具体减电目标的方式推行建筑物节能。2003～2014 年，政府实现了 15%的能源削减，并且承诺到 2020 年再减 5%，并已于 2018 年实现 4.9%的能耗降低。这一目标于《香港都市节能蓝图 2015～2025+》被提出。政府以其所有的包括公共住房的新建筑采用绿色建筑作为模范，展示特区政府在绿色建筑中的承诺和绿色建筑的优点。

---

①香港推行绿色建筑物认证始于 1995 年的“绿建环评”，也就是如今“绿建环评+”的前身，而香港绿色建筑议会在发展局及建筑业的专业团体的共同支持下，于 1996 年成立并颁布了香港第一个绿色建筑标准。创始成员包括建造业议会、商界环保协会、建筑环保评估协会及环保建筑专业议会，见 https://www.beamsociety.org.hk/en_about_us_2.php。

②资料来源：建筑环保评估协会，个人访谈，2019 年 6 月 26 日。

相较香港，澳门在绿色建筑方面的努力相对较少。因节能与博彩及度假建筑物运营者利益相符，澳门的绿色建筑主要集中于此①。2016 年，澳门博彩业的碳排放占澳门总碳排放的 62.46%，而住宅和商业建筑的排放分别仅占 20.51%和 17.03%（Zhao S. *et al.*, 2019）。尽管如此，能源业发展办公室在 2009 年发布了《澳门建筑物能耗优化技术指引》，确定了建筑节能的六个主要领域：①建筑材料；②照明系统；③空调及通风系统；④电力系统；⑤升降机及自动梯系统；⑥再生能源系统。能源业发展办公室发布的其他政策更侧重于使用节能的电子产品和再生能源（例如太阳能）。此外，澳门环境保护局于 2007 年起颁发“澳门环保酒店奖”以促进旅游业绿色建筑的推广和实施②。

2. 其他领域的节能措施

（1）节能设备

2008 年颁布的《能源效益（产品标签）条例》为“强制性能源效益标签计划”在香港的实施提供了基础。该计划要求在规定的产品上标明能源等级标签，以告知消费者产品的能源性能。强制性标签计划首阶段涵盖三类产品，即空调机、冷冻器具和紧凑型荧光灯（省电灯泡），于 2009 年 11 月 9 日起全面实施。第二阶段将涵盖范围扩大至另外两类电器产品，即洗衣机和烘干机，于 2011 年 9 月 19 日起全面实施。2015 年 11 月，室内空调、制冷器具、洗衣机的能效分级标准被进一步收紧。第三阶段于 2019 年 12 月起开始实施，规定产品覆盖室内空调的制冷及制热功能、大容量洗衣机、电视机、储水式电热水器和电磁炉。迄今为止，三个阶段的计划覆盖的产品的用电量约占住宅用电量的 70%。据机电工程署估计，由此每年可节约 6 亿千瓦时电。此外，“自愿性能源效益标签计划”还覆盖了供家庭、办公室以及车辆使用的 22 种

①Nelson Moura, 本地绿色建筑可持续发展主要由博彩运营商的倡议推动——环境顾问，澳门新闻通讯社，2017 年 11 月 30 日。http://www.macaubusiness.com/macau-local-green-building-sustainability-mainly-driven-gaming-operators-initiatives-environmental-advisor/。

②澳门环境保护局：“澳门环保酒店奖”，2016 年。http://www.dspa.gov.mo/h_index.aspx。

电器和燃气产品（见下文），以帮助公众选择节能产品。目前，澳门尚未制定电器能源标签计划。

（2）交通运输节能

报告第三章已详细介绍了交通运输中的能源使用及碳排放问题，运输部门的碳排放与其终端所使用的能源相关。铁路和有轨电车系统由电力驱动（约占公共交通总量的43%），道路运输车辆主要由柴油、汽油和液化石油气提供动力，而本地船舶则由船用柴油提供动力。如本章前文指出，道路车辆中，出租车使用液化石油气，公共小型巴士使用液化石油气或柴油，卡车等柴油商用车由柴油驱动，私家车主要使用汽油，少量使用柴油。在此，报告主要关注载客量最大的公共交通道路车辆及一部分柴油商用车。

香港特区政府通过以下两种途径来节约能源使用减少运输方面碳排放：第一，确保公共交通仍然是日常出行的首选。这项措施要求继续发展"以铁路为骨干的运输系统"，改善与其他公共交通网络（巴士及小巴）的连接，并协调新界新区的房屋及交通规划，减少私家车的使用。多年来，香港特区政府一直在不断扩大地铁网络，以便在2021年前为70%的人口提供连接地铁站的便捷通道，并计划于同年将地铁出行在公共交通出行中的占比从40%提高到43%。预计至2031年，香港进一步的扩建计划将使地铁服务覆盖75%的人口居住区，使铁路在公共交通出行中所占的比例达到50%[①]。第二，推进零碳运输模式。这项措施要求改善城市步行的通过性、易行度，缩短人们步行距离，使人们能相对轻松地步行。由于香港坐拥群山，在步道建设有独特的创新，其中包括建造人行天桥和电梯/自动扶梯以方便人们更轻松地在城市中步行、穿梭。提高行走便利度的措施使步行成为大量香港居民日常生活的一部分。调查发现香港是46个被调查地区中人均每日行走步数最多的地区。

---

①《铁路发展策略2014》，2014年9月，见 https://www.thb.gov.hk/sc/psp/publications/transport/publications/rds2014.pdf。

此外自行车道在新界新城中也得到推广和普及[①]。

澳门交通事务局制定了《澳门陆路整体交通运输政策（2010-2020）》，鼓励公众将公共交通作为其出行首选[②]。公共巴士的月度载客量从2012年的1 330万人次增加到2014年的1 550万人次[③]。该政策旨在实现2020年公共交通（公共巴士、轻轨和出租车）占所有出行交通工具使用的50%的目标。

（3）私人汽车自愿性能源效益标签计划

自2002年起，上述的香港自愿性能源效益标签计划开始适用于8座私家汽油车，其目的是让购车者在标准测试条件下比较不同车型的燃油消耗情况[④]。由于目前为止没有任何车型被列出，该计划尚未达到预期目的。如国际评论所述，由于来自日本、欧盟和美国等不同地区的检测标准不同，汽车贸易一直担心不同标准带来的燃料消耗量对比的公平性问题[⑤]。在协调不同的检测方法后，欧盟颁布了一项新的测试，并于2018年9月生效。根据现有方案，香港特区政府需要考虑是否以及如何推进香港本地的汽车检测。

---

①斯坦福大学，运动不均：通过大规模体育运动数据揭示的全球运动不均等，2017，http://activityinequality.stanford.edu/。

②澳门交通事务局：《澳门陆路整体交通运输政策（2010-2020）》，2010年。

③澳门交通事务局：《澳门陆路整体交通运输政策2014年度总结》，2014年。

④香港机电工程署："香港自愿性能源效益标签计划——汽油载客车辆"，2017年1月，https://www.emsd.gov.hk/filemanager/sc/content_358/2017_Petrol%20Passenger%20Cars_chi.pdf。

⑤亚太经合组织："汽车燃油效率标签和消费者信息计划审查和评估"，2015年，https://apec.org/Publications/2015/12/A-Review-and-Evaluation-of-Vehicle-Fuel-Efficiency-Labeling-and-Consumer-Information-Programs。

三、气候变化减缓政策评估

（一）电力和天然气部门的碳减排

由于发电造成的碳排放在香港和澳门的总碳排放中占比均排第一，因此对发电量的关注至关重要。就香港特区而言，1997年起特区政府决定停止增加燃煤发电并改用天然气。从1996年第一座燃气发电厂投产以来，香港从煤炭发电向天然气发电转型迅速。2015年香港已有10家天然气发电厂；到2020年，天然气将提供约50%的香港电力，而煤炭将下降到25%左右。《管制计划协议》为中电和港灯提供了一个合约框架。在此框架下，中电和港灯可以就其转型所需投资获得合理回报。此外，香港特区政府会审核投资立项，以确保完成相关发电厂转型的顺利进行。

事实上，使用《管制计划协议》来规管电力行业是香港能源系统的一个独特方式，为香港提供了世界上最安全和稳定的电力，并确保了合理的定价，也最大限度地减少了发电对环境的影响，同时让中电和港灯公司可就其供电投资获得合理回报。最新的《管制计划协议》中的新条款为电力公司及其客户生产接驳至电网的可再生能源和消费者节能提供了有吸引力的奖励。可以看出，政府与两家电力公司之间已经通过谈判有效地执行了政策。然而《管制计划协议》的一个疏漏是没有要求电力公司向当局提供详细的电力生产和消费数据，或进行公开披露。这些数据现在被视为决策的关键，尤其是在需求侧管理方面。更全面的信息披露可能涉及隐私问题，然而这一观点有待商榷，因为数据透明度对于任何辖区实现更好的政策（包括智慧城市的实现）都是至关重要的。

然而，《管制计划协议》在允许投资水平和回报率方面受到了批评①。进一步减少电力储备容量的强化机制将成为下次谈判的重点。在过去的两版协议中，港灯和中电的储备容量已从50%以上减少到30%左右。此外，一直有意见要求加强中电和港灯之间的电网连接，以便实现一定程度的竞争从而降低香港的整体蓄电能力，并舒缓电费加价压力。同时，未来需要进口更多的天然气，需要建设新的基础设施以接收天然气。目前，两家电力公司正在香港水域共同建造一个海上液化天然气接收站，预计于2022年启用②。

从竞争的角度看，只要发电仍需更多土地和投资成本，其他的私人投资者投资香港或澳门的发电公司就是难以实现的。就香港而言，如果假设项目本身就具有足够的商业吸引力，且政府计划建设上述的两个海上风力发电厂，那么发电厂之间的竞争关系是有望形成的。但最终，投资者仍需将电力出售给港灯及中电进行电力输送。目前，政府和公共部门是香港第三大拥有各种可再生能源（包括垃圾发电）项目的能源生产商。如果生产可再生能源的政府部门无法直接使用所有能源，则该部门仍须将所生产的能源输送给港灯、中电或中华煤气公司的输电系统。也许将来，电力输送可以发展至建筑物可以通过包括外墙的现场光伏装置自己生产电力的情况。然而，由于港灯、中电等电力生产商具有良好的业绩记录，且提供了可信赖的优秀的服务，公众可能仍更偏好现有公司投资这些新技术。这将是未来政策重点关注的方向。

中电及港灯之前都没有节能奖励。对世界各国政府来说，从奖励销售向奖励节约转变仍是一个普遍挑战。上一版《管制计划协议》(2008/09-18)为节约能源提供了适度的奖励，而在当前的《管制计划协议》(2018/19-33)中，

---

①香港特别行政区立法会："数据摘要：香港电力市场"，2006年，https://www.legco.gov.hk/yr05-06/chinese/sec/library/0506in26c.pdf。

②香港环境局："两家电力公司二零一四至一八年发展计划和二零一四年电费检讨"，https://www.legco.gov.hk/yr13-14/chinese/panels/edev/papers/edev1210-enbcr145760813pt14_enbcr245760813pt10-c.pdf。

这些奖励变得更具吸引力。此外，由于早前的《管制计划协议》行动不足，因此可再生能源的使用没有得到实质性的推广。但目前，由于光伏技术的进步，光伏安装和运营成本的下降，以及公众环保意识的提高，《管制计划协议》在香港得到积极响应。此外，2015 年以来香港特区政府致力以额外的资金和项目奖励措施推广公共基建、建筑节能和可再生能源项目，政府部门、公共部门、学校和私营部门节能及可再生能源推广的吸引力大大增加。同时，为支持《管制计划协议》的节能措施，港灯和中电两家电力供应商都将用带有后端设施的智能电表取代原有的机电式电表。这些智能电表将为个人客户提供即时耗电信息，帮助个人客户实现节能，并实施需求响应计划，配合香港的“智慧城市”规划。香港特区政府已与两家电力公司达成共识，逐步发放相关数据，推动智慧城市发展。

目前，澳电在主变电站的屋顶上安装了并网光伏系统，并研究如何在澳门扩建光伏项目①。2014 年，澳门特区政府颁布了《太阳能光伏并网安全和安装规章》，于 2015 年开始生效，并于同年发布了太阳能光伏并网的上网电价措施。直到 2018 年底，共有三个并网光伏发电系统，装机容共约 22.9 千瓦。虽然太阳能发电的总发电量占比较小，但它被视为澳门电力的重要组成部分。

本地发电与进口电力

根据一项长期供应合约，香港特区自 1994 年起从广东省大亚湾核电站购入核电。目前，中电集团进口的电力约占广东大亚湾核电站总发电量的 80%，其中 70%是根据 2034 年到期的 20 年期合同购买的，另外 10%是根据 2023 年到期的另一份合同购买的。进口的核能约占本港燃料组合的四分之一。大亚湾核电站由广东核电合营有限公司持有，中电控股通过全资子公司香港核电投资有限公司持有其 25%的股份。作为投资者，中电以出售大量的电力的

---

①澳门能源发展办公室：“澳门拥有开发太阳能的巨大潜力”，2014 年，http://www.gdse.gov.mo/gdse_gb/newsDetails.asp?newID=433。

方式为核电站提供了资金。这种方式取得了巨大的成功——它掀起了中国商用核电的发展浪潮，并在较长时间内以合理的价格为香港提供了大量安全且可靠的电力供应。因此，中央政府在其他地区也开发了核电站。

21 世纪初，澳门能够从南方电网获取电力以应对迅速增长的需求。2004 年，澳门可满足自身 93%的电力供应，只有 7%电能来自内地。而到 2014 年，从内地进口的电力占比达到 88%。2018 年这一比例达到 88.8%，其余部分有 3.1%的电力来自澳门垃圾焚烧厂。2018 年，澳门特区政府批准在澳门新建一座燃气发电厂，预计将于 2024 年投入使用。澳门电力股份有限公司认为，扩大本地发电能力以提高电力供应的稳定性和可靠性符合澳门的利益。

对于香港和澳门特区而言，从内地进口更多的电力可以作为一种选择，例如中电从具备专用输送能力的大亚湾核电站购买电能并直接输送电力至香港，又或者像澳电与南方电网签订合同，直接从内地电网购买电力。在 20 世纪 90 年代初大亚湾核电站建成时，由于内地自身缺乏足够的电力供应，香港无法从内地电网购买电力。但是，随着近年来基础设施完善，电力供应大幅增加，可靠性持续改善，现在南方电网可以为两个特别行政区提供更多电力。香港的用电量仅占南方电网在广东、广西、云南、贵州和海南供电总量的 5%左右[①]，澳门的份额更小。

两个特别行政区对于供电可靠性的顾虑是可以理解的，尤其是在恶劣天气或系统故障期间，毕竟他们没有发电厂（特别是核电站）的直接控制权。自 2010 年以来，香港特区政府已就是否向内地增加购电进行了讨论，然而由于公众对供电可靠性的担忧，尽管展开了相关研究，政府还没有采取实质性行动。与此同时，中电还将与南方电网、大亚湾核电站一道加强其清洁能源

①电力市场未来发展公众咨询文件，2015，6—7 页，https://www.enb.gov.hk/en/resources_publications/policy_consultation/public_consultation_future_development_electricity_market.html。

输电系统的升级，为中电使用更多“零碳能源”（包括核能和可再生能源）方面提供灵活性。该计划在 2025 年完成后可使香港提前 5 年达到其 2030 年的碳强度减排目标（即与 2005 年基准年的水平相比，将碳强度削减 65% 至 70%）[①]。此外，通过强化输电系统，可以延迟或减少建设新天然气发电机组的资本投资。

就澳门而言，其电力政策已确定为以输入为主、本地发电为辅，本地发电可调整的空间不大。本地大部分电力需求将持续由南方电网提供，且主要使用天然气，但在建设新的燃气发电厂过程中，澳电将能够更好地确保可靠性，特别是在有严重风暴的情况下。此外，在可再生能源应用方面，受环境资源的制约，相关潜力仍需进行评估，参考其他国家及地区的经验，相信在较长时间内，可再生资源占供电组合的比例难以显著增加。长远而言，在应用可再生能源和其他节能减排技术方面，考虑到澳门的地理条件和经济体量，区域合作将会是实现达峰和碳中和政策目标的重要手段。

这里需要注意的一点是，碳排放分布的变化取决于发电的地点。因此，从内地进口电力的非核能发电所产生的碳排放，会计算在内地的排放清单内，而不会计算在特区的排放清单内。因此，内地生产并销售给 CEM 的非核电产生的碳排放在内地计算。最终，重要的是全面减少电力消耗，而不是增加发电容量——特别是在使用化石燃料的情况下——当现有的发电能力已足够使用的时候。

（二）需求侧的能源节约

成功的需求侧能源管理需要结合明确的政策目标，并通过治理和技术参

①香港环境局：“两间电力公司二零一八至二三年发展计划和二零一九年电费检讨”，https://www.legco.gov.hk/yr17-18/chinese/panels/edev/papers/edev20180704-enbcr145760818pt28_enbcr245760818pt27-c.pdf。

与得到持续不断的加强。环境局清楚地认识到，必须首先在公共部门内协调政策目标，然后激励私营部门采取行动。

2017 年 6 月发表的政策文件《透过 4T 合作伙伴加强在香港现有建筑物节约能源》对香港具有开创性意义，因为它表明香港特别行政区政府始终发挥着领导作用，持续与公共和私营部门的建筑业主、物业经理和房地产开发商开展合作。环境局召集商界领袖，向他们解释香港特区在减少耗电量方面的使命，并寻求他们的合作。机电工程署与香港绿色建筑议会合作，致力向在公司工作的管理及工程专业人士推广节能建筑。政府希望通过接触这些公司的管理层使他们相信节能的益处。

鼓励利益相关者和公众的参与是必须持续进行的艰巨工作。在政府依赖自愿行动的环境中，还需要特定的政策才能将合适的人员聚集在一起共同学习和协作，这是因为假如香港没有达到目标，特区政府无法对能源使用者进行惩罚。随着投资者和公众对企业环境绩效的关注，领先企业已有意识地准备企业的可持续发展报告，但仍存在许多落后者。2013 年，香港证券交易所出台了一项新政策，要求企业在自愿的基础上编制环境、社会和治理报告。2016 年，它又向前迈出了一步，要求上市公司“报告或解释”相关信息。2019 年 5 月，香港证券交易所提议在 2020 年生效的咨询文件中引入强制信息披露作为下一步举措[①]。逐步收紧的信息披露要求正在提高企业对自然资源尤其是能源和水的使用和管理。

由于澳门正处于实现适度多元化和经济转型的阶段，包括博彩牌照的重新竞投可能会导致新一波大型娱乐项目的投资，这不仅引领其他新兴产业的发展，而且将影响澳门的电力需求。未来的能源利用对经济发展的边际效应

---

①香港交易及结算所有限公司：“咨询文件：检讨《环境、社会及管治报告指引》及相关《上市规则》条文”，2019 年 5 月，https://www.hkex.com.hk/-/media/HKEX-Market/News/Market-Consultations/2016-Present/May-2019-Review-of-ESG-Guide/Consultation-Paper/cp201905.pdf?la=en。

将有所变化，关键要在继续推动社会经济活动和限制用电增长这两方面取得平衡，以实现澳门可持续发展。

总体而言，香港特别行政区和澳门特别行政区的政策发展缓慢。其中一个原因可能是，在现实中，当真正来自社会的压力很小时，政策制定者花了相当长的时间来优先考虑节能，而对于一个行业，比如建筑部门，需要花费时间才能真正为节能投入承诺和行动。现在港澳地区已经奠定了一定的节能基础，贯彻政策重点和利益相关者的参与仍然至关重要。

（三）运输中的碳减排总量

香港有明确的政策支持公共交通。经济实惠、高效的公共交通系统有助于减少对私家车的依赖。在香港特别行政区，私家车非常昂贵，首次登记税最高可达车辆售价的100%至110%，停车费也非常昂贵。与伦敦和纽约等城市相比，这些政策导致了相对于香港经济发展程度较低的汽车拥有率，较高的公共交通使用率。90%以上的乘客每日出行采用的是公共交通工具，从结果来看可谓非常成功。香港一直被评为城市公共交通领域的领导者。

本报告的第三章提供了运输部门碳排放的详细信息。需要指出的是，车辆相关的政策主要是为减少路边空气污染物而设计的。总体而言，这些政策对减少碳排放没有太大影响。

正如上文第二节所述，香港特别行政区政府为专营巴士运营商提供了慷慨资助，以试验新技术巴士，改善路边空气质量。具体来说，从2014年开始六辆双层欧六混合动力柴电巴士被采购以进行为期两年的试验。试验结果表明，新车型在燃料使用方面没有总体效益，而在当地丘陵地形和炎热月份大量使用空调的工况下进行测试时，污染物排放和碳排放也没有总体效益（Keramydas *et al.*, 2018）。

此外，特区政府还资助了36辆单层电动巴士（包括28辆电动巴士及8辆超级电容巴士）进行为期两年的试验，以测试其在本地状况下的性能、可靠性及经济可行性。截至 2020 年底，已有 26 辆电动巴士和 7 辆超级电容巴士投入运营，剩余的公交车将很快投入运营。

初步结果显示，香港能否更广泛地使用单层电动巴士，将取决于单层电池电动巴士的电池电量日后能否大幅提升，如在充满电后可每日行驶超过300千米，或是否有空间及电力容量在现有的巴士总站或公共交通交会处加装充电设施，可在日间为单层电池电动巴士补充电力。

至于混合动力车辆，在绿色运输试验基金支持下试用的商用混合动力小巴，与传统的同类车辆相比，迄今只节省不多于 4%的燃料开支。政府会继续鼓励汽车供应商推出适合本地运输行业使用的电动商用车和混合动力商用车，并邀请这些行业在该基金下申请绿色创新运输技术的应用试验。

此外，环保署于 2018 年完成了四个政府露天停车场（分别位于机电工程署总部、香港湿地公园、蕙荃体育馆及石硖尾公园）推进的先导计划，安装了 11 个户外中速充电器以评估其可靠性。结果显示，户外中速充电器运作良好。

主要的问题仍是香港的车辆总数一直处于上升阶段。2014 年，香港的汽车保有量不到 70 万辆。到 2018 年底，这一数字已增至 78.4 万多。最令人担忧的是私家车数量从 49.5 万辆增加到 56.52 万辆。车辆保有量的改变是香港客运能源使用量增加的最大原因，因此当局认同需要控制私家车的增加[①]。运输造成的日益增长的能源使用和外部性需要采取协调一致的政策行动。政策制定者将继续密切留意私家车保有量，以适时制定解决方案。

①可参考 Transport Advisory Committee, Report on Study of Road Traffic Congestion in Hong Kong 2014, https://www.thb.gov.hk/eng/boards/transport/land/Full_Eng_C_cover.pdf；及 Environment Bureau, Hong Kong's Climate Action Plan 2030+, January 2017, https://www.enb.gov.hk/sites/default/files/pdf/ClimateActionPlanEng.pdf。

第三节　港澳气候变化适应与恢复政策的综述与评估

恶劣天气事件造成的破坏程度是由物理威胁（恶劣天气事件），受到威胁（可能受到不利影响的在场人员、资源和资产）和脆弱性（产生不利影响的倾向）所构成的函数。根据第一章展示的港澳气候变化的预测和第二章所讨论的在水管理、洪水和山体滑坡预防、海平面上升和公共卫生等领域所需要的适应性，香港和澳门未来仍然面临着人身和财产的高度威胁。

香港作为重要的商业和金融中心，必须确保有充足的淡水供应，香港的基础设施需要能够抵御暴雨、热带气旋等恶劣天气及极端温度。由于香港多山且沿海，海平面上升、极端降雨越趋频繁、热带气旋强度增加和与其有关的降雨及风暴潮风险增加会带来多重挑战，尤其是山体滑坡、洪水和风暴潮。澳门的滑坡风险虽然远低于香港，但也需要应对这些挑战。港澳分别总结了应对 2017 年超强台风“天鸽”和 2018 年“山竹”的经验，对气候变化的适应和恢复系统提供了有益的反思。2017 年的经验教训在 2018 年得到了应用。然而，为了做得更好，当局必须在未来扩大私营部门和社区的参与。可以通过各种手段减少脆弱性，例如良好的预警系统、应急准备以及适当的防御工程。这些措施和公众对风险的认识对改进应急准备工作至关重要。

本节将对香港特别行政区和澳门特别行政区针对气候变化的适应能力和应对能力进行概述和评估，包括近年超强台风的经验和教训。值得注意的是，经济发展增加了系统失败的可能性。

一、管理过剩水量与防洪

香港特区渠务署提供污水和雨水的处理排放服务。澳门特区市政署（由原民政总署变更而来）排水科负责排水网络的运作。两个特区都有强降雨，因此良好的排水是必不可少的。即便是在适应气候变化成为重要政策之前，香港就已付出了巨大的努力预防洪水，并取得了良好的效果。澳门仍然需要更多努力，其规划涉及与内地珠海市等的密切合作。

第二章已经指出，直至20世纪90年代，香港北部的低洼乡郊地区、天然洪泛区以及一些较老的城区，在特别严重的暴雨期间仍经常发生洪水。快速城镇化导致新界地区包括洪泛平原的密集开发，这些开发项目将天然土地变成建成区，令原本由土地存留的雨水变成地表径流。接近主要水道的建成区也被加以扩建，同样削弱了这些水道的排洪能力①。旧城区的问题是由于发展速度超过了以前的设计标准及现有系统的老化②。

1994～2010年，渠务署完成了一系列研究，被称为“雨水排放整体计划及雨水排放研究”，涵盖香港的集水区，并为水浸/内涝问题提供了全面的解决方案。这些研究全面检视了现有排水系统是否满足需要，并建议在短期及长远改善措施时考虑技术限制、成本效益和对环境的影响③。新界区必须兴建新的排水系统、设定乡村防洪计划，而市区必须采取各种措施，例如拦截雨水、管道升级，以及蓄洪抽水等。自2008年以来，有关雨水排放整体计划的修订工作已经逐步展开，修订过程正在进行中。自1989年渠务署成立至2019年，政府共投资约300亿港元在铺设排水渠（约100千米），治理河道（约110

①关于历史讨论，可参见 Luk, K. K. C., A. K. F. Kwan and J. K.Y. Leung 2001. Drainage Master Planning for Land Drainage Flood Control in the Northern New Territories, *HKIE, Proceedings*, 3(1). https://www.dsd.gov.hk/EN/Files/Technical_Manual/technical_papers/ LD0102.pdf。

②关于历史讨论，可参见 Chan, K. K., K. W. Mak, A. S. K. Tong and K. P. Ip 2011. Flood Relief Solutions in Metropolitan Cities the Hong Kong Experience Theme: Planning for Sustainable Water Solution Oral Presentation. *Water Practice & Technology*, 6(4)。

③改善措施的类别载于：https://www.dsd.gov.hk/SC/Flood_Prevention/Long_Term_Improvement_Measures/Categories_of_Long-Term_Improvement_Measures/index.html。

千米），修筑四条雨水排放隧道，以及开展 27 个乡村防洪计划。由 1995 年至今，渠务署合共消除了 126 个水浸黑点，现在仅剩 5 个水浸黑点[①]。

多年来，渠务署对排水基础设施的规划和设计策略已逐渐由抵御转变为适应。香港推行“蓝绿建设”，利用自然环境和自然过程来增强韧性。传统工程旨在通过阻止、击退或控制自然力来保护资产，而新方法要求基础设施的设计规划可以在增强抗洪能力的同时提高水资源的利用率。这与内地正在实施的“海绵城市”理念相似。已经实施的多种方案被证明是成功的，生态上合理且具有视觉吸引力[②]。它们还提供了除工程之外的许多学科的融合，包括城市设计和生态系统弹性。

虽然香港的水浸风险已大大降低，但仍不能完全消除。当发生特大暴雨而降雨量超过设计标准时，便可能发生水浸。如果排水系统的某处尤其是集水井入口被垃圾、倒下的树木或山泥倾泻带来的泥石阻塞，排水系统便不能发挥功能。当台风过境时，风暴潮与波浪作用相结合也造成异常高水位，沿岸低洼地带也可能发生水浸。

展望未来，渠务署应更加注重协调发展，特别是在新发展区，政府需要提供污水及雨水基础设施并采用“蓝绿建设”方案。此外，除了大型防洪基础设施外，还将探索防洪措施和非结构性措施，以便在极端天气事件发生前有效减轻洪水风险。

由于暴风雨和风暴潮，澳门仍时有洪水发生。为此，澳门特区政府已投资当地的设备，如采购更多排水泵、建设沿岸的防洪设施，并且一直在与中

---

①可参考渠务署网站：https://www.dsd.gov.hk/SC/Flood_Prevention/Our_Flooding_Situation/Flooding_Blackspots/index.html。

②何耀光、香港渠务署：“香港的可持续排水系统”，2017 年 4 月 1 日，https://www.hk2030plus.hk/document/KSS_PPT/5th_KSS/Sustainable_Drainage_System_in_HK.pdf。这些工程包括林村河及元朗排水绕道。正在建设的新项目包括“安达臣道石矿场用地发展”中的堰塞湖。

央政府讨论建立有利于澳门及邻近地区的防洪系统。2017 年 8 月 23 日，台风“天鸽”过境中国南部，澳门经历了自 1925 年以来最严重的洪水，澳门半岛的一半被淹没，其中内港区受影响最大且沿岸最高淹没深度达 2.38 米[①]。台风“天鸽”促使中央政府加速决定推进珠海湾仔防洪体系建设，包括防潮屏障的研究和规划，该体系将有助于澳门、珠海和中山[②]。

（一）斜坡管理

由于多山、开发范围广及人口密集等现状，暴雨给香港带来的威胁非常大。20 世纪 60 年代和 70 年代初期进行的香港地盘平整工程，甚少有专业地质工程人员的参与。当时香港发展迅速，产生了大量可能不合标准的人造斜坡。再加上山坡上及附近有很多房屋和其他开发项目，形成了生命和财产的损失风险。土力工程处/土木工程拓展署是负责斜坡管理的部门。多年来，它已创建了世界上令人印象最深刻的斜坡管理系统之一。

土力工程处成立于 1977 年，负责管制新建或重建项目的岩土工程，并制定策略以处理了大量可能不合标准的人造斜坡。自此以后建成的新斜坡，其设计已普遍经由土力工程处审核，以确保符合安全标准。在 2010 年前，土力工程处按防止山泥倾泻计划，已经有系统地巩固了不合标准的政府人造斜坡。从 2010 年开始，土力工程处推行长远的防治山泥倾泻计划，把工作扩展至天然山坡的山泥倾泻风险减缓工程。

1. 防止山泥倾泻计划

防止山泥倾泻计划于 1976 年开始推行，该计划根据风险评级系统，处理大批可能不合标准并影响现有发展项目的斜坡。防止山泥倾泻计划的工程和

①数据来源：https://www.smg.gov.mo/zh/subpage/355/page/38。

②广东省珠江水利委员会珠江水利科学研究院 2016 至 2017 年已完成相关研究。另见澳门新闻：“澳门与广东加强防洪合作”，2017 年 9 月 12 日，https://macaunews.mo/macau-guangdong-strengthening-flood-prevention-cooperation/。

研究，最初由土力工程处的部门员工负责。1991 年土力工程处精简程序，以提高斜坡研究的数目，并于 1992 年开始运用私营机构的资源，以加速防止山泥倾泻计划的进度。

为执行 1995 年通过的《斜坡安全检讨报告书》的部分建议，土力工程处获增拨资源，进一步加快防止山泥倾泻计划的工作。在 1995 年 4 月 1 日，土力工程处展开了“五年加速防止山泥倾泻计划”，巩固了约 800 个政府人造斜坡，并就约 1 500 个私人人造斜坡进行了安全调研。

作为政府改善斜坡安全的长远策略，“延续十年的防止山泥倾泻计划”在 2000 年开始，即在“五年加速防止山泥倾泻计划”完成后展开。“延续十年的防止山泥倾泻计划”提升了防止山泥倾泻的成果，改善了更多会影响发展的项目和需要优先处理的不合标准人造斜坡。在“延续十年的防止山泥倾泻计划”下，土力工程处巩固了约 3 100 个不合标准的政府斜坡，并就约 3 300 个私人人造斜坡进行了安全调研。这个数量是“五年加速防止山泥倾泻计划”展开前的 4 倍。

2. 长远防治山泥倾泻计划

在 2010 年防止山泥倾泻计划完成时，人造斜坡的整体山泥倾泻风险已大幅减少至低于 1977 年风险的 25%，达到国际认可的最佳风险管理水平。不过，其余的山泥倾泻风险仍对现有发展和重要交通廊道构成潜在危险。由于斜坡的状态会日渐变差，人口增长与越来越多都市发展或重建靠近陡峭的山坡地带，以及气候变化可能引致更频密和严重的暴雨等极端天气，如果不继续在斜坡安全方面投放资源，山泥倾泻的风险便会日趋增加。除了人身安全构成威胁外，也可能因堵塞道路或危机时居民需撤离楼宇，带来重大的经济损失、干扰社会秩序、危害公众安全，危及可持续发展的能力，进而影响香港作为一个现代化都会和旅游枢纽的声誉。

2007 年年底，香港当局展开了长远防治山泥倾泻计划以衔接防止山泥倾泻计划，以处理人造斜坡和天然山坡引起的山泥倾泻风险。“长远防治山泥倾泻计划”的每年目标如下：①巩固 150 个政府人造斜坡；②为 100 个私人人造斜坡进行安全风险调研；③为 30 个天然山坡推进风险缓减工程。“长远防治山泥倾泻计划”的防治山泥倾泻工程已于 2010 年展开。

自 1997 年以来，政府在斜坡安全方面投入了约 246 亿港元[①]。值得一提的是，香港的斜坡安全系统已获联合国认可为管理滑坡风险的良好做法[②]。土力工程处/土木工程拓展署定期会发布和更新报告，作为政府承诺的证明。

尽管香港斜坡安全系统大幅降低了山泥倾泻风险，但斜坡老化、人口增长和气候变化令山泥倾泻的风险与日俱增，更频繁、更强烈的极端暴雨会增加发生山泥倾泻的可能性和后果。

为了加强社区应对及防范山泥倾泻灾害的能力，土力工程处/土木工程拓展署也通过各类项目及活动（如传媒简介会、展览、学校讲座、电视节目）推广全年的公共教育计划及宣传活动，以提高公众对渐增的滑坡风险的警惕，并教育公众在暴雨期间采取预防措施。

有关澳门特区政府斜坡管理是由土地工务运输局牵头的“斜坡小组”为主导[③]，据资料显示，澳门有 214 个危险斜坡，其中只有 2 个属于高风险。土地工务运输局和澳门特区市政署共同负责维护路旁及公园的斜坡，并签订合约以降低山泥倾泻的风险[④]。

（二）水资源

第二章所述的香港三种主要水资源，即东江水、本地淡水及冲厕用海水，

①香港土力工程处：“防治山泥倾泻研究和工程第 1/2020 号报告”，https://www.cedd.gov.hk/sc/our-projects/landslip/land_quar/index.html。

②联合国减灾办公室：“滑坡危险和风险评估”，https://www.preventionweb.net/files/52828_03landslidehazardandriskassessment.pdf。

③相关详细资讯建议参考澳门特区政府斜坡安全资讯网。

④《2017 澳门年鉴》，396 页。

均由发展局辖下的水务署管理。澳门自来水公司拥有在澳门提供供水服务的独家特许经营权，而澳门市政署辖下化验处则负责监督公共供水网的水质。供水是一项核心公共服务，即使由私营企业管理也有必要受到政府监管。

香港和澳门人口稠密，经济发展迅速，耗水量大。香港目前每年消耗约10.1亿立方米淡水[①]；澳门耗水约0.88亿立方米。广东省提供香港70%～80%的原水、澳门约95%的原水（具体数字每年根据当地的产量变化）用于淡水供应。可以说，两地的淡水供应严重依赖广东省。2018年，香港为从广东购买原水支付了48亿港元，而澳门在2017年的供水成本为1.846亿澳门元[②]。

第二章已经指出，由于20世纪60年代初严重缺水，香港和澳门开始从广东进口原水。虽然预计未来总降雨量不会减少，但仍有可能出现偶尔的干旱时期。事实上，这种降雨预测影响了整个华南地区。因此，长期节约用水和调高用水效率对整个粤港澳大湾区至关重要，也是区域合作的一个重要部分。

1. 香港特别行政区和澳门特别行政区的水资源挑战

两个行政特区面临相似的供水挑战：①年产量不稳定：本地年产量波动较大，没有天然湖泊、河流或主要的地下水资源蓄水。香港的记录显示其本地产量波动可能超过2亿立方米[③]。②经济快速发展：两个特区均面对人口增长和持续经济扩张，这都需要更多的淡水支持，未来也可能有更高的用水需求。在缺乏水资源管理措施的情况下，预计到2040年，香港每年的淡水需求量约为11亿立方米。③公众对可用资源的错判：由于有能力从广东引进水资

---

①香港大部分人口使用海水冲厕以节省淡水资源。现时每年的冲厕用水量约为3亿立方米。

②Macao Water 2019. Water Supply Statistics, https://www.macaowater.com/frontend/web/index.php?r=statisticaldata%2Findex.

③在过去10年里，香港本地的年产量在1.03亿至3.85亿立方米之间波动。见：https://www.wsd.gov.hk/en/core-businesses/total-water-management-strategy/local-yield/index.html。

源，而即使广东面临干旱，也会优先保证两个特区的用水需求，这使得公众认为没有需要以各种迫切方式减少用水和加速实施节约用水的必要性。

水务署于 2008 年提出的“香港全面水资源管理策略”（Total Water Management Strategy, TWMS 2008），阐述了香港特区在水资源方面所面对的挑战。TWMS 2008 旨在通过需求侧和供应侧手段，实现 2030 年之前每年减少 2.36 亿立方米淡水消耗量的目标（相当于预计用水量的五分之一）。2019 年完成的 TWMS 回顾降低了减排目标。新修订的 TWMS 2019 预计，到 2030 年，年节约量为 1.1 亿立方米，到 2040 年为 1.2 亿立方米[①]。

表 4–4　TWMS 2019 总览

控制食水需求增长	管理用水流失
	扩大使用次阶水
	加强节约用水的宣传
提升食水供应的应变能力	第一阶段将军澳海水淡化厂
新增的本地集水量	水塘间转运隧道计划（由渠务署进行）
后备选项	第二阶段将军澳海水淡化厂
	重启香港岛已停运的滤水厂
	在船湾淡水湖内的较高位置兴建一个水塘以提升食水储存量
	扩建船湾淡水湖的集水区
	推行其他海水淡化工程项目
	增加东江水供应

为了能在 2030 年和 2040 年完成着重控制饮用水需求增长的需求，以及利用多元化水资源提升饮用水的应对能力，水务署表示，在采取节约用水、用水流失管理和扩大使用次阶水（即海水和再造水）作非饮用用途等方面的

① https://www.wsd.gov.hk/en/core-businesses/total-water-management-strategy/twm-review/containing-fresh-water-demand-growth/ index.html.

需求管理措施下，现有的食水供应将可应对预期的用水需求。在加强本港淡水供应弹性方面，根据表 4–4 提供的备选方案，政府现正兴建海水淡化厂，以便在出现偏离水供应项目的情况下提供弹性方案（例如人口超过预期、气候变化、控制需求增长）[①]。

香港的水资源管理有两个长期存在的问题特别值得关注：

（1）提高水价的障碍

香港特区政府曾于 1996 年及 1999 年建议提高水费，但香港特区立法会并不支持[②]。香港特区政府自 1995 年起便没有调高水费，使香港的水价成为发达地区中最便宜的，且人均每日用水量约 130 升，而世界平均用水量为 110 升。政府指出，2015 年对世界各地城市的调查证明，提高水价对降低用水量的影响并不显著[③]，重点应该放在公共节水教育上。不过，值得注意的是，在其他许多发达城市，例如新加坡、东京、首尔、伦敦、巴黎、日内瓦和纽约，水价比香港高得多。事实上，价格在用水管理中并非无足轻重，非政府专家不断呼吁当局对水价进行现代化管理和重新评估[④]。此外，公众教育需要更有针对性，特别应面向家庭展开香港市民饮用水消耗量、能源和用水之间的相互关系的宣传教育，并就香港的水费补贴等问题进行沟通，鼓励他们改变用水习惯（Zhao Y. *et al.*, 2019）。

---

①https://www.wsd.gov.hk/en/core-businesses/total-water-management-strategy/index. html.

②*The Illusion of Plenty: Hong Kong's Water Security, Working Towards Regional Water Harmony*, 2019, p.73.

③除了 TWMS 2019，政府还提到了这个调查：https://www.wsd.gov.hk/en/faqs/index.html#41。根据作者与水务署的信件，上述调查是香港大学于 2015 年受政府委托进行的“其他主要城市的水费结构及水费水平研究”的一部分。

④Frederick Lee, Hong Kong Water – Agenda and Goals, China Water Risk, 10 October 2013, http://www.chinawaterrisk.org/opinions/hong-kong-water-agenda-goals-2/; “Extreme shortage of domestic freshwater”, China Daily, 11 June 2017; Civic Exchange, “Modernising Hong Kong's water Management Policy, Part I”, June 2019, https://civic-exchange.org/wp-content/uploads/2019/06/Conservation-and-Consumption_ExSum-EN.pdf.

根据香港特别行政区政府 2017～2018 年度水务设施运作账目，只有 26% 的供水成本是通过水价直接回收的，而 46%的成本是由利率贡献的收入支付的，还有 10%来自政府对消费者的补贴，其余主要来自一般收入。政府审定水费时，除了考虑饮用水生产成本（包括海水淡化）外，还会考虑多项因素，包括市民的负担能力、水务的财政状况、社会的经济形势、立法会议员的意见。政府目前没有计划提高水费。

鉴于澳门在博彩、酒店和酒店设施方面的庞大扩张，其用水总量有所增加。澳门的用水量在 2008 年为 6 746 万立方米，而到 2017 年就已增加到 8 844 万立方米。与 MSWC 签订的特许经营合同允许每年根据原水、能源、劳动力的实际成本，维修与维护费用以及所需投资额来调整水价，并对低收入家庭为主的低用量用户提供补贴，因此整体来看，水价是可以负担得起的。

（2）私人水管渗漏造成的水流失

现在，香港每年的饮用水流失约 8%来自私人水管渗漏。水务署一直致力于采取多手段处理私人水管渗漏问题，包括在屋内安装监测私人水管漏水的主水表，为业主及住宅管理人员提供技术意见及支援以监察渗漏情况，现行政策仍相对温和，水务署协助业主，而不是进行处罚。水务署于 2000 年展开了大型的“更换及修复水管计划”，更换及修复了约 3 000 千米的老化水管。随着计划于 2015 年大致完工，供水管网的状况已大为改善。现在，水务署正逐步在全港的饮用水分配管网建立“智管网”[①]以监测用水流失的情况，从而制定最有效的管网管理措施以维持管网健康。与此同时，水务署也推行风险为本的水管资产管理策略，以维持供水管网的健康状况，减少水管爆裂或渗漏的风险。此外，水务署正在探讨通过立法修正案向未能修复私人公用渗漏管道的业主或楼宇管理商征收费用。

---

①关于水智能网络的更多细节，见 https://www.wsd.gov.hk/en/core-businesses/operation-and-maintenance-of-waterworks/reliable-distribution-network/index.html。

由于地形崎岖、道路工程繁忙、交通繁忙、地下公用设施拥挤水管，令香港面临水管水压过高的问题，进而使得水管更容易渗漏，但是在管道渗漏率方面，香港仍需要做出巨大努力。香港水务署的目标是到2030年将公共水管的泄漏降至10%以下。

2. 海水冲厕

香港限制淡水用量的一种独特方法是在靠近大海的区域使用海水冲厕。因此，另一项重要的水务工程是海水供应系统，其独立的配水干管、抽水站和配水库网络覆盖了约85%的居民。2018年度，约2.8亿立方米的海水供冲厕使用。除了节约淡水外，使用海水还有很多积极影响，但也不乏缺点。从积极的方面来看，最近的研究表明，经氯化处理的含盐污水对微生物的毒性通常低于经氯化处理的淡水（Yang *et al.*, 2015）。通过香港科技大学开发的杀泥污水处理技术（SANI）[①]工艺，使用海水冲厕还能减少污水污泥。负面影响是，海水腐蚀性极强，需要使用耐腐蚀的管道材料。

3. 海水淡化

全面的水资源管理策略包括在香港发展海水淡化，并作为水资源开发的一种新方式。将军澳第137区被论证是建造海水淡化厂的合适地点，年产水量为5 000万立方米并有望增加至最终1亿立方米，这足以满足香港5%～10%的淡水需求量。海水淡化厂在香港建造的预计成本，包括生产、分销及客户服务为每立方米13港元左右。而东江水成本为每立方米10.5港元，本地集水为每立方米4.7港元。该海水淡化厂预计在2023年落成。

4. 循环再用水

全面水资源管理策略包括使用循环再用水，其中包含再造水、中水重用及雨水回收。水务署现正推行多项计划，详见第二章。水务署已在2018年年

①杀泥工艺: SANI is the acronym of the technology of Sulphate Reduction, Autotrophic De-nitrification and Nitrification Integrated to reduce the volume of sewage sludge。

底完成公众咨询，现正就循环再用水的供应制订法例。此外，水务署一直提倡在新的政府工程项目中采用中水重用系统或雨水回收系统，作冲厕、灌溉等非饮用用途。对于私人住宅，设有中水重用系统或雨水回收系统的住宅可获绿建环评分数，以此鼓励开发商提供这些设施从而代替使用淡水作非饮用用途。

二、风暴、风暴潮和海平面上升

根据第一章中的预测，未来整个区域的热带气旋将更加强烈。此外基于IPCC 第五次评估报告的结果和《巴黎协定》的相关情景，香港特区政府预计在中等浓度温室气体的情况下，到 21 世纪末，平均海平面将上升 0.49 米[①]。

随着海平面上升，风暴潮和洪水的组合风险超过以前，并将随海平面继续上升而加剧。2018 年的超强台风“山竹”又一次展示了前所未有的风险。风暴潮导致香港维多利亚港的潮位上升至高于天文潮 2.35 米，其中鲗鱼涌的潮位最高升至 3.88 米（基准面以上），升幅相当可观。如果“山竹”没有经过吕宋岛并被削弱，并且它的风暴潮如果正好与当天天文高潮的时间相一致，海平面将会高出更多，可能达到 5 米以上（Choy *et al.*, 2020）。

以澳门为例，超强台风“天鸽”期间的最大潮位和最大洪水位分别达到5.58 米和 2.38 米，创下澳门的历史纪录[②]。如防洪章节所述，洪水、严重的风暴、风暴潮和海平面上升的组合将对整个大湾区构成更大的威胁。因此，明智的做法是各地方政府携手合作以降低这些风险，共享气候数据并制定对策来解决问题。

---

①所述海平面预测并未考虑香港天文台在大老山观测到的可能的向下垂直地壳运动。当有足够的数据包括地质证据来证实这一发现时，它的影响可能会被纳入未来的海平面上升。

②风暴潮影响澳门期间最高潮水高度及最高水浸高度，https://www.smg.gov.mo/zh/subpage/355/page/38。

对港澳地区来说，除了使港口工程和海堤等防御结构标准更加坚固以外，为基础设施制定适应、应变和应急准备措施也是重要的。投放资源研究以识别高危地区，研究不同类型风险的非线性和局部特殊性（例如，地下停车场是香港和澳门防洪的优先事项）是有价值的。争议更多的是划定发展区和限制开发区，以及某些资产在远期看来是否需要迁移。有关限制发展和重新安置资产的问题肯定会引起争议，将需要广泛协商。进行研究、咨询、讨论以及决策需要时间和跨政府管理（特区实行五年任期）。目前的问题是，现任政府打算如何改善面向气候变化和海平面上升风险的规划？

此外，鉴于香港、澳门和广东（即总体上大湾区）之间的社会经济联系非常紧密，三地对全面合作与提高应对能力具有浓厚的兴趣。

三、沿海水域和保护生物多样性

第二章已描述了香港沿海水域上层的富营养化和底层缺氧的增加趋势，其严重程度将导致水体酸化。造成这一问题的主要原因是珠江的营养负荷增加以及气候变化，如风速减弱和气温升高。

香港特别丰富的生物多样性是由气候条件、地理位置和地质因素共同作用的结果。作为一个小区域，这个城市有无数的陆地和海洋栖息地，他们都是本地动植物的家园，其中一些物种具有代表性或特别珍贵。2016 年 12 月发布的香港首份城市级生物多样性策略及行动计划描述了香港独特的生物多样性。它是根据《联合国生物多样性公约》制订的一项多方面的五年保护计划。作为公约缔约方，中国的《中国生物多样性保护战略与行动计划》将涵盖 2011～2030 年[①]。

---

①中国生物多样性保护战略与行动计划（2011～2030），http://www.biodiv.gov.cn/lygz/xdjh/201601/P020160107513841377238.pdf。

香港特别行政区政府的《香港气候行动蓝图2030+》包含一章关于保护生态系统的内容，提及了生物多样性保护计划的重要性[①]。2021年在云南昆明举行的联合国《生物多样性公约》第十五次缔约方会议将会决定全球生物多样性的新框架，中国很可能将更新其计划，香港也将随之修订生物多样性工作计划。大湾区合作的一个重要领域是澳门和广东是否也可以考虑开发自己的生物多样性保护策略及行动计划，以适应2021年后的生物多样性框架，方便整个大湾区统筹保护管理，并作为加强应对气候变化的弹性的一部分。

生物多样性策略及行动计划的设计和实施需要多个政府部门的共同努力。在香港，环境局局长主持一个跨部门工作小组，负责统筹各部门的工作，并监察工作的进展。香港特别行政区政府特别拨款2.5亿元，在五年计划期间推行各项措施。粤港澳大湾区在保护管理方面的区域合作也需要跨机构协调和更多的资源协同。

四、气温变化与公共卫生

第二章提及，气候变化正在提高与天气状况相关的多种风险，香港及与其情况相似的澳门面临严重的公共卫生和安全问题。此外，气候变化对人口健康影响的研究还处于起步阶段，相应的与气候变化和公共卫生相关的政策制定也如此。气候政策的制定不应与科学研究分开，因为极端温度和更高的降雨量可能导致新的健康与安全风险。同时，为做好迎接更大风险的准备，应加强卫生保健专业人员的培训，完善社会服务[②]。

---

①中国生物多样性保护战略与行动计划（2011～2030）。

②见《2018年香港CARe会议总结报告和政策建议》，2018年12月，第六章改变风险和公共卫生（http://care2018.ust.hk/index.php/report01/）。

（一）炎热和严寒

如第一章和第二章所述，未来高温天气及热浪将更加频繁和严重，并且它们会在一年中的较早时期开始，个别情况的持续时间也可能加长。香港和澳门密集的高层城市环境加剧了热浪。因此，城市规划和设计需要考虑到不断攀升的气温。

香港特别行政区政府利用其规划标准努力减少城市热岛效应并改善城市建成环境的微气候。政府在新发展区域遵循关于城市设计和空气流通指引，对于现有的建成区，鼓励项目发起人在规划和设计其再开发项目时采纳这些设计原则，以便逐步改善城市风环境。自2006年以来，政府要求对所有大型政府项目进行空气通风评估，以便评估结果、改进设计，促进风向周围地区的渗透，并鼓励私营部门参考这种做法。香港特别行政区和澳门特别行政区都需要考虑在不断变化的天气（见本书第一章）和日益增加的健康风险（见本书第二章）的情况下，为密集的城市环境降温。

本书第二章指出，酷热会导致人体脱水、中暑甚至死亡。最危险的人群包括儿童、生活在通风不良的低标准环境中的人群及老年人。即使气候正在整体变暖，非常寒冷的天气仍有机会发生。由于香港特区和澳门特区的建筑物和房屋通常没有暖气，因此低温可能导致人体体温过低，尤其是对于老年人。因此需要开展更多相关研究，而香港专家的工作对整个大湾区也将有所帮助。

（二）天气警告、卫生服务和卫生工作人员意识

香港中文大学在 2017 年进行的一项调查评估了新预警系统的有效性（Chan *et al.*, 2018）。其中，87%的受访者表示收到天文台发出的酷热警告；97.5%的人表示，他们会在酷暑期间增加饮水量作为预防措施，而37.2%的人会选择涂防晒霜。然而，有45.3%的人认为高温根本不会影响到他们的健康，

在65岁以上的人群中，28.8%同样忽视高温带来风险。因此，需要提高公众对相关风险的理解，特别是与高温有关的健康影响的认识。

第四节　科学研究与跨部门合作支撑的气候变化政策

科学研究和跨部门合作是香港和澳门两个特区气候变化政策的重要支撑，科学研究对于决策的方向和内容起关键作用，而跨部门合作则是实现这些决策的途径。

一、气候变化科学研究

正如第一章和第二章提到的，香港和澳门两地长期以来一直向公众提供气象服务，香港天文台在其漫长的发展过程中收集了丰富的气象数据，对于理解和研究气候变化科学极其珍贵。香港天文台成立于1883年，在发展和提供气象服务方面表现卓越，澳门的一些气象数据则可以追溯到1861年。香港天文台和澳门地球物理暨气象局提供的气象服务及研究结果奠定了本地气候变化决策的基础，并逐渐发展成为适应政策的一部分，例如上文提到的热带气旋警告、暴雨及山泥倾泻警告、寒冷及酷热天气警告等。

同时，港、澳特区政府设立了专门的环保基金，鼓励市民改变个人行为及生活方式，达到可持续发展的目的；开发或引入创新科技及实践方法，改善环境，促进节能减排，进而支持及推动环保产业的发展。香港特区政府设有“环境及自然保育基金”（下称“环保基金”）。环保基金于1994年6月成立，同年8月开始运作。环保基金设有“环保研究、技术示范和会议项目”，该项目的优先研究主题包括“气候变化——减缓、适应及应变”“生物多样性、保育及地质保育”等，协助气候变化应对项目的开发。环保基金于2019年共批出约2 950万港币资助有关研究项目，而2020年该项目的资助预算为3 000万港币。香港特区政府拨款港币2亿元成立的“低碳绿色科研基金”于2020年12月开始接受申请，为有助减碳和加强环保的科研项目提供更充裕到位的资助。澳门特区政府也设有“环保与节能基金”，该基金下设“环保、节能产品和设备资助计划”，旨在资助购买或更换有助改善环境质量、更具能源效益或节水效果的产品和设备。特区政府所设环保基金是促进本地气候变化科研转化的重要支撑。

在发展本地气候适应政策的同时，两个特区与国家、国际气候变化科学研究服务机构接轨交流，互通有无，反哺气候决策。香港和澳门分别于1948年和1996年加入世界气象组织，并由香港天文台和澳门地球物理暨气象局代表①。同时，香港天文台和澳门地球物理暨气象局与中国气象局、国家海洋局和广东省气象局等部门有广泛合作项目，长期保持着良好的工作关系。

二、跨部门合作实施气候变化政策

在实际实施气候变化政策的过程中，不仅政府部门间需要建立紧密合作，政府与社会的良好沟通与配合也必不可少。跨部门合作是保证气候政策顺利实施的条件。

第二章所述的天气警告系统始于香港天文台于2000年首次建立的酷热天气警告系统。2014年，香港天文台又推出了“炎热天气特别提示”服务，

①根据香港特别行政区和澳门特别行政区各自的基本法，港澳两个特区可进行“对外事务”，例如保留它们在回归前拥有的国际组织（如世界气象组织）成员资格。

提醒市民在天气炎热但又未达酷热水平时保持警惕，并采取适当的预防措施，慎防中暑的风险。该系统在运作过程中，首先由香港天文台发出警告，随后通过媒体进行资讯传播。相关政府部门也会展开行动，例如民政事务总署会在酷热天气警告期间开放临时避暑中心；香港天文台网站也列出炎热或酷热天气下的注意事项，包括对户外工作人员、老年人、慢性病患者和其他弱势人群的警告和建议。如酷热天气持续，香港天文台会向市民持续发出酷热天气警告，并在有需要时更新警告内容。卫生署卫生防护中心也会通过不同的途径，向公众通告最新情况、传达相关的预防措施及健康信息。居民在收到这一系列通告时可及时采取必要措施，以规避相应气候风险。

热带气旋产生的波浪和风暴潮或会导致香港和澳门沿海低洼地区受到越堤浪的冲击。两个特区政府需要尽力减少资产暴露，而确定低洼地区尤为重要。为达成这一目标，香港特区政府基于 IPCC 第五次评估报告及《巴黎协定》完成了修订气候变化如何影响沿岸构筑物设计的研究，并于 2018 年根据海平面上升及风速增加，更新了政府基础设施的设计标准中的海平面及雨量等参数。

在斜坡管理方面，政府官员和其他专家共同合作开展了许多研究，以填补知识空白从而指导政策制定和项目开发。要预测可能发生滑坡的地点和时间非常具有挑战性，因此除香港天文台、土力工程处/土木工程拓展署和渠务署间的日常合作外，工程学术界以及其他地区的专业人士也建立了合作关系。这些跨界合作最近的研究结果包括发表的研究论文（Kwan *et al.*, 2018）和联合项目，以及开发“智能泥石坝”系统实时监测泥石坝的状况。当发生泥石堆积时，系统会发送实时警报以触发及时的应急行动。

第五节 区域合作

从第一章、第二章和第三章中可以清楚看到，要保持自然环境和资源的完整，不仅需要香港和澳门的共同努力，更需要大湾区作为一个整体来共同应对气候变化。

一、协作框架

香港与广东在环境方面的合作始于20世纪80年代。20世纪80年代初，深圳和香港开始在跨境污染控制方面展开合作。粤港环保机构的合作始于20世纪80年代中期，粤港环境保护联络小组于1990年成立。1993年，广东、香港和澳门联合举办了珠江三角洲环境污染防治研讨会，开启了三方合作的新纪元[①]。

1995年，广东在《珠江三角洲经济区现代化建设规划纲要（1996-2010）》中考虑到区域合作，提出了许多雄心勃勃的设想。该规划的成果之一是1998年成立的粤港合作联席会议（联席会议），目的在于加强香港特区与广东省之间的联系。粤港环境保护联络小组于1999年更名为粤港可持续发展与环保合作小组，作为联席会议的一部分就环境事宜建立合作交流框架及运作机制。合作小组每年召开一次会议，审议商定的工作。

粤港可持续发展与环保合作小组由多个专家组组成，负责讨论和协调工作计划并提出建议。目前，该合作小组共有七个专题小组，分别针对空气质

①见"Unity call over delta pollution"和"Joint bid to fight water pollution", reported by Kathy Griffin on 19 and 20 February 1993, *South China Morning Post*。

量、水质、林业、海洋资源护理、大鹏湾及后海湾（深圳湾）保护、东江水质及海洋环境管理等具体领域进行专题研究[①]。粤港双方在 2011 年的联席会议上签署了粤港应对气候变化合作协议，并于同年成立了粤港应对气候变化联络协调小组[②③]。此外，《珠江三角洲 2008-2020 改革计划纲要》中有更为雄心勃勃的改革计划，如广东、香港和澳门逐步采用统一的汽车燃料、船舶燃料及排放标准，这将是全国领先的计划；再如，广东和香港合作进行清洁能源和可再生资源的研发，在清洁生产领域合作，构建能源供应和市场营销的有益网络，目前还有待实现。

由于缺乏可比较数据，以及法律和行政制度的差异，粤港澳区域环境合作最初面临困难。尽管如此，在各地的积极配合下，相关联动仍出现了重要创新，尤其是在空气质量管理方面，如粤港澳珠江三角洲区域空气监测网络的建立[④]。另一创新是 2008 年 4 月推出的清洁生产伙伴计划（CP3），旨在帮助区域港资工厂，特别是中小型企业。清洁生产伙伴计划由香港环境保护署与当时的广东省经济和信息化委员会（现更名为工业和信息化厅）共同发起，旨在鼓励和促进粤港两地的港资工厂采用更清洁的生产技术和作业方式，从而为改善环境作出贡献。节能节水、减少污染物排放以及降低生产成本是该项目重点关注的领域。对工厂进行现场环境改善评估以识别和

---

①有关背景资料，可以参阅香港环境保护署与广东省环境保护局《粤港合作无间，共享天清海澄》，https://www.enb.gov.hk/sites/default/files/zh-hant/booklet_jwgsdep.pdf。

②粤港应对气候变化联络协调小组，创建于 2011 年，由香港环境局局长和广东省发展和改革委员会主任共同领导，目的是就共同适应和减轻气候变化的议题进行合作和交流。

③粤港合作联席会议于 1998 年成立，目标是加强香港与广东在特定专家小组所涵盖的多个政策范畴内的横向协调。

④创新一直被特别重视，见“Hong Kong—Mainland Innovations in Environmental Protection since 1980”, *Asian Survey*, Vol 51, No. 4, July/August 2011。

分析它们所面对的问题，并提出实用的改进方案，建设示范项目展示新清洁生产技术的有效性，由行业协会和在其他相关工厂推进示范工程，以及行业协会举办的行业清洁能源使用推广活动，香港特区政府均对其部分成本进行资助。该计划被认为是成功的，并已延长了三次。目前的阶段将持续到2025年3月[①]。

二、未来前景

2019年2月18日发布的最新合作框架《粤港澳大湾区发展规划纲要》包含了气候变化内容。它比《珠江三角洲地区改革发展规划纲要（2008-2020）》更具雄心和方向性。新的大湾区规划明确指出，该区域应追求“绿色发展和生态保护”，包括低碳的生产生活方式、能源系统和水资源保护方案，并改善防洪条件。大湾区规划要求有关部门在2019～2020年进行短期合作，并有望将这种合作机制延长至2035年。在编写本报告的同时，特区政府尚未公布它们将如何在现有机制之外加强和扩大合作结构与细节。

可以看出，随着时间的推移，区域绿色发展与合作变得越来越重要。已经建立的体制机制为更密切的合作奠定了基础，但它必须继续改变，以满足区域社会经济和多学科政策需求，从而更快变革并实现广泛的绿色发展成果。

---

①香港特区政府在2019年《施政报告》中宣布延长服务期限，详情请参阅https://www.policyaddress.gov.hk/2019/eng/highlights.html,《施政报告》第七章也有提及，请参阅https://www.policyaddress.gov.hk/2019/eng/pdf/supplement_7.pdf。

第六节　港澳应对气候变化方面的独特之处与政策建议

香港和澳门由于是高度城市化的沿海亚热带经济体，受风暴、暴雨和其他相关自然灾害的影响较大，对市民和城市资产构成较高的风险，因此两地在抵御天气相关风险方面有着悠久历史和丰富经验。随着未来极端天气事件可能会更加频繁和严重，它们在应对气候变化方面仍有进步空间。但不可否认的是，两地政府在天气相关的风险管理的一些方法是独特而卓越的，这可以为其他城市提供借鉴。分享经验的同时也能够给特区政府提供继续改进的机会。本节总结了港澳在应对气候变化方面的独特之处，并在此基础上提出了可行的改进建议。

一、在城市适应和应对能力方面的独特经验

香港的热带气旋警告信号系统、山泥倾泻警告、斜坡安全管理和防洪系统，都是气候适应和应对能力的杰出典范，在这方面的努力也得到国际认同。热带气旋警告信号系统已有超过 100 年的历史，是保护市民生命和财产的第一道防火墙。预警系统发出每个信号后，特区政府和社会各界应当采取的措施和行动都被清楚列明，并付诸实践。这为城市应对不同紧急情况提供了工作范本。例如，香港的斜坡安全管理系统按不同风险等级进行了明确界定，并定期进行维修和管理。防洪策略和计划则兼顾了人口稠密地区的预防，应对洪灾的短期和长期需求，以确保将对市民的日常生活工作的影响降到最低。在大湾区的背景下，各城市应当携手提升气候适应能力，一些能够影响整个地区的应对战略和项目需要逐步落实，香港可以在知识分享和经验交流方面

同其他城市开展紧密合作。

以下是本报告的分析及建议：

（1）持续升级天气警告系统，加入健康提醒。现有的天气警告系统发挥了很大的作用，但仍可以进一步升级成为包含更多针对性信息的综合系统，以促进极端天气期间社区互助和护理服务，从而降低人口风险。这一系统还可以帮助相关机构审查在极端天气下服务需求激增时的应对能力，以便在社区层面重构服务体系，开展相关的公共卫生、医疗、教育和社会服务工作，提高服务的相关性、合理性、公平性和成本效益。为了使卫生机构能够应对日益频繁和严峻的极端天气条件，如持续高温、寒潮以及可能带来人员伤亡及疾病风险的暴雨，与气温有关的死亡率/发病率/住院率，以及其他天气事件的相关数据需要被深入研究和分析。特区政府还需要与医院、医学院、医护人员及社会福利专业人士合作，积极开展培训，以提高各机构对于气候变化给健康带来的影响等问题的认识水平，这些培训需要涵盖技术技能提升，制定气候变化的应对计划。

（2）加强斜坡管理和滑坡的原型设计。香港在滑坡预警、预防和应急准备方面表现卓越。对于内地多山地的部分大城市来说，山体滑坡风险较大，威胁生命财产安全，学习香港的先进经验显得至关重要。粤港澳大湾区可以考虑在未来几年内提升其滑坡和防洪系统的应对能力：①支持跨学科研究和实验，处理过剩的水资源问题，更好地应对气候变化。建立一个基于科学、数据驱动的滑坡预测系统，能够精准预测在什么地方以及在什么情况下可能发生危险，分析隔水、过水、土壤和植被之间的相互作用。②在基础架构设计上执行“蓝绿部署”。③完善风险沟通和具体行动的设计，提高社区的抗灾能力。④充分利用新的公共教育工具，如大规模开放在线课程，这些课程可以配备可视化和交互式组件，鼓励公众无限制参与并开放访问。⑤对斜坡管理和滑坡进行设计提升，推广至整个粤港澳大湾区。

（3）积极保护海岸线和区域水域。粤港澳大湾区拥有漫长的海岸线，加强对海岸线和区域水域的保护将提升整个区域的气候应对能力。此外，鼓励相关的海洋学研究可以指导大湾区管理机构审慎考虑包括填海造陆在内的各种陆地开发活动，以确保这些活动不会影响区域的水流动，及随之而来的诸如海水缺氧等负面影响。大湾区还可以通过软硬结合的工程手段来加强海防结构，为填海工程中的生态海岸线建设制定指导性设计原则。此外，大湾区应积极开展与海洋问题有关的联合研究，为控制污染负荷和保护海洋生态系统提供有益指导。

（4）深度加强气候抗御合作。在大湾区范围内开展恶劣天气事件的应对方面的宣传合作，将增强社区对恶劣天气的抵御能力和准备能力，也将促进大湾区社区的绿色和低碳生活方式转型。香港和澳门特区可以考虑将以下内容纳入增强大湾区内合作时的重要方面：①由于大湾区面临类似的天气风险，合作应从整体上改善区域内的城市规划和设计原则，并探讨该区域是否有必要采用类似的基础设施设计原则来应对潜在危害（如风，雨和海平面上升等）。②为大湾区建筑绿色化设定时间表，以便各地区能够合力完成高节能和节水的目标。③鼓励在海绵城市发展方面进行合作，特别是对整体防洪和复原的计划（蓝绿建设），活化河道以恢复其自然排水和生态系统。④将数据共享扩大到气象学以外的领域，以便区域和国家主管部门加强有关单位开展研究和共享信息，以应对极端天气风险。⑤交换新技术和商业模式，改变运输系统和服务，以实现低碳或零碳排放。⑥考虑在污水处理方面采取大湾区统筹设计，以便大湾区各地政府可以提高排污质量，效率和成本效益，采用最适当的污水处理技术。⑦进一步协调内地与港澳地区之间的能源供应系统，加强能源供应，进而高效减缓气候变化，提高生态恢复力。⑧在大湾区中强制推广能源标签，并对电器降低能耗及车辆的燃油效率提出更高要求。

二、在治理、金融、科研等方面具的独特优势，可提高区域效率和竞争力

作为特别行政区，香港特别行政区和澳门特别行政区的行政体制不同于内地其他地区，这为特区政府及国家探索、试验政策提供了的灵活性和机会。例如，香港几十年来一直是全球领先的金融中心；香港大学的科研质量在全球名列前茅；综合法律制度吸引了香港的商业和人才，使香港成为国际上最具活力的城市之一。这些优势应继续保留。

以下是本报告的分析及建议：

①加快绿色金融的发展。香港已经具备了成为中国绿色金融中心的能力。香港已为本地、内地及其他地方的项目筹集了大量绿色债券，可为大湾区及全国其他地区的一揽子绿色项目筹集资金，这一作用还可以进一步扩展到其他筹资方式。随着香港和内地上市公司提供环境、社会和治理报告的重要性日益提高，香港有关机构可以提供高报告质量的范本。此外，随着内地继续发展其碳排放交易体系，内地和香港特区政府应考虑香港如何为中国的碳排放交易提供国际市场。此外，香港和深圳可以考虑建立一个大型建筑物减排量交易系统（限额交易方案）。②在技术应用和大数据监管方面进行创新。在大数据和人工智能时代，每个司法管辖区都必须不断更新其政策。此外，大湾区可以考虑联合开展一项“大环境数据”项目，以此平台发布该地区的相关数据。③发挥专才优势，促政策革新。香港特别行政区在专业人才培养、认定方面独树一帜，为学界、业界培育了大量包括环境领域在内的优秀专才。面对日益复杂的气候变化和环境问题，进一步发挥专才的优势，促进政府政策的前沿发展和科学制定，对香港自身变革、保持竞争力意义重大。④加强培养跨学科跨部门的开放和协作文化。应对气候变化的复杂性需要采取多学科交叉方法，并需要加强政府、大学、行业和专业人士之间的协作。他们可

以一起将理论研究和应用研究纳入实施范围，包括医疗专业培训，这都需要适当的平台和资金来支持积极和定期地参与。更重要的是，由于区域空气质量管理是大湾区的重要问题，主管部门应确保将碳排放结果也纳入考量范围，以使空气质量与碳排放都能获得最佳结果。⑤设计和开发区域合作机制。根据大湾区规划，粤港澳相关主管部门，及相关机构、大学、企业需要建立适当的合作机制以实现大湾区的绿色发展转型。

参考文献

Chan, E. Y. Y., A. Y. T. Man, H. C. Y. Lam, 2018. *Personal Heat Protective Measures During the 2017 Heatwave in Hong Kong: A Telephone Survey Study*. First Global Forum for Heat and Health, Hong Kong, December 2018.

Choy, C. W., D. S. Lau, Y. He, 2020. Super Typhoons Hato (1713) and Mangkhut (1822). Part Ⅱ: Challenges in Forecasting and Early Warnings. *Weather*.

Keramydas, C., G. Papadopoulos, L. Ntziachristos, *et al.*, 2018. Real-World Measurements of Hybrid Buses' Fuel Consumption and Pollutant Emissions in a Metropolitan Urban Road Network. *Energies*,11(10).

Kwan, J. S. H., H. W. K. Lam, C. W. W. Ng, *et al.*, 2018. Recent Technical Advancement in Natural Terrain Landslide Risk Mitigation Measures in Hong Kong. *HKIE Transactions*, 25(2).

Song, Q. B., J. H. Li, H. B. Duan, *et al.*, 2017. Towards to Sustainable Energy-Efficient City: A Case Study of Macau. *Renewable and Sustainable Energy Reviews*, 75.

Yang, M. T., J. Q. Liu, X. R. Zhang, *et al.*, 2015. Comparative Toxicity of Chlorinated Saline and Freshwater Wastewater Effluents to Marine Organisms. *Environmental Science & Technology*, 49(24).

Zhao, S. J., Q. B. Song, H. B. Duan, *et al.*, 2019. Uncovering the Lifecycle GHG Emissions and Its Reduction Opportunities from the Urban Buildings: A Case Study of Macau. *Resources, Conservation and Recycling,* 147.

Zhao, Y., Y. Bo, W. L. Lee, 2019. Barriers to Adoption of Water-saving Habbits in Residential Buildings in Hong Kong. *Sustainability*, 11(7).

英文版

Contents

Chapter 1　Observed Climate Change

Since the industrial revolution, with the rapid development of human society and enhancement of production efficiency, the environment on Earth, on which human survival depends, is suffering from unprecedented pollution and destruction. At the same time, the global climate has changed, exacerbating the deterioration of the environment. In particular, greenhouse gas emissions and anthropogenic pollutants have increased rapidly in recent decades, leading to more and more climate and environmental problems (e.g., snow and ice melting, sea level rising, more frequent extreme weather events, urban pollution, etc.) and threatening human survival. In the face of such arduous climate and environmental problems, climate change mitigation is urgently needed. Anthropogenic carbon emissions and pollution have crucial impacts on climate (Chen *et al.*, 2004) and human health. There are two types of anthropogenic emissions contributing significantly to global warming: the first one is long-lived greenhouse gases such as carbon dioxide while the second one is short-lived climate pollutants including tropospheric ozone, methane, black carbon, freon and its alternatives (Xie, 2016). The abundance of these emissions not only raises global temperatures but also has a crucial impact on regional climates such as accelerating the melting of glaciers and snow cover, affecting the amount of solar radiation reaching the surface and even monsoon systems and typhoon related precipitation.

This chapter describes the observed climate change and climate projection in Hong Kong and Macao.

1.1 Observed Changes

1.1.1 Changes in basic climate elements

Hong Kong

Situated in Tsim Sha Tsui, Kowloon, the Hong Kong Observatory is a centennial observing station recognized by the World Meteorological Organization. Daily weather observation at the Observatory began in 1884 but was interrupted during 1940～1946 due to the Second World War. More than 130 years of observations provide valuable data for climate change research. Previous studies used the Observatory data to analyze long-term changes in several basic climate elements in Hong Kong (Ginn *et al.*, 2010; Chan *et al.*, 2012; He *et al.*, 2016). This section extends the relevant analysis to 2018.

(1) Changes in temperature

Figure 1–1 shows the annual mean temperature of the Hong Kong Observatory from 1885 to 2018. There is a significant increasing trend with a rate of 0.13℃ per decade, partly reflecting the impact of global warming and partly reflecting the impact of local urbanization. Ginn *et al.* (2010) analyzed the temperature time series and pointed out that the warming in Hong Kong accelerated in the 1970s and 1980s. In addition to global warming, the acceleration was also related to the rapid urbanization during that period. The

Observatory used two different methods to assess the impact of urbanization. In one of the methods, Macao was considered as a rural station and the difference in warming trends between Hong Kong and Macao was analyzed (Lee *et al.*, 2011; Chan *et al.*, 2012). In the other method, the 850 hPa temperature of the upper air sounding data taken at the King's Park meteorological station was compared with the NCEP reanalysis data to obtain the background warming trend (Chan *et al.*, 2012). Based on these two different methods, it was estimated that the effects of urbanization accounted for about 40%~50% of the overall warming trend (Chan *et al.*, 2012).

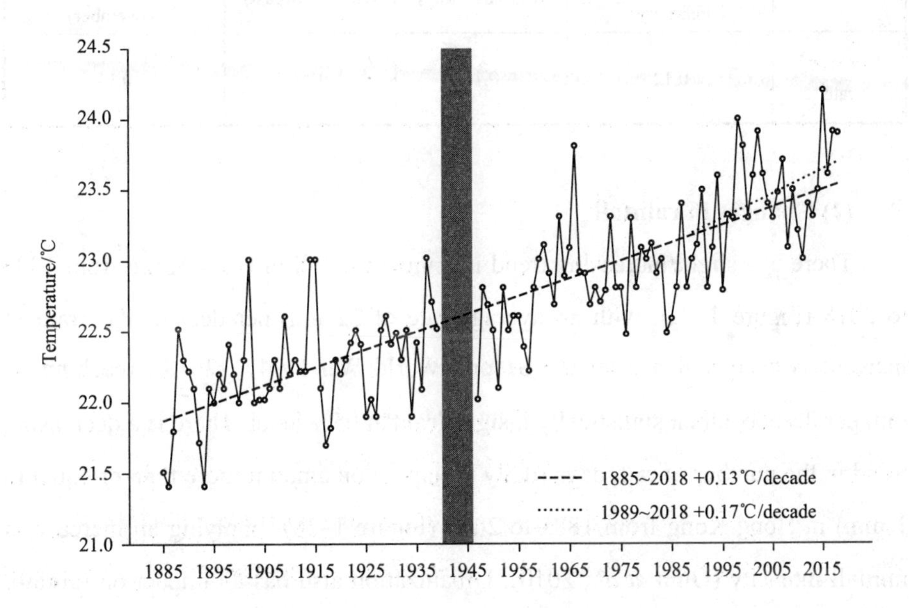

Figure 1–1 Annual mean temperature recorded at the Hong Kong Observatory (1885 ~ 2018) (No data in 1940 ~ 1946, same in Figure 1–2)

There are significant warming trends in all seasons in Hong Kong (Table 1–1), with the highest trend in spring, reaching 0.16℃ per decade, and smaller trends in summer and autumn. According to the method developed by Chan *et al.* (2012), the 25th and 75th percentiles (i.e., 19.1℃ and 27.7℃) of the daily average temperature from 1981 to 2010 are defined as the thresholds of cool and warm days, respectively. The number of cool days is decreasing, while the number of warm days is increasing. Meanwhile, warm days tend to start earlier and finish later.

Table 1–1 Increasing trend of seasonal average temperature in Hong Kong from 1885 to 2018 (℃ per decade)

	Winter (December～February)	Spring (March～May)	Summer (June～August)	Autumn (September～November)
Increasing rate	0.12	0.16	0.11	0.11

(2) Changes in rainfall

There is a slight increasing trend in annual rainfall in Hong Kong from 1885 to 2018 (Figure 1–2a), with an average rate of 22 mm per decade. The rate of increase is even higher after the Second World War (1947～2018), reaching 37 mm per decade, albeit statistically insignificant at 95% level. There is a decreasing trend in the number of rain days (daily precipitation amount more than or equal to 1 mm) in Hong Kong from 1885 to 2018 (Figure 1–2b), implying an increase in rainfall intensity (Ginn *et al.*, 2010). Urbanization also has an impact on rainfall. Mok *et al.* (2006) found that the increase rate of rainfall in urban areas in Hong Kong is higher than those in other regions (e.g., the New Territories, high grounds and offshore areas).

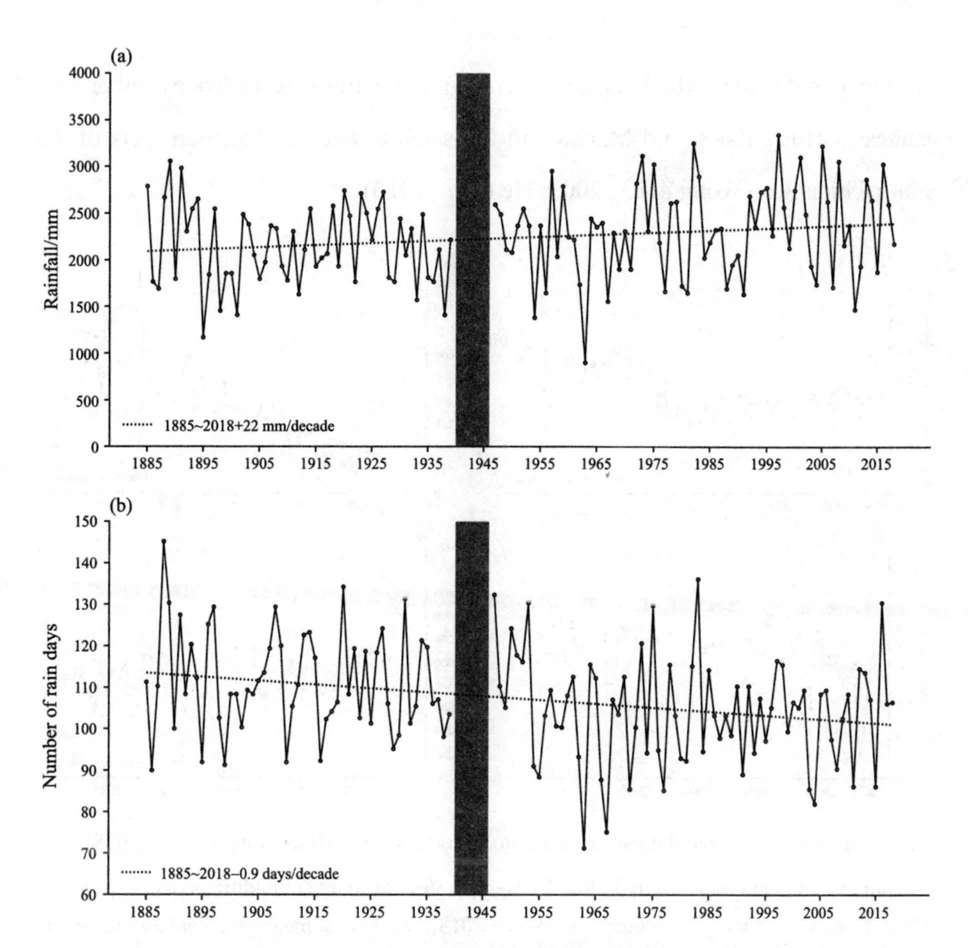

Figure 1–2 (a) Annual rainfall recorded at the Hong Kong Observatory Headquarters (1885 ~ 2018); (b) Annual number of rain days recorded at the Hong Kong Observatory Headquarters (1885 ~ 2018)

(3) Changes in sea level

Under global warming, thermal expansion of seawater and melting of land-based ice and snow lead to a rise in global sea level. Since 1954, Hong Kong set up an instrument to measure sea level in Victoria Harbor (Quarry Bay/North Point). Figure 1–3a shows a clear rising trend of annual mean sea level at a rate of

31 mm per decade, which is consistent with the trend recorded by other tide gauges in Hong Kong and Macao and the satellite over the northern part of the South China Sea (Wong *et al.*, 2003; He *et al.*, 2016).

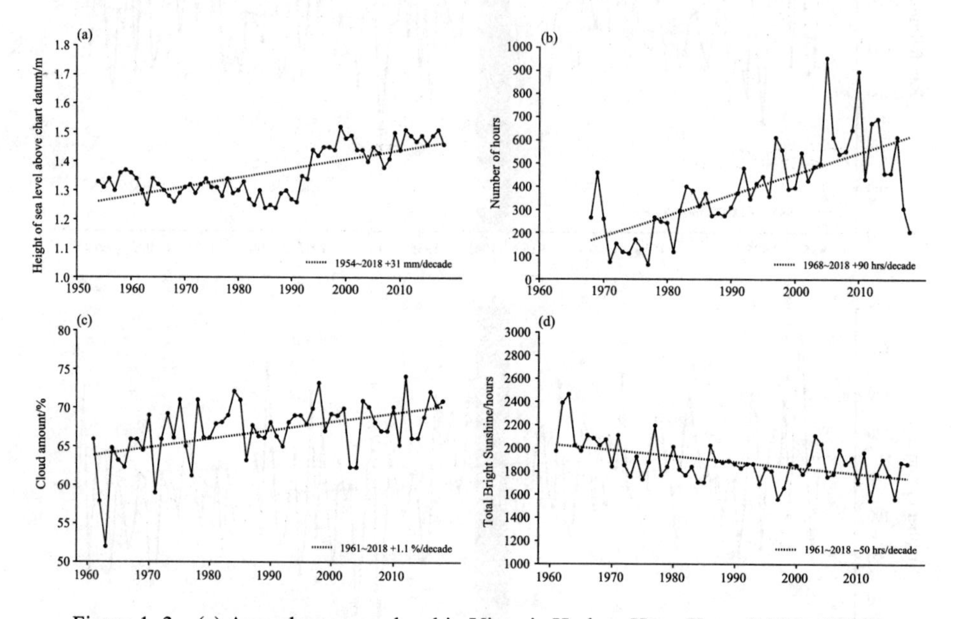

Figure 1–3 (a) Annual mean sea level in Victoria Harbor, Hong Kong (1954 ~ 2018); (b) Number of hours of visibility below 5 km (regardless of weather conditions) recorded at the Hong Kong Observatory Headquarters (1968 ~ 2018); (c) Annual mean cloud amount recorded at the Hong Kong Observatory Headquarters (1961 ~ 2018); (d)Annual total bright sunshine hours recorded at King's Park Meteorological Station (1961 ~ 2018)

(4) Changes in visibility, cloud amount and sunshine duration

Visibility at the Hong Kong Observatory was taken hourly by trained observers. The number of hours of visibility below 5 km (regardless of weather conditions) between 1968 and 2018 shows a clear increasing trend from the 1970s to the early 2010s (Figure 1–3b) but decreases in recent years. Leung *et al.* (2008) pointed out that the changes in the number of hours of low visibility in Hong

Kong may be related to the concentration of $PM_{2.5}$, which is mainly due to anthropogenic activities such as construction, emissions from vehicles and coal-fired power generation.

Cloud amount in Hong Kong is also reported in oktas hourly by trained observers. Between 1961 and 2018, the annual mean cloud amount increases by 1.1% per decade (Figure 1–3c). One of the potential reasons for the increase in cloud amount is the increase in the concentration of atmospheric condensation nuclei generated by urban activities, which is favorable to cloud formation (Ginn *et al.*, 2010). Sunshine duration decreases as a result of the increase in cloud cover (Figure 1–3d).

(5) Others

One study examined the sounding data of King's Park Meteorological Station and found that the autumn, winter, and annual mean temperature at 850hPa show significant long-term increasing trends. The long-term variations of some atmospheric stability indices (e.g., K index, CAPE index) indicate that the atmosphere over Hong Kong has become more unstable in summer in recent decades (Tong and Lee, 2014).

Macao

The meteorological observation in Macao began in 1861. The Portuguese Navy mainly carried out the observation at the beginning. The basic meteorological elements (e.g., temperature, precipitation, pressure, wind speed, wind direction, evaporation, sunshine duration, radiation) that can be queried at present can be started as early as 1901. The current meteorological stations include Sun Yat-Sen Municipal Garden station, Outer Harbor Ferry Terminal station, Maritime Museum station, Monte Fort station, Sai Van Bridge station,

Governador Nobre de Carvalho Bridge station, Amizade Bridge station, Taipa Grande station, Ka-Ho station, and Coloane station.

(1) Changes in temperature

According to the data provided by Macao Meteorological and Geophysical Bureau, the time series of the annual mean temperature and anomaly (relative to the average of 1901～2018) are plotted (Figures 1–4). Annual mean temperature in Macao increased by 0.71℃ from 1901 to 2007, with a rate of 0.066℃ per decade. The temperature increase rate is 0.056℃ per decade from 1901 to 2008, that is, the temperature increase rate is slower on a longer time scale. Analysis of the annual average temperature from 1998 to 2018 found a cooling rate of -0.14℃ per decade. Macao is of small size and faces to the ocean, and the impact of urbanization on local climate is smaller, which may lead to the temperature increase rate in Macao is slower than that in the global and surrounding regions.

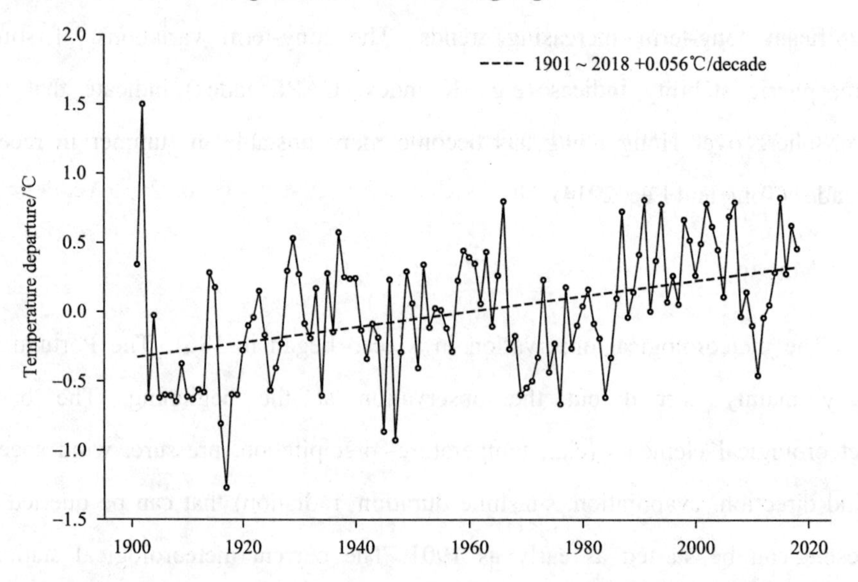

Figure 1–4 Annual mean temperature at Macao from 1901 to 2018 (data source: Macao Meteorological and Geophysical Bureau)

Fung *et al.* (2010) analyzed the annual average temperature in Macao and found that the interdecadal variations of temperature were obvious, two cold periods occurred in the first 20 years of the 20th century and the period form the late 1960s to early 1980s, respectively. During the period from the 1930s to 1940s, and the late 1950s to the 1960s, the temperature increased. The other warm period occurred after the mid-1980s. Therefore, the annual average temperature change in Macao has an oscillation of several decades. In the background of this slow climate fluctuation, there are obvious characteristics of interannual variation. From Figure 1–4, during the periods from 1903 to 1928, from 1940 to 1957, and from 1967 and 1985, the temperature was lower. There was also a decrease in temperature from 2008 to 2014. After these four periods, the temperature increased significantly, and this warming trend lasted longer after 1985.

A study based on a small wave band analysis found that the annual average temperature change in Macao has a relatively significant period of 2~5 years oscillations, and has a strong locality in the time domain, that is, in the early 20th century, the mid-1910s, mid-1930s, mid-1940s, mid-1960s, mid-1980s~mid-1990s, and early 21st century, the oscillations with a period of 2~5 years were relatively strong. The oscillations around 1910, 1921, 1941, 1957, 1981 and 2002 were relatively weak. Therefore, the oscillation intensity of the period of 2~5 years had an obvious interdecadal variation, and the time scale is about 10~20 years (Fung *et al.*, 2010).

Fung *et al.* (2010) also explored the relationship between climate change in Macao and Atlantic Multidecadal Oscillation (AMO) and found that the low-frequency changes in annual mean temperature in Macao are closely related to AMO on a decadal timescale. The smoothed Macao annual average temperature and AMO share the same phase changes, and the oscillation cycle is about 60 years, with a correlation coefficient of as high as 0.67 (significant at 99% confidence level). Meanwhile, the correlation between the average temperature of each season in Macao and AMO is significantly different, with the highest correlation coefficient in winter (0.47) and the lowest in summer (0.21). South China is mainly dominated by subtropical and tropical weather systems in summer, and changes in the sea surface temperature in the Pacific may have a more direct impact on these systems.

The number of days with maximum temperature over 33℃ from 1901 to 2018 does not increase significantly (-0.02 days per decade). During the same period, the number of cold days (daily minimum temperature less than 13℃) has decreased, with a decline rate of -0.75 days per decade.

Qian *et al.* (2015) used the continuous daily temperature data from 1912 to 2012 to test the normality of two indicators of a hot day (> 33℃) and hot night (> 28℃) in Macao and found that the hot day and night index are related to the interannual and interdecadal variations of the El Niño-Southern Oscillation (ENSO)–East Asian Summer Monsoon coupling system.

(2) Changes in precipitation

The annual average precipitation in Macao is 1 873 mm. Figure 1–5 shows the annual precipitation anomaly in Macao, there is an overall increasing trend for average precipitation during 1901 to 1982, with the peak in 1982 of 3 041 mm. The anomaly from 1982 to 1991 decreased greatly and the anomaly in the recent 20 years showed an overall decrease trend. The annual precipitation increasing rate is 41.4 mm per decade, which is slightly lower than the global average, and in the recent 20 years, the precipitation in Macao decreased.

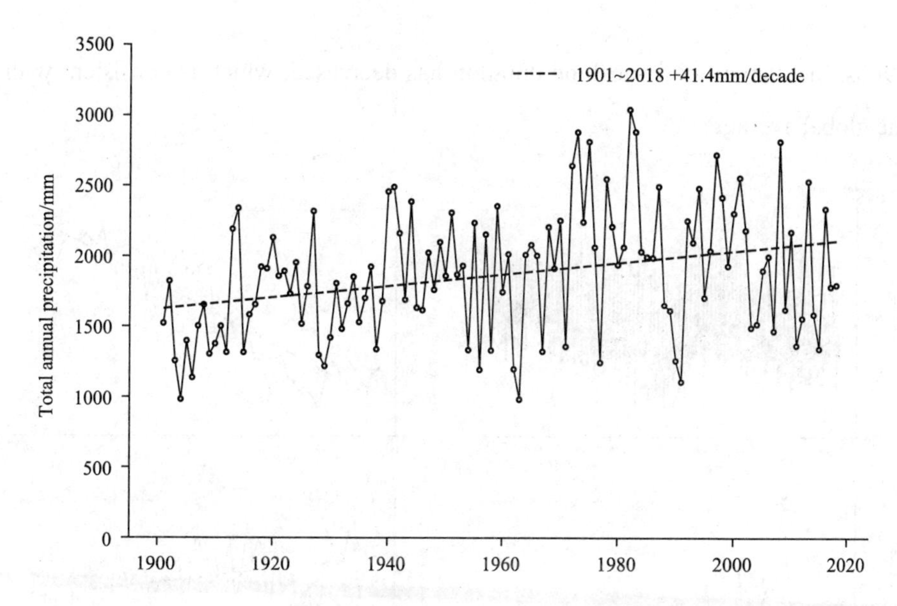

Figure 1–5 Annual precipitation in Macao from 1901 to 2018 (data source: Macao Meteorological and Geophysical Bureau)

(3) Changes in sea level

During 1925~2010, the sea level in Macao has increased at a rate of 1.35 mm per year, and after 1970, the sea level has increased at a rate of 4.2 mm per year. Meanwhile, the sea level rising rate in Macao during the period of 1993 to 2012 was 1.1 times that of the global average (Wang L. *et al.*, 2016). Figure 1–6a shows the average sea level change in Macao from 1925 to 2010.

(4) Changes in visibility, cloud cover and sunshine duration

The sunshine duration from 1952 to 2018 is shown in Figure 1–6c. The annual average total sunshine duration of Macao from 1952 to 2018 is 1 861.5 hour, and the total sunshine duration has increased from 1952 to 1963, reaching the peak in 1963 with 7.2 hour per day. The annual average sunshine duration is 5.1 hour per day over the recent 68 years. The linear decline rate of annual total sunshine duration from 1952 to 2018 is 51.4 hour per decade. Since the early

1960s, the annual total sunshine duration has decreased, which is consistent with the global average.

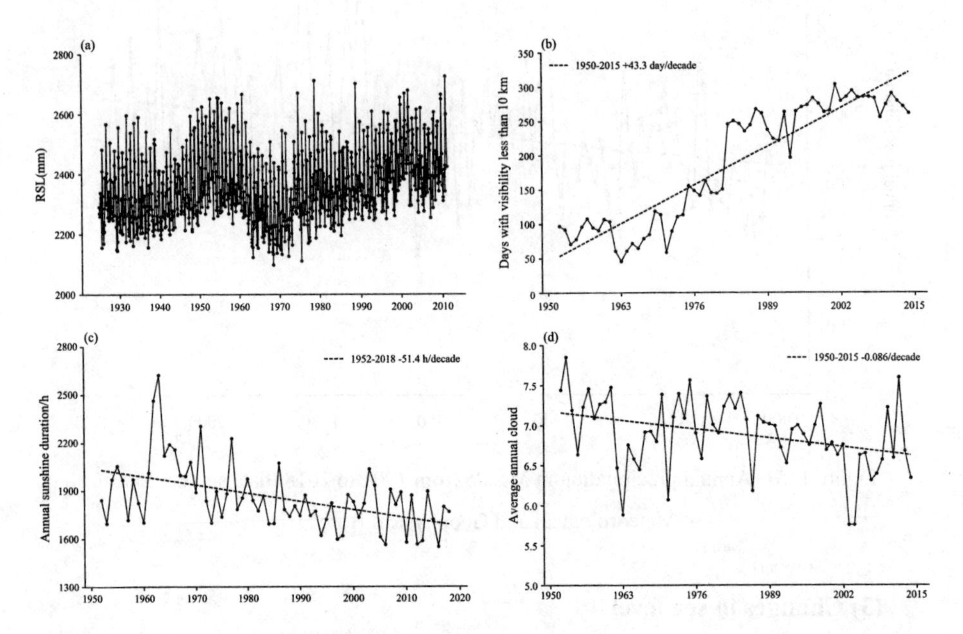

Figure 1–6 (a) Temporal evaluation of relative sea level at Macao (units: mm) in reference to local chart datum (Wang *et al.*, 2016); (b)The number of days with visibility less than 10 km in Macao from 1950 to 2015 (data source: Macao Meteorological and Geophysical Bureau); (c) Annual total sunshine duration in Macao from 1952 to 2018 (data source: Macao Meteorological and Geophysical Bureau); (d) Annual average cloud over in Macao from 1950 to 2015 (data source: Macao Meteorological and Geophysical Bureau)

The annual cloud cover in Macao during 1952～2014 shows a declining trend with a rate of 0.86% per decade (Figure 1–6d), which is different from the trend of cloud cover in the adjacent Guangdong province (Wu *et al.*, 2011). The number of days with visibility less than 10 km during 1952～2014 is increasing with a rate of 43.3 day per decade (Figure 1–6b).

(5) Others

Mi (2010) found that the heat island effect areas of Macao in 1998 and 2004 were 4.07 km^2 and 2.83 km^2, respectively, indicating that the heat island effect of Macao during that period has weakened, however, taking the small size of Macao into consideration, the heat island effect of Macao is still relatively strong.

1.1.2 Changes in extreme weather and climate events

Hong Kong

Changes in climate mean state also alter the frequency of extreme weather events. Taking temperature as an example, if the average temperature rises, the frequency of extremely high temperature will increase and the frequency of extremely low temperature will decrease. The numbers of very hot days (daily maximum temperature more than or equal to 33℃) and hot nights (daily minimum temperature more than or equal to 28℃) exhibit significant increasing trends while the number of cold days (daily minimum temperature less than or equal to 12℃) shows a decreasing trend (Wong *et al.*, 2011). The average annual numbers of hot nights, very hot days and cold days for the periods of 1885～1914 and 1989～2018 (Figure 1–7a) exhibit a sharp contrast. The number of hot nights increased by dozens of times. Other extreme temperature indicators such as annual absolute maximum temperature and annual absolute minimum temperature also increased significantly (Wong *et al.*, 2011).

Wang W. *et al.* (2016) defined extreme urban heat island in Hong Kong based on extreme value theory (EVT). The thresholds for winter and summer were 7.8℃ and 4.8℃, respectively. They introduced parameter changes in the EVT model and pointed out that there was a significant increasing trend in extreme urban heat

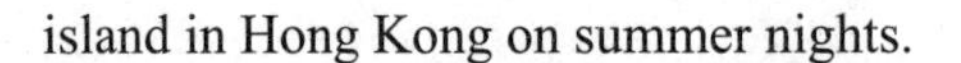

island in Hong Kong on summer nights.

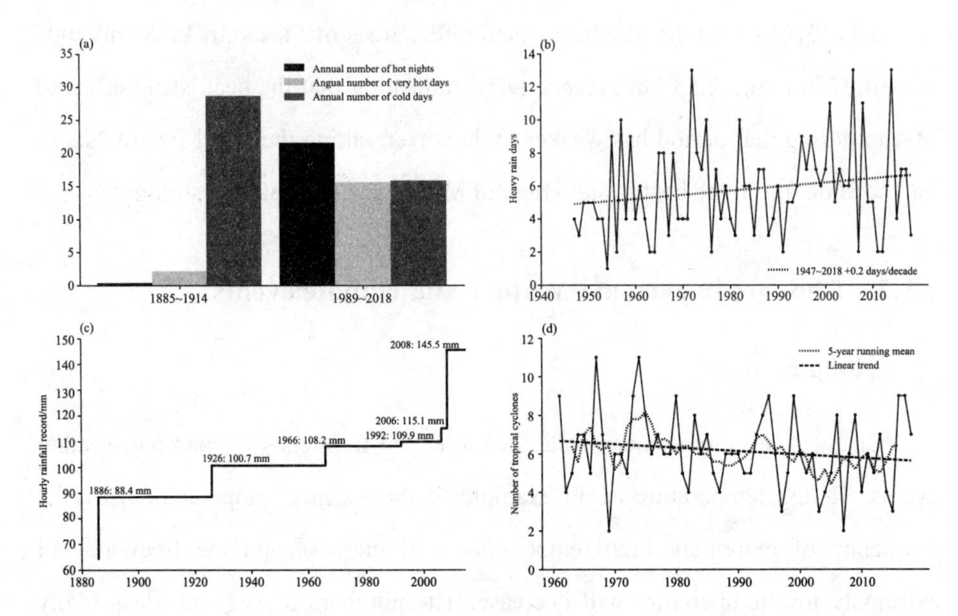

Figure 1–7 (a) Comparison of the average annual numbers of hot nights, very hot days and cold days recorded at the Hong Kong Observatory Headquarters for 1885 ~ 1914 and 1989 ~ 2018; (b) The number of heavy rain days (hourly rainfall over 30 mm) at the Hong Kong Observatory Headquarters (1947 ~ 2018); (c) Record of maximum hourly rainfall at the Hong Kong Observatory Headquarters; (d) Number of tropical cyclones entering 500 km of Hong Kong (1961 ~ 2018)

Ginn *et al.* (2010) pointed out that there was an increasing trend in the number of heavy rain days (hourly rainfall over 30 mm) (Figure 1–7b). Wong *et al.* (2011) used the non-stationary time-dependent generalized extreme value distribution model to analyze changes in extreme rainfall events. It was found that the return period of extreme rainfall events of hourly rainfall equal to or above 100 mm was shortened from 37 years in 1900 to 18 years in 2000, indicating the extreme rainfall event becoming more frequent. The record of maximum hourly rainfall at the Hong Kong Observatory has been broken more and more frequently (Figure 1–7c), reflecting the increasing occurrence of extreme rainfall events to a certain extent. On the other hand, the maximum number of consecutive dry days between April and September in Hong Kong also shows a clear increasing trend, indicating an increasingly uneven distribution in rainfall.

Tropical cyclones entering the northern part of the South China Sea can bring extreme weather to Hong Kong. Lee *et al.* (2012a; 2012b) used the tropical cyclone best tracks from different meteorological agencies to analyze the changes in frequency and intensity of tropical cyclones in the South China Sea and the vicinity of Hong Kong. They found a long-term decreasing trend in the numbers of tropical cyclones entering the South China Sea and within 500 km of Hong Kong (Figure 1–7d). However, this decreasing trend is not statistically significant. In terms of impacts on Hong Kong, the rainfall caused by tropical cyclones within 500 km of Hong Kong has a decreasing trend, but the trend is not statistically significant. The number of days with daily rainfall equal to or above 100 mm (R100) caused by tropical cyclones within 300 km of Hong Kong was found to decrease significantly (Li *et al.*, 2015). The changes in atmospheric circulation in the mid-troposphere led to more tropical cyclones moving towards Taiwan and Japan rather than entering the South China Sea, which is the main reason for the decrease in the numbers of tropical cyclones in the South China Sea and the vicinity of Hong Kong (Lee *et al.*, 2012a; Li *et al.*, 2015). The trend of tropical cyclone track changes coincides with the poleward shift of the location of the maximum intensity of tropical cyclones (Kossin *et al.*, 2014; 2016). Some studies (Kossin *et al.*, 2014; Liu and Chan, 2017; Kossin, 2018) suggested that the trend is quite robust in the western North Pacific and is unlikely to be caused by natural variability. According to the latest assessment report of UNESCAP/WMO

Typhoon Committee, the reliability level of the northward shift trend of the maximum intensity of tropical cyclones in the northwest Pacific since the 1940s is low to medium (Knutson *et al.*, 2019; Lee *et al.*, 2020).

In terms of the extreme wind speed induced by tropical cyclones, the maximum 10-minute mean wind speed and 1-second gust at Waglan Island associated with tropical cyclones within 500 km of Hong Kong decreased insignificantly. However, wind speed at urban stations decreased significantly, probably due to local urbanization (Lee *et al.*, 2012a; Peng *et al.*, 2018). In addition to strong winds and heavy precipitation, tropical cyclones also bring storm surge to Hong Kong. In the past 100 years, several typhoons (e.g., in 1874, 1937 and 1962) brought severe storm surge to Hong Kong, causing major casualties and damage. For example, during the passage of Super Typhoon Wanda in 1962, a maximum sea level of 3.96 m was recorded at Victoria Harbor (Table 1–2) while the maximum sea level recorded at Tai Po Kau in Tolo Harbor reached 5.03 m (Table 1–3). Based on observations available at the time, Lee *et al.* (2010) estimated the return value of a 50-year extreme sea level event at Vitoria Harbor to be 3.5 m. However, Hato in 2017 and Mangkhut in 2018 caused the occurrences of sea level exceeding 3.5 m at Victoria Harbor within just two years (Table 1–2) and the storm surges due to Mangkhut set a record in Hong Kong. After the passage of Mangkhut, the return period of the extreme sea level event of 3.5 m at Victoria Harbor has shortened by half (less than 25 years).

Macao

Wang *et al.* (2015) homogenized long-term observations in Macao and found that the frequency of extreme precipitation in Macao significantly increased, however, the increasing trend of the extreme precipitation intensity was not significant. In terms of summer heat waves, the increasing trend of intensity and duration was not significant.

Table 1–2 Maximum sea level and storm surge recorded at Vitoria Harbor during the passage of tropical cyclones (1954～2018)

Rank	Year	Name of tropical cyclone	Maximum sea level recorded at Victoria Harbor (m)	Year	Name of tropical cyclone	Maximum storm surge recorded at Victoria Harbor (m, above astronomical tide)
1	1962	Wanda	3.96	2018	Mangkhut	2.35
2	2018	Mangkhut	3.88	1962	Wanda	1.77
3	2017	Hato	3.57	1954	Ida	1.68
4	2008	Hagupit	3.53	1964	Ruby	1.49
5	2001	Utor	3.38	1979	Hope	1.45

Table 1–3 Maximum sea level and storm surge recorded at Tai Po Kau during the passage of tropical cyclones (1962～2018)

Rank	Year	Name of tropical cyclone	Maximum sea level recorded at Tai Po Kau (m)	Year	Name of tropical cyclone	Maximum storm surge recorded at Tai Po Kau (m, above astronomical tide)
1	1962	Wanda	5.03	2018	Mangkhut	3.40
2	2018	Mangkhut	4.71	1979	Hope	3.23
3	1979	Hope	4.33	1962	Wanda	3.20
4	2017	Hato	4.09	1964	Ruby	2.96
5	2008	Hagupit	3.77	1964	Ida	2.16

From 1971 to 2010, the number of typhoons that required issuing No. 8 signal decreased insignificantly. These phenomena are consistent with the changes in the number of tropical cyclones making landfall in Guangdong province (Li and Duan, 2010; Zhang *et al.*, 2011).

1.1.3 Changes in greenhouse gases and aerosols

Hong Kong

(1) Greenhous gas

Since May 2009, the Hong Kong Observatory has set up an instrument at the King's Park Meteorological Station to measure outdoor carbon dioxide concentration in urban areas. In addition, the Observatory has collaborated with the Hong Kong Polytechnic University (PolyU) to measure carbon dioxide concentration in rural areas at PolyU's background atmospheric monitoring station at Cape D'Aguilar, the southeast tip of Hong Kong Island since October 2010 (Feng *et al.*, 2011). In January 2020, the Observatory replaced the measuring instrument with a newer model and relocated it to the Hong Kong Environmental Protection Department's Cape D'Aguilar Supersite Air Quality Monitoring Station. Since the monitoring station at Cape D'Aguilar is far away from urban areas and is less directly affected by local carbon dioxide sources, the measured carbon dioxide concentration can represent the background carbon dioxide concentration in Hong Kong for most of the time. Figure 1–8 shows the variation in carbon dioxide concentration at King's Park and Cape D'Aguilar. Although the observation period is only about ten years, the rise in carbon dioxide concentration in urban and rural areas is clear and the increasing trend is consistent with that of global atmospheric carbon dioxide concentration.

(2) Aerosols

According to the emissions inventory of 2015, the emission of major air pollutants in Hong Kong decreased by 14% to 45% compared to 2010. From 2010 to 2015, (1) the emission of sulfur dioxide decreased by 78%, mainly due to the

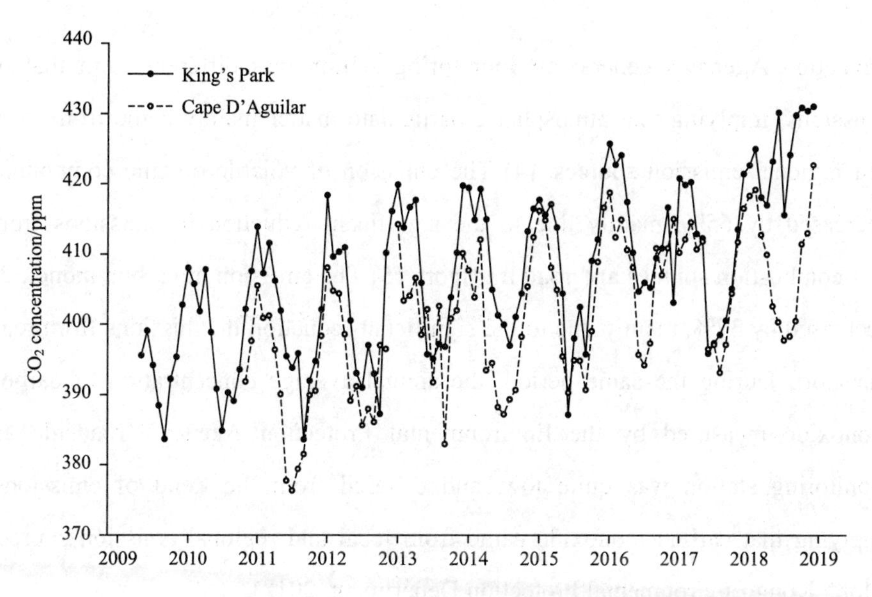

Figure 1–8 Carbon dioxide concentration in Hong Kong (2009 ~ 2018)

significant reduction in emissions from public electricity generation. During the same period, the average annual concentration of sulfur dioxide measured by the Environmental Protection Agency's general air monitoring station was consistent with the changes in emissions, indicating that the sulfur dioxide in the atmosphere mainly comes from local sources, in addition to being affected by meteorological factors. (2) The emission of nitrogen oxides decreased by 39%, during the same period, the average annual concentration of nitrogen oxides measured by the Environmental Protection Agency's roadside air monitoring stations has a similar trend with the emission, indicating that nitrogen oxides on the roadside mainly came from local emission sources. (3) Both PM_{10} and $PM_{2.5}$ decreased by 69%, mainly due to the significant reduction in emissions from road transport and public electricity generation. During the same period, the trend of the annual average concentration of PM_{10} and $PM_{2.5}$ measured by the Environmental

Protection Agency's general air monitoring station was different from that of emissions, implying that atmospheric particulate matter mainly came from local and regional emission sources. (4) The emission of volatile organic compounds decreased by 65%, mainly due to the significant reduction in emissions from non-combustion sources and road transport. (5) The emission of carbon monoxide decreased by 37%, mainly due to the significant reduction in emissions from road transport. During the same period, the annual average concentration of carbon monoxide measured by the Environmental Protection Agency's roadside air monitoring station was quite low, and differed from the trend of emissions, implying that carbon monoxide came from local and regional emission sources (Hong Kong Environmental Protection Department, 2018).

The annual average concentration of air pollutants, except for ozone decreased year by year from 1990 to 2017 (Hong Kong Environmental Protection Department, 2018). The emissions of ozone precursors, i.e., NMVOCs, CO and NOx decreased gradually, but the ozone concentration continued to rise (urban and new towns) or remained stable at a relatively high concentration level (suburbs). Since 1990, the Hong Kong Environmental Protection Agency has begun to monitor atmospheric ozone. The annual average ozone concentration at the general monitoring station has increased from less than 20 $\mu g/m^3$ initially to 52 $\mu g/m^3$ in 2018, breaking the record. According to the Environment Bureau expectation, the atmospheric ozone concentration will exceed the relevant level of Air Quality Objectives in 2025 and will be higher than the current level. Lam *et al.* (2005) used the PATH model and found that the pollution transport in the Pearl River Delta led to an increase in ozone concentration in Hong Kong. Before the arrival of the typhoons, pollutants were imported through northwestern Hong Kong and contributed significantly to the ozone pollution in Hong Kong, which accounted for 60%～90% of ozone concentration. Huang *et al.* (2005) suggested that about 30% of the ozone in Hong Kong was produced by local photochemical reactions, and about 70% came from transportation from South China.

The analysis of the chemical composition of atmospheric PM_{10} in Hong Kong from 1995 to 2017 found that the variation in ambient concentration was different from that in primary emission inventory, and the percentage of secondary inorganic ions in PM_{10} increased, indicating the complexity of the secondary reaction and the important influence of the secondary reaction on air quality.

Macao

(1) Greenhouse gas

The estimated emissions of greenhouse gases in Macao during 2007～2014 were generally decreasing, with a significant rebound in 2015 and 2016. The main sources of greenhouse gas emissions are local electricity generation, transportation (land transportation, maritime transportation and air transportation) and waste incineration (Macao Environmental Protection Agency, 2018). If Macao's GDP, population, and passengers continue to increase, Macao's greenhouse gas emission are expected to continue to increase.

(2) Aerosol

The overall trend of estimated emissions of air pollutants in Macao from 2007 to 2016 indicates that SOx generally decreased, NH_3 remained stable, however, the estimated emissions of CO, NOx, and NMVOC increased significantly. PM_{10} and $PM_{2.5}$ decreased from 2007 to 2010 and increased since 2011. The use of fossil fuels is a major source of air pollutants and greenhouse gases in Macao.

In terms of the overall trend of the annual average concentration of air pollutants, SO_2, PM_{10} and $PM_{2.5}$ decreased. The particulate matter concentration was already lower than the standard value after 2014. The SO_2 concentration was at a relatively low level. In the recent five years, the average concentration of $PM_{2.5}$ decreased significantly. In the recent decade, the annual average concentration of NO_2 and CO had a relatively flat trend, with NO_2 at a relatively high level and local NO_2 concentration exceeding the standard. It is worth noting that the average concentration of O_3 in the past decade has increased, and O_3 has become the major air pollutant in Macao with a higher concentration in September.

Wang *et al.* (2019) collected filter samples from Outer Harbor Ferry Terminal (PE) and Taipa Grande Country Park (TG) for chemical analysis and applied PMF to analyze source apportionment of $PM_{2.5}$ in Macao. Sulfate and organic carbon were the two dominant components with the highest concentration. Element carbon concentration and percentage were also high (2.20 and 4.43 μg C/m^3, 8% and 15% TG and PE), and higher than those measured at Foshan in 2011～2012 (2.95 μg C/m^3, 5%) and Guangzhou (1.92 μg C/m^3, 4%) (Wang, 2016). The results are like those in Hong Kong, indicating that the secondary reaction has an important impact on the concentration of atmospheric fine particles in Macao.

1.2 Changes in Atmospheric Circulation and Climate Phenomenon

1.2.1 Internal variability of the climate system

Regional decadal climate change may be contributed by both internal variability of the climate system and anthropogenic forcing. The attribution of decadal climate change relies on accurate data and climate model, but the resolution of current global climate models may not be able to capture the climate change over Hongkong and Macao. Up to now, very few attribution study focuses on Hong Kong and Macao, but a plenty of studies have focused on the mechanism of climate change over southern China including Hongkong and Macao. Based on these previous studies, the relative contribution from internal variability of the climate system and anthropogenic forcing on Hongkong and Macao will be addressed in this section.

Decadal variability

The most favorable modes of the decadal internal variability are the Pacific Decadal Oscillation (PDO) and Atlantic Multi-decadal Oscillation (AMO). The PDO is featured by the decadal oscillation of sea surface temperature (SST) over the Pacific Ocean, and the positive (negative) phase of PDO is associated with positive (negative) SST anomaly over tropical Pacific (mid-latitude North Pacific) Ocean. The PDO has shifted from a negative phase to a positive phase in the late 1970s, and back to a negative phase in the late 1990s. The PDO has an impact on the climate over South China, and deficient summer rainfall and excessive winter rainfall are seen during the positive phase of PDO, while hotter summer and colder winter are seen during the positive phase of PDO (Lu *et al.*, 2013; Xu *et al.*, 2017). A positive phase of PDO also favors a weaker East Asian summer monsoon (Cheng *et al.*, 2017). According to the periodicity of the PDO, a shift of PDO to a positive phase may occur in the recent future.

AMO is the dominant mode of decadal SST variability over the North Atlantic Ocean. The positive (negative) phase for the AMO is characterized by a positive (negative) SST anomaly over North Atlantic at a multi-decadal time scale. The AMO shifts from a negative to a positive phase in the late 1920s, from a positive to a negative phase in the 1960s, and from a negative to positive phase in the 1990s. During the positive phase of AMO, the temperature is higher over East Asia including Hong Kong and Macao (Wang *et al.*, 2013; Han *et al.*, 2016). The South China Sea summer monsoon onsets earlier after the early 1990s (Kajikawa and Wang, 2012; Zhang *et al.*, 2014; Jian *et al.*, 2018), coincident with the phase shift of AMO. But it is unclear whether the phase shift of AMO caused the earlier onset of the South China Sea summer monsoon.

Monsoon

East Asian climate is featured by monsoon. Cold and dry northerly wind prevails in winter, whereas warm and moist southerly wind prevails in summer. Summer is the major rainy season for most areas in China, and major meteorological disasters occur during the summer monsoon season. The South China Sea summer monsoon (SCSSM) usually onsets and arrives at South China in mid-May, which is substantially earlier than the onset time of summer monsoon in other regions of East Asia. The SCSSM usually onsets abruptly within a few days when the low-level prevailing southwesterly wind outbursts. The onset of SCSSM is the precursor of the onset of the East Asian summer monsoon, and the date of onset of SCSSM is a predictor for the climate prediction in East Asia. Overall, the onset of SCSSM is associated with abundant moist transport from tropical ocean to South China and an abrupt increase of heavy rainfall over South China. The time for the onset of SCSSM is monitored by the National Climate Center and Guangdong Provincial Meteorological Administration. The intensity of SCSSM is closely related to the onset date of SCSSM, an earlier (late) onset of

SCSSM often indicates a weak (strong) SCSSM (Zhang *et al.*, 2014).

The date of onset of SCSSM is modulated by the climate variability at multiple time scales. El Niño-Southern Oscillation (ENSO) and other modes of the tropical SST modulate the onset of SCSSM (Lin *et al.*, 2013; He and Zhu, 2015; Shao *et al.*, 2015; Gu *et al.*, 2018). Currently, the climate prediction can only tell whether the SCSSM onsets earlier or later than the climatology, but it is hard to accurately predict the date of SCSSM onset (Shao *et al.*, 2015). Besides interannual SST variability, the tropical intraseasonal oscillation also modulates the time for the onset of SCSSM (Shao *et al.*, 2014; Li *et al.*, 2017). The onset date of SCSSM is subject to great interannual variability, and it ranges from late April to late June in recent decades. It has also experienced a decadal change, and the onset of SCSSM is advanced after 1994 than before 1993 (Kajikawa and Wang, 2012; Zhang *et al.*, 2014; Jian *et al.*, 2018). Indeed, the onset date of SCSSM is subject to decadal oscillations since 1945, without a significant linear trend (Kajikawa and Wang, 2012).

Southern China cold spell

Cold spells in Southern China mainly flows from Siberia and Mongolia to northern China, and reaches Southern China through eastern China, middle China or southwestern China. The influence of cold spells is mainly modulated by the location of the East Asian trough, the shortwave trough near Lake Baikal and the Indo-Burma trough. Before the onset of a cold spell, there is always a blocking high near the Ural Mountains (Zhou *et al.*, 2009; Park *et al.*, 2011; Cheung *et al.*, 2015), which is related to the eastward extension of Rossby wave over North Pacific (Takaya and Nakamura, 2005), and also associated with the interaction between the troposphere and stratosphere (Cheung *et al.*, 2016). The intensity and location of East Asia through is mainly associated with the signal over the equatorial Pacific (such as ENSO) (Leung and Zhou, 2016; Leung *et al.*, 2017). When the East Asian trough deepened, the pathway of cold spell located more southward, the impact on Southern China was enhanced (Leung *et al.*, 2015). In addition, the recent studies also found that the MJO low-frequency signal over tropical regions affected the convective activity over the Maritime Continent, thus affecting the intensity of northerly winds and the cold spells (Jeong *et al.*, 2005; Hong and Li, 2009).

From the end of 1980 to the early 2000s, the Ural blocking high occurred relatively less, Siberia cold high was relatively weaker, and the South cold spell was relatively weaker (Wu and Leung, 2009; Wang and Chen, 2014). Since the mid-2000s, the Ural blocking high occurred relatively more, Siberia cold high became stronger, and decades-return-period strong cold spell reached Southern China, such as January in 2008 and January in 2016 (Zhou *et al.*, 2009; Cheung *et al.*, 2016). The variations in the recent Southern China cold wave have been modulated by low-frequency oscillations such as Pacific Decadal Oscillation and North Atlantic Oscillation (Zhou *et al.*, 2007; Hong *et al.*, 2008; Luo *et al.*, 2016), and maybe affected by Arctic sea ice loss, Arctic warming, and weakening of the polar vortex (Wu B. *et al.*, 2011; Wu Z. *et al.*, 2011; Kim *et al.*, 2014; Mori *et al.*, 2014; Luo *et al.*, 2018). Using EOF to analyze the leading mode of the winter cold spell duration in China from 1957 to 2010, they found that the center of the leading mode was in the north before 1980, and after 1980, the center of the leading mode moved to Southern China, which may be related to the southward shift of cyclone activities in East Asia (Ma *et al.*, 2012). Based on the climate model prediction, the leading mode of the cold wave also exhibits a southward shift to Southern China, indicating the global warming may strengthen the intensity and duration of cold spells in Southern China (Ma *et al.*, 2012). In addition, the frequency of the Ural blocking high will not change significantly in the future, but its relationship with the southward extension of Siberia clod high become stronger on the interannual scale, indicating that blocking high will have a stronger impact on the interannual variation of the Southern China cold spell (Cheung and Zhou, 2015).

Typhoon

The formation, intensity, and structural changes and movement of tropical cyclones are controlled by their external environment. There are two sources of tropical cyclones affecting Hong Kong and Macao, one in the seas east of the Philippines and the other in the South China Sea. Tropical cyclones formed in the seas east of the Philippines generally pass through the Philippines and enter the South China Sea. They will weaken after landing in the Philippines but could intensify again after entering the South China Sea. However, most of them will not become too strong. On the contrary, some tropical cyclones that cross the Bashi Strait directly into the South China Sea will not weaken much, so they have the chance to become super typhoons. In addition, the sea surface temperature on the coast of Guangdong is quite high, and the continental shelf is shallow. When the tropical cyclone passes, there is very little upwelling of cold water. Therefore, if the atmospheric conditions (such as vertical wind shear) are favorable, a tropical cyclone affecting Hong Kong and Macao can continue to intensify before making landfall.

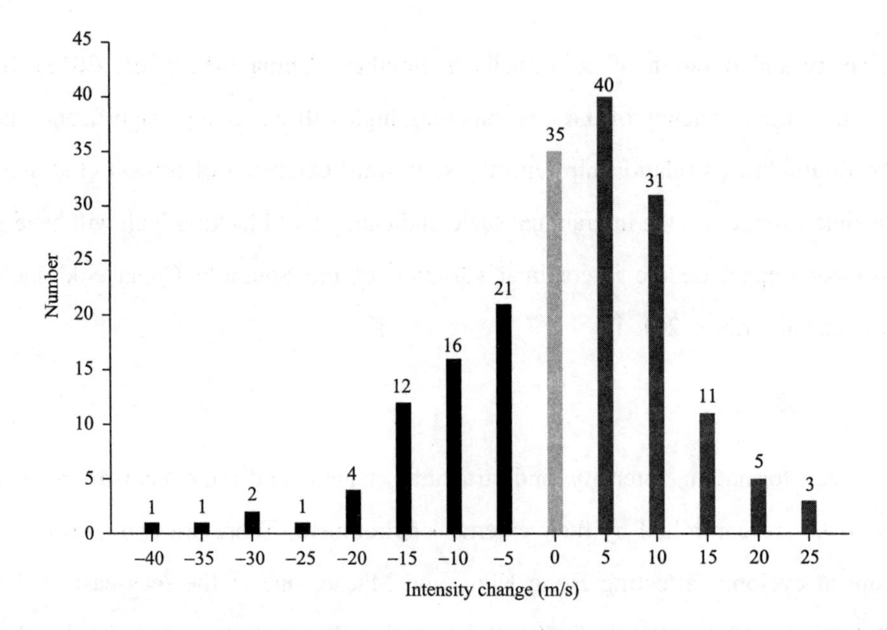

Figure 1–9 The comparison of tropical cyclone intensity landing in Hong Kong and Macao and 48 hours ahead of landing. The positive values indicate enhancement. The data is from Joint Typhoon Warming Center 1960 ~ 2013 best track dataset.

The movement of the tropical cyclone is mainly driven by the surrounding flow (steering flow) (Chan and Gray, 1982), and the change in the direction of steering flow is related to the changes in the western North Pacific Subtropical High (WNPSH). The interdecadal variation in the number of tropical cyclones affecting Guangdong province is due to the changes in its location, strength and westward extension, and these changes are associated with ENSO and PDO (Liu and Chan, 2017). Another factor is the formation location of the tropical cyclones. In the years with a low number, the average location is more to the north so that after the tropical cyclones form, the steering flow will push them northward, and they will not likely enter the South China Sea. In addition, if the number of tropical cyclones in the WNP decreases, it will also lead to a reduction in the number of tropical cyclones affecting Guangdong province (Zhang *et al.*, 2011).

1.2.2 Anthropogenic forcing

Large-scale circulation

The Western North Pacific Subtropical High (WNPSH) is a key large-scale atmospheric circulation system affecting the East Asian climate. The variation of WNPSH modulates the location of the East Asian rain belt (Zhu *et al.*, 2003; Ren *et al.*, 2013) and the track of tropical cyclones (Wu *et al.*, 2005; Kim *et al.*, 2012) at the intraseasonal and interannual time scales. Given the importance of WNPSH to the climate of South China, the area, location and intensity of the WNPSH draw great attention from the community.

The most commonly used metric for monitoring the WNPSH is geopotential height. However, recent studies suggest that global geopotential height is increasing along with global warming due to the thermal expansion of the air column (He *et al.*, 2015; 2018). The systematic increase of geopotential height makes it difficult to measure the change of atmospheric circulation associated with WNPSH, and it requires comparing the change of geopotential height with multiple variables associated with atmospheric circulation. Reanalysis datasets consistently show that the zero contours of eddy geopotential height at 850 hPa shifted eastward after the late 1970s, suggesting that the WNPSH at the lower troposphere has weakened and retreated eastward after the late 1970s (Huang *et al.*, 2015). The WNPSH at the mid troposphere has also weakened and retreated eastward after the late 1970s (Wu and Wang, 2015).

The moving of the tropical cyclones over the Western North Pacific (WNP) is mediated by the steering flow of WNPSH. The decadal weakening and eastward retreat of WNPSH may contribute to the change of tropical cyclones over WNP (Lee *et al.*, 2020). The landing location of the tropical cyclones has generally shifted northward (Liu and Chan, 2017), featured by an upward (downward) decadal trend of the numbers of tropical cyclones landing to the north (south) of Xiamen city (Wang *et al.*, 2009). Although the annual total number of tropical cyclones making landfall in China decreases in recent decades, the number of severe tropical cyclones follows a rising trend (Yang *et al.*, 2009). The average intensity of tropical cyclones making landfall in China is increasing, especially for South China (Yang *et al.*, 2009; Mei and Xie, 2016; Li *et al.*, 2017). In recent decades, the number of TC affecting Hong Kong (1961-2018) and Macao (1953-2017) has no significant decadal trend (Lee *et al.*, 2020). But it is still hard to detect whether the averaged moving speed of tropical cyclones over WNP is decreasing or not (Kossin, 2018; Moon *et al.*, 2019; Lai *et al.*, 2020), due to the quality of data in the pre-satellite era (Moon *et al.*, 2019). It is difficult to determine whether the average speed of tropical cyclones in the northwest Pacific has a decadal deceleration trend at present (Kossin, 2018; Moon *et al.*, 2019; Lai *et al.*, 2020).

The observed climate change in recent decades may result from either the internal variability of the climate system or its response to global warming. Based on the ensemble of 30 coupled climate models under the forcing of global warming scenario, the internal variability of the climate system can be suppressed and its response to global warming may be obtained. Although the geopotential height rises substantially under a warmer climate over the entire WNP, the anticyclone circulation associated with WNPSH is weakened. A comparison of the zero contours of the eddy geopotential height between the late 21st century under the RCP8.5 scenario andthe late 20th century suggests that the area of the WNPSH contracts and its western boundary of the WNPSH retreats eastward for about 10 degrees in longitude. Scaled by the amplitude of warming under RCP8.5, the western boundary of the WNPSH retreats for about 3 degrees in longitude per degree of warming (He *et al.*, 2015).

Despite a weakened WNPSH, climate projections and paleoclimate evidence show an enhanced East Asian summer monsoon (EASM) circulation in response to global warming. The strength of EASM circulation is usually measured by the low-level southerly wind over East Asia. As shown by the coupled climate model projections, the low-level southerly wind over entire eastern China in summer is enhanced under a warmer climate, and the response of EASM circulation to global warming is associated with an anomalous cyclone around the Tibetan Plateau. Multiple pieces of evidence suggest that the thermal forcing of the Tibetan Plateau on the atmosphere plays a key role in the response of EASM circulation to global warming (He *et al.*, 2019; He and Zhou, 2020). Climate models participating CMIP3 and CMIP5 consistently show an enhanced EASM circulation under a warmer climate (Ding *et al.*, 2013; Jiang and Tian, 2013; Kamae *et al.*, 2014; Kitoh, 2017). An enhanced EASM circulation is associated with a northward shift of East Asian major rain belt, and excessive (deficient) summer rainfall in North China (southern China). Paleoclimate evidence also shows that the EASM circulation is stronger and the East Asian rain belt shifts northward in warmer periods (Man *et al.*, 2012; Yang *et al.*, 2015). The enhanced EASM circulation and weakened WNPSH under a warmer climate do not contradict each other, since EASM circulation is also modulated by other factors (such as Tibetan Plateau forcing) besides WNPSH.

Heavy precipitation

The urban heat island effect would increase the storm rainfall over the Guangzhou megacity area (Meng *et al.*, 2007). Due to the anthropogenic activities, the air pollutants concentrations increased in the Pearl River Delta region, which not only increased the precipitation rate, but also enhanced the area of heavy precipitation (Wang Y. *et al.*, 2011). Holst *et al.* (2016) used the WRF model to simulate a heavy precipitation process and added the anthropogenic heat (AH) into the surface sensible heat flux function. They found when AH reached a certain large value, the precipitation over the city, especially the heavy precipitation would increase. In a city with huge size, the increase in heavy precipitation will be more obvious (Holst *et al.*, 2017).

1.3 Future Projection of Climate Change

1.3.1 Future projection of basic climate elements

Hong Kong

The Hong Kong Observatory mainly uses statistical downscaling methods to estimate future climate change (except for sea level projection). For details in the climate projection methods, please refer to previous studies (Lee *et al.*, 2011; Chan *et al.*, 2014; Cheung M. *et al.*, 2015; Chan *et al.*, 2016; Tong *et al.*, 2017).

Figure 1–10a shows the projection of annual average temperature in Hong Kong under the medium-low greenhouse gas concentration (RCP4.5) and high greenhouse gas concentration (RCP8.5) scenarios. Under the RCP8.5 scenario, it is expected that by the end of the century (2091～2100), the annual average temperature in Hong Kong will be about 3℃～6℃ higher than the average of 1986～2005. The increases in annual average temperature under the RCP4.5 and RCP6.0 scenarios are smaller (Chan *et al.*, 2014).

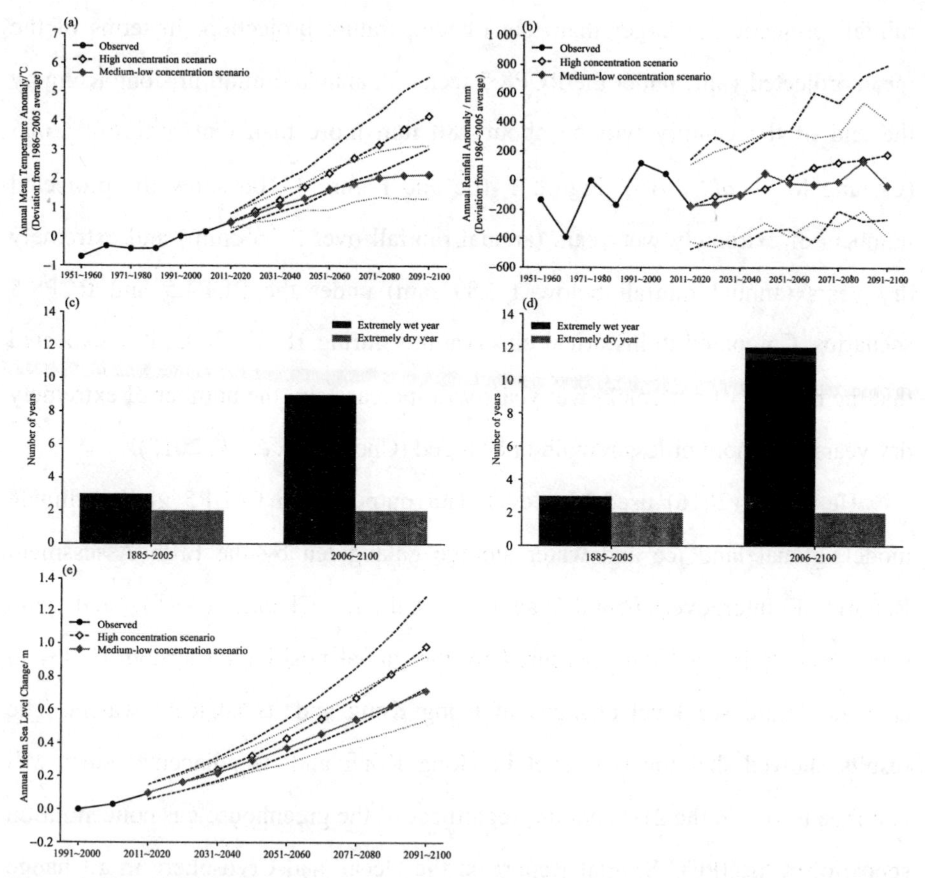

Figure 1–10 (a) Projections of annual average temperature anomaly in Hong Kong (relative to the 1986 ~ 2005 average) under RCP8.5 and RCP4.5 scenarios (lines with symbols are mean values, dashed lines show the likely range of projections). Historical observations are shown in black line with solid dots; (b) Projections of annual rainfall anomaly in Hong Kong (relative to the 1986 ~ 2005 average) under RCP8.5 and RCP4.5 scenarios. Future changes in extremely wet years and extremely dry years under RCP4.5; (c) and RCP8.5; (d) scenarios; (e) Projected changes in the mean sea level in Hong Kong and its adjacent waters (relative to the 1986 ~ 2005 average) under RCP8.5 and RCP4.5 scenarios

Figure 1–10b shows the projections of Hong Kong annual rainfall under the RCP4.5 and RCP8.5 scenarios. As shown by the likely range of projections, there are considerable inter-model differences in the projection. The uncertainty of rainfall projection is larger than that of temperature projection. In terms of the mean projected value under the RCP8.5 scenario, annual rainfall in Hong Kong by the end of the century will be about 180 mm more than that in 1986～2005 (Cheung M. *et al.*, 2015). Figure 1–10c and Figure 1–10d show the projected numbers of extremely wet years (annual rainfall over 3 168 mm) and extremely dry years (annual rainfall below 1 289 mm) under the RCP4.5 and RCP8.5 scenarios. Compared to historical observations during 1885～2005, it is expected that the number of extremely wet years will increase and the number of extremely dry years will more or less remain unchanged (Cheung M. *et al.*, 2015).

He *et al.* (2016) used sea level data output from CMIP5 global climate model, global land ice and water storage data given by the Fifth Assessment Report of Intergovernmental Panel on Climate Change (AR5), and land subsidence in Hong Kong measured by the global positioning system (GPS) to estimate future sea level changes in Hong Kong and its adjacent waters. The results showed that the sea level in Hong Kong and its adjacent waters will continue to rise in the 21st century, regardless of the greenhouse gas concentration scenario. In the IPCC Special Report on the Ocean and Cryosphere in a Change Climate (SROCC) published in September 2019, a higher projection of future global sea level rise was issued. The projection of future sea level rise in Hong Kong and its adjacent waters was updated accordingly based on the latest data provided by SROCC. Under the RCP8.5 scenario, the mean sea level in Hong Kong and its adjacent waters in 2091～2100 (with the effect of local vertical land displacement incorporated) will be 0.73 m～1.28 m higher than the 1986～2005 average (Figure 1–10e). As the mean sea level rises, the threat of storm surges from tropical cyclones will increase accordingly.

Macao

Tang (2008) used the different greenhouse gas emission scenarios adopted by the IPCC Fourth Assessment Report (emission scenarios from low to high: B1, A1B, A2) and grid data from climate model prediction, and combined with historical temperature data in Macao and Southern China. He predicted that the average temperature in Macao at the end of the 21st century (2091～2100) is likely to be 1.9℃～3.4℃ higher than the average of 1971～2000, and the average increase in temperature in multi-model and multi-scenario is 2.7℃. In terms of seasonal changes, four seasons shows a warming trend, the average increase in temperature in multi-model and multi-scenario is 2.9℃, the minimum increase was in summer about 1.9℃. The city-scale temperature projection for Macao indicates that by the end of the 21st century, with the significant reduction in cold days, the temperature increases by approximately 2.7℃ (Tang *et al.*, 2008).

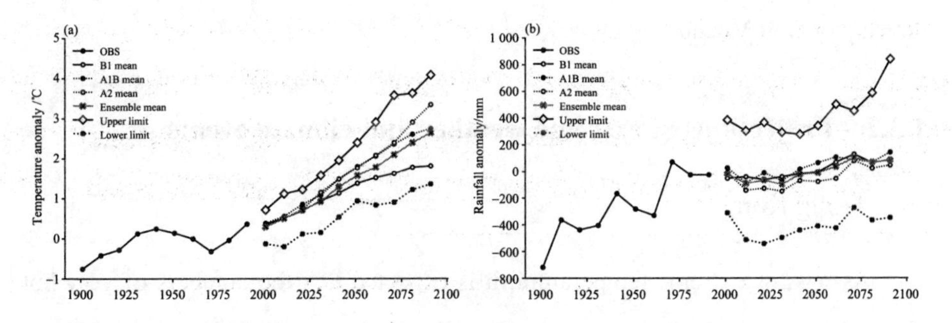

Figure 1–11 Future projections of 21st century annual (a) surface air temperature and (b) precipitation (relative to 1971 ~ 2000) in Macao (Tang, 2008)

Tang (2008) predicted that the annual precipitation in Macao (multi-model and multi-scenario average, emission scenarios from low to high: B1, A1B, A2) showed a decreasing trend before 2040, and increase after 2040. However, the results of different models are different, and the uncertainty is high. From the average of multi-model and multi-scenarios, the precipitation is less than the average of 1971～2000 in spring and winter, especially for winter, the decrease in precipitation is significant. The precipitation in summer and autumn is increasing after 2040 and 2050.

It is predicted that the rate of increase in sea level in Macao will be 20% faster than the global average in the future, mainly because the oceans near Macao are warming faster than the global average, and the enhancement of southerly winds in the northern part of the South China Sea lead to the seawater accumulation. By 2020, 2060, and 2100, sea level rise can reach 8 cm～12 cm, 22 cm～51 cm, and 35 cm～118 cm, respectively, depending on greenhouse gas concentration scenario and climate sensitivity (Wang L. *et al.*, 2016). The high sea level has intensified disasters such as storm surges, floods, coastal erosion, and seawater intrusion, which have had a certain impact on the social and economic development in Macao.

1.3.2 Projections of extreme weather and climate events

Hong Kong

As regards extreme temperature, it is expected that the numbers of very hot days and hot nights will increase significantly. Under the RCP8.5 scenario and by the end of the century (2091～2100), the average annual number of very hot days will exceed 100 and the average annual number of hot nights will be about 150 while the average annual number of cold days will decrease significantly (Figure 1–12a and 12b). Human comfort is related to temperature and humidity. Wet bulb temperature is a basic indicator of heat stress. High wet bulb temperature indicates that the environment is warm and humid, and heat dissipation of the human body is difficult. The projection of daily maximum wet bulb temperature shows that regardless of the greenhouse gas concentration scenario, the annual number of extremely warm-and-humid days (daily maximum wet bulb temperature at 28.2℃ or above) and the annual maximum number of consecutive extremely warm-and-humid days will increase in Hong Kong in the 21st century. The increase is most pronounced under the RCP8.5 scenario (Tong *et al.*, 2017) (Figure 1–12c and 12d).

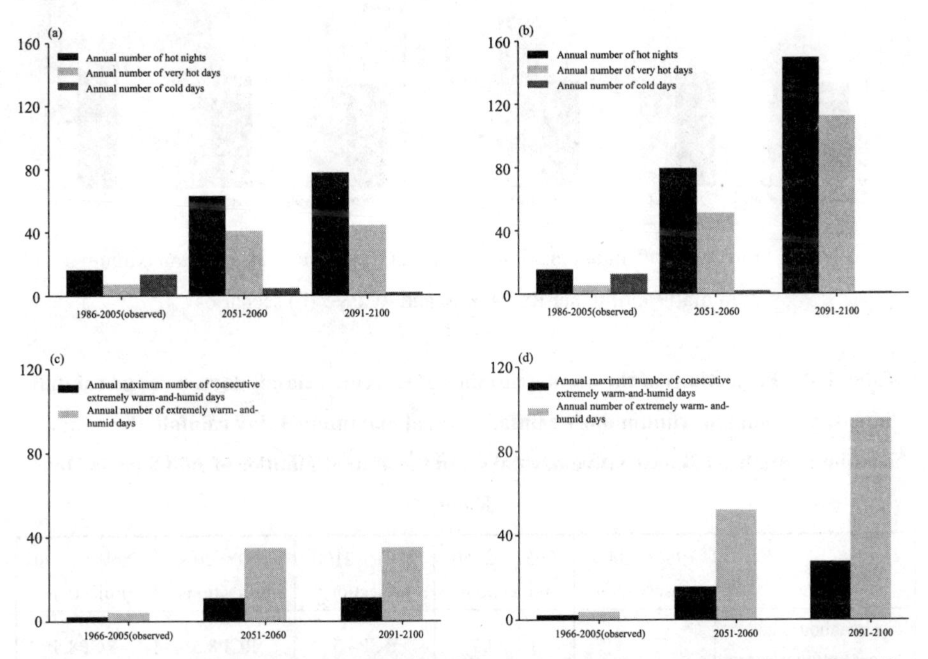

Figure 1–12 Projected changes in the annual number of very hot days, hot nights and cold days in Hong Kong under RCP4.5 (a) and RCP8.5 (b) scenarios. Projected annual number of extremely warm-and-humid days and annual maximum number of consecutive extremely warm-and-humid days in Hong Kong under RCP4.5 (c) and RCP8.5 (d) scenarios

As regards extreme rainfall, under the RCP8.5 scenario, the number of extreme rainfall days (daily rainfall at 100 mm or above) will increase from an average of 4.2 days per year in 1986～2005 to about 5.1 days per year by the end of the century (2091～2100). The average rainfall intensity (i.e., annual rainfall divided by the number of rain days), the annual maximum daily rainfall, the annual maximum 3-day rainfall, and the annual maximum number of consecutive dry days will increase, but the number of rain days will decrease (Chan *et al.*, 2016) (Figure 1–13 and Table 1–4).

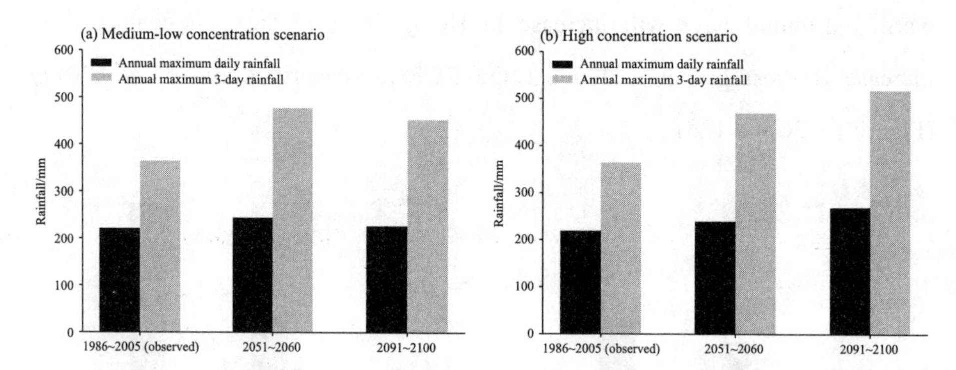

Figure 1–13 Projections of annual maximum daily rainfall (mm) and annual maximum 3-day rainfall (mm) under RCP4.5 (a) and RCP8.5 (b) scenarios

Table 1–4 Projections of the annual number of extreme rainfall days, average rainfall intensity, annual maximum daily rainfall, annual maximum 3-day rainfall, the annual maximum number of consecutive dry days and the annual number of rain days in Hong Kong

	1986～2005 observation	2051～2060 projections	2091～2100 projections	2051～2060 projections	2091～2100 projections
Greenhouse gas concentration scenario	-	RCP4.5	RCP4.5	RCP8.5	RCP8.5
The annual number of extreme rainfall days	4.2	4.5	4.4	5.0	5.1

Continued

	1986～2005 observation	2051～2060 projections	2091～2100 projections	2051～2060 projections	2091～2100 projections
Average rainfall intensity/mm/day	23.4	25.0	24.0	25.4	26.7
Annual maximum daily rainfall/ mm	221	246	228	243	273
Annual maximum 3-day rainfall/ mm	367	482	454	476	523
Annual maximum number of consecutive dry days	46	49	52	54	59
The annual number of rain days	102	103	102	100	97

The resolution of most climate models is not enough to predict the number of tropical cyclones affecting Hong Kong and Macao in the future. The latest assessment from the expert team of the UNESCAP/WMO Typhoon Committee indicated that the intensity, proportion of intense tropical cyclone and rainfall rate of tropical cyclones in the western North Pacific will increase in the future (Cha *et al.*, 2020). Yukoi and Takayabu (2009) analyzed five CMIP3 models and found that the frequency of typhoons in the South China Sea will decrease significantly under different scenarios. Another study has similar results (Wang R. *et al.*, 2011).

One downscale simulation system was used to make projections for three 10-year periods: 2030～2039, 2060～2060, and 2090～2099. It is found that the frequency of tropical cyclone landing at South China will likely decrease (Figure 1–14a), which is consistent with the conclusion of previous studies on tropical cyclone activity in the South China Sea. The probability density function of intensity (represented by APDI) shows that the closer to the end of the century, the greater the probability that tropical cyclones will become stronger (Figure 1–14b)

(Lok and Chan, 2018). With sea level rise and increase in tropical cyclone intensity in the future, the risk of storm surge and coastal inundation due to tropical cyclones will increase (Cha *et al.*, 2020; Chen *et al.*, 2020).

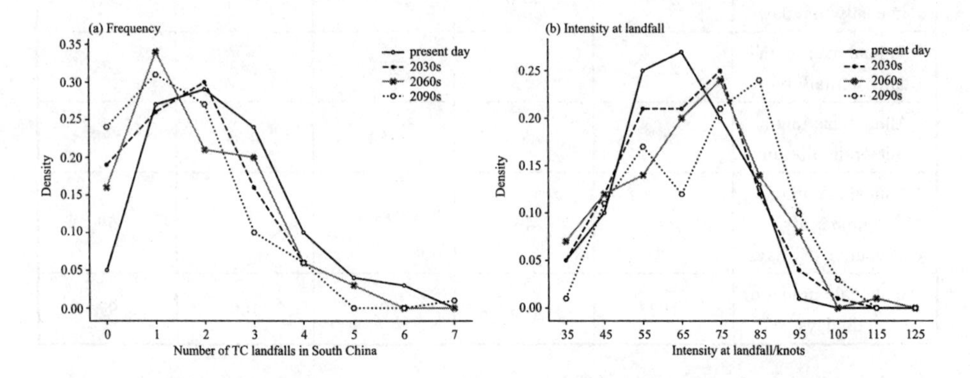

Figure 1–14 Probability density functions of (a) number of tropical cyclones landfalls in Southern China, (b) tropical cyclone maximum sustained wind speeds (knots) at the time of landfall in Southern China (Lok and Chan, 2018)

References

Cha, E. J., T. R. Knutson, T. C. Lee, *et al.*, 2020. Third Assessment on Impacts of Climate Change on Tropical Cyclones in the Typhoon Committee Region – Part Ⅱ: Future Projections. *Tropical Cyclone Research and Review*, 9(2).

Chan, H. S., M. H. Kok, T. C. Lee, 2012. Temperature Trends in Hong Kong from a Seasonal Perspective. *Climate Research*, 55(1).

Chan, H. S., H. W. Tong, S. M. Lee, 2014. Temperature Projection for Hong Kong in the 21st Century Using CMIP5 Models. *Hong Kong Meteorological Society Bulletin*, 24.

Chan, H. S., H. W. Tong, S. M. Lee, 2016. Extreme Rainfall Projection for Hong Kong in the 21st Century Using CMIP5 Models. 30th Guangdong-Hong Kong-Macao Seminar on Meteorological Science and Technology, Guangzhou, April 20-22, HKO Reprint No. 1222.

Chan, J. C. L., W. M. Gray, 1982. Tropical Cyclone Movement and Surrounding Flow Relationships. *Monthly Weather Review*, 110(10).

Chen, J. L., Z. Q. Wang, C. Y. Tam, *et al.,* 2020. Investigating Climate Change on Tropical Cyclones and Induced Storm Surges in the Pearl River Delta Region using Pseudo-Global Warming Method. *Scientific Reports*, 10.

Chen, L. X., X. J. Zhou, W. L. Li, *et al.*, 2004. Characteristics of Climate Change and Its Formation Mechanism in China in the Last 80 Years. *Acta MeteorologicaSinica*, 62(5) (in Chinese).

Cheng, C., Y. M. Zhu, H. X. Ding, *et al.*, 2017. The Interdecadal Shift of Summer Precipitation and Atmospheric Circulation over East China and Its Relationship with PDO. *Journal of the Meteorological Sciences*, 37(4) (in Chinese).

Cheung, H. H. N., W. Zhou, 2015. Implications of Ural Blocking for East Asian Winter Climate in CMIP5 GCMs. Part II: Projection and Uncertainty in Future Climate Conditions. *Journal of Climate*, 28(6).

Cheung, H. H. N., W. Zhou, S. M. Lee, *et al.*, 2015. Interannual and Interdecadal Variability of the Number of Cold Days in Hong Kong and Their Relationship with Large-Scale Circulation. *Monthly Weather Review*, 143(4).

Cheung, H. H. N., W. Zhou, M. Y. T. Leung, *et al.*, 2016. A Strong Phase Reversal of the Arctic Oscillation in Midwinter 2015/2016: Role of the Stratospheric Polar Vortex and Tropospheric Blocking. *Journal of Geophysical Research-Atmospheres*, 121(22).

Cheung, M. S., H. S. Chan, H. W. Tong, 2015. Rainfall Projection for Southern China in the 21st Century Using CMIP5 Models. 29th Guangdong-Hong Kong-Macao Seminar on Meteorological Science and Technology, Macao, January 20-22, HKO Reprint No. 1165.

Ding, Y. H., Y. Sun, Y. Y. Liu, *et al.*, 2013. Interdacadal and Interannual Variabilitites of the Asian Summer Monsoon and Its Projection of Future Change. *Chinese Journal of Atmospheric Sciences*, 37(2) (in Chinese).

Fung, S. K., C. S. Wu, T. Wang, *et al.*, 2010. Multiple Time Scales Analysis of Climate Variation in Macao During the Last 100 Years. *Journal of Tropical Meteorology*, 26(4) (in Chinese).

Fung, W. Y., S. W. Chan, K. H. Tam, *et al.*, 2011. Analysis and Measurement of Outdoor CO_2 Concentration in Hong Kong. 25th Guangdong-Hong Kong-Macao Seminar on Meteorological Science and Technology, Hong Kong, January 26-28, HKO Reprint No. 952 (in Chinese).

Ginn, W. L., T. C. Lee, K. Y. Chan, 2010. Past and Future Changes in the Climate of Hong Kong. *Acta Meteorologica Sinica*, 24(2).

Gu, D. J., Z. P. Ji, A. L. Lin, 2018. The Key Oceanic Regions Affecting Interannual Variation of Onset Date of SCSSM and Their Preliminary Mechanisms. *Journal of Tropical Meteorology*, 34(1) (in Chinese).

Han, Z., F. F. Luo, S. L. Li, *et al.*, 2016. Simulation by CMIP5 Models of the Atlantic Multidecadal Oscillation and Its Climate Impacts. *Advances in Atmospheric Sciences*, 33(12).

He, C., A. L. Lin, D. J. Gu, *et al.*, 2018. Using Eddy Geopotential Height to Measure the Western North Pacific Subtropical High in a Warming Climate. *Theoretical and Applied Climatology*, 131.

He, C., T. J. Zhou, A. L. Lin, *et al.*, 2015. Enhanced or Weakened Western North Pacific Subtropical High under Global Warming? *Scientific Reports*, 5.

He, C., Z. Q. Wang, T. J. Zhou, *et al.*, 2019. Enhanced Latent Heating over the Tibetan Plateau as a Key to the Enhanced East Asian Summer Monsoon Circulation under a Warming Climate. *Journal of Climate*, 32(11).

He, C., W. Zhou, 2020. Different Enhancement of the East Asian Summer Monsoon under Global Warming and Interglacial Epochs Simulated by CMIP6 Models: Role of the Subtropical High. *Journal of Climate*, 33(22).

He, J. H., Z. W. Zhu, 2015. The Relation of South China Sea Monsoon Onset with the Subsequent Rainfall over the Subtropical East Asia. *International Journal of Climatology*, 35(15).

He, Y. H., H. Y. Mok, E. S. T. Lai, 2016. Projection of Sea-Level Change in the Vicinity of Hong Kong in the 21st Century. *International Journal of Climatology*, 36(9).

Holst, C. C., J. C. L. Chan, C. Y. Tam, 2017. Sensitivity of Precipitation Statistics to Urban Growth in a Subtropical Coastal Megacity Cluster. *Journal of Environmental Sciences*, 59.

Holst, C. C., C. Y. Tam, J. C. L. Chan, 2016. Sensitivity of Urban Rainfall to Anthropogenic Heat Flux: A Numerical Experiment. *Geophysical Research Letters*, 43(5).

Hong, C. C., H. H. Hsu, H. H. Chia, *et al.*, 2008. Decadal Relationship Between the North Atlantic Oscillation and Cold Surge Frequency in Taiwan. *Geophysical Research Letters*, 35.

Hong, C. C., T. Li, 2009. The Extreme Cold Anomaly over Southeast Asia in February 2008: Roles of ISO and ENSO. *Journal of Climate*, 22(13).

Hong Kong Environmental Protection Department, 2018. *Emission Trends*. https://www.epd.gov.hk/epd/english/environmentinhk/air/data/emission_inve.html#1.

Huang, J. P., J. C. H. Fung, A. K. H. Lau, *et al.*, 2005. Numerical Simulation and Process Analysis of Typhoon-Related Ozone Episodes in Hong Kong. *Journal of Geophysical Research-Atmospheres*, 110.

Huang, Y. Y., H. J. Wang, K. Fan, *et al.*, 2015. The Western Pacific Subtropical High after the 1970s: Westward or Eastward Shift? *Climate Dynamics*, 44.

Jeong, J. H., C. H. Ho, B. M. Kim, *et al.*, 2005. Influence of the Madden-Julian Oscillation on Wintertime Surface Air Temperature and Cold Surges in East Asia. *Journal of Geophysical*

Research-Atmospheres, 110.

Jian, M. Q., M. Peng, X. Luo, 2018. Coheren Leading Modes of Precipitation Variation in Late Spring and Early Summer over the South China Sea and Surrounding Area and Their Mechanisms. *Acta ScientiarumNaturalium Universitatis Sunyatseni*, 57(5) (in Chinese).

Jiang, D. B., Z. P. Tian, 2013. East Asian Monsoon Change for the 21st Century: Results of CMIP3 and CMIP5 Models. *Chinese Science Bulletin*, 58(12).

Kajikawa, Y., B. Wang, 2012. Interdecadal Change of the South China Sea Summer Monsoon Onset. *Journal of Climate*, 25(9).

Kamae, Y., M. Watanabe, M. Kimoto, *et al.*, 2014. Summertime Land-Sea Thermal Contrast and Atmospheric Circulation over East Asia in a Warming Climate. Part I: Past Changes and Future Projections. *Climate Dynamics*, 43(9).

Kim, B. M., S. W. Son, S. K. Min, *et al.*, 2014. Weakening of the Stratospheric Polar Vortex by Arctic Sea-Ice Loss. *Nature Communications*, 5.

Kim, J. H., C. H. Ho, H. S. Kim, *et al.*, 2012. 2010 Western North Pacific Typhoon Season: Seasonal Overview and Forecast Using a Track-Pattern-Based Model. *Weather and Forecasting*, 27(3).

Kitoh, A., 2017. The Asian monsoon and Its Future Change in Climate Models: A Review. *Journal of the Meteorological Society of Japan*, 95(1).

Knutson, T., S. J. Camargo, J. C. L. Chan, *et al.*, 2019. Tropical Cyclones and Climate Change Assessment. Part I: Detection and Attribution. *Bulletin of the American Meteorological Society*, 100(10).

Kossin, J. P., 2018. A Global Slowdown of Tropical-Cyclone Translation Speed. *Nature*, 558(7708).

Kossin, J. P., K. A. Emanuel, S. J. Camargo, 2016. Past and Projected Changes in Western North Pacific Tropical Cyclone Exposure. *Journal of Climate*, 29(16).

Kossin, J. P., K. A. Emanuel, G. A. Vecchi, 2014. The Poleward Migration of the Location of Tropical Cyclone Maximum Intensity. *Nature*, 509(7500).

Lai, Y. C., J. F. Li, X. Gu, *et al.*, 2020. Greater Flood Risks in Response to Slowdown of Tropical Cyclones over the Coast of China. *Proceedings of the National Academy of Sciences of the United States of America*, 117(26).

Lam, K. S., T. J. Wang, C. L. Wu, *et al.*, 2005. Study on an Ozone Episode in Hot Season in Hong Kong and Transboundary Air Pollution over Pearl River Delta Region of China. *Atmospheric Environment*, 39(11).

Lee, B. Y., W. T. Wong, W. C. Woo, 2010. Sea-level Rise and Storm Surge: Impacts of Climate Change on Hong Kong. HKIE Civil Division Conference, Hong Kong, April 12-14, HKO Reprint No. 915.

Lee, T. C., K. Y. Chan, W. L. Gin, 2011. Projection of Extreme Temperature in Hong Kong in the 21st Century. *Acta Meteorologica Sinica*, 25(1).

Lee, T. C., T. R. Knutson, H. Kamahori, *et al.*, 2012a. Impacts of Climate Change on Tropical Cyclones in the Western North Pacific Basin. Part I: Past Observations. *Tropical Cyclone Research and Review*, 1(2).

Lee, T. C., T. R. Knutson, T. Nakaegawa, *et al.*, 2020. Third Assessment on Impacts of Climate Change on Tropical Cyclones in the Typhoon Committee Region. Part I: Observed Changes, Detection and Attribution. *Tropical Cyclone Research and Review*, 9(1).

Lee, T. C., C. Y. Y. Leung, M. H. Kok, *et al.*, 2012b. The Long Term Variations of Tropical Cyclone Activity in the South China Sea and the Vicinity of Hong Kong. *Tropical Cyclone Research and Review*, 1(3).

Leung, M. Y. T., H. H. N. Cheung, W. Zhou, 2015. Energetics and Dynamics Associated with Two Typical Mobile trough Pathways over East Asia in Boreal Winter. *Climate Dynamics*, 44.

Leung, M. Y. T., H. H. N. Cheung, W. Zhou, 2017. Meridional Displacement of the East Asian Trough and its Response to the ENSO Forcing. *Climate Dynamics*, 48.

Leung, M. Y. T., W. Zhou, 2016. Direct and Indirect ENSO Modulation of Winter Temperature over the Asian-Pacific-American Region. *Scientific Reports*, 6.

Leung, Y. K., M. C. Wu, K. K. Yeung, 2008. A Study on the Relationship Among Visibility, Atmospheric Suspended Particulate Concentration and Meteorological Conditions in Hong Kong. *Acta MeteorologicaSinica*, 66(3) (in Chinese).

Li, C. H., W. J. Pan, X. Li, *et al.*, 2017. The Variation Characteristics of 10～30 Days Oscillation Intensity of Spring over the South China Sea-Western Pacific Ocean and Their Effects on the South China Sea Summer Monsoon Onset. *Journal of Tropical Meteorology*, 33(1) (in Chinese).

Li, Q. Q., Y. H. Duan, 2010. Tropical Cyclone Strikes at the Coastal Cities of China from 1949 to 2008. *Meteorology and Atmospheric Physics*, 107.

Li, R. C. Y., W. Zhou, T. C. Lee, 2015. Climatological Characteristics and Observed Trends of Tropical Cyclone-Induced Rainfall and Their Influences on Long-Term Rainfall Variations in Hong Kong. *Monthly Weather Review*, 143(6).

Li, R. C. Y., W. Zhou, C. M. Shun, *et al.*, 2017. Change in Destructiveness of Landfalling Tropical Cyclones over China in Recent Decades. *Journal of Climate*, 30(9).

Lin, A. L., D. J. Gu, B. Zheng, *et al.*, 2013. Relationship Between South China Sea Summer Monsoon Onset and Southern Ocean Sea Surface Temperature Variation. *Chinese Journal of Geophysics*, 56(2) (in Chinese).

Liu, K. S., J. C. L. Chan, 2017. Variations in the Power Dissipation Index in the East Asia Region. *Climate Dynamics*, 48.

Lok, C. C. F., J. C. L. Chan, 2018. Changes of Tropical Cyclone Landfalls in South China Throughout the Twenty-First Century. *Climate Dynamics*, 51.

Lu, C. H., Z. Y. Guan, Y. H. Li, *et al.*, 2013. Interdecadal Linkages Between Pacific Decadal Oscillation and Interhemispheric Air Mass Oscillation and Their Possible Connections with East Asian Monsoon. *Chinese Journal of Geophysics*, 56(4) (in Chinese).

Luo, D. H., X. D. Chen, A. G. Dai, *et al.*, 2018. Changes in Atmospheric Blocking Circulations Linked with Winter Arctic Warming: A New Perspective. *Journal of Climate*, 31(18).

Luo, D. H., Y. Q. Xiao, Y. N. Diao, *et al.*, 2016. Impact of Ural Blocking on Winter Warm Arctic-Cold Eurasian Anomalies. Part II: The Link to the North Atlantic Oscillation. *Journal of Climate*, 29(11).

Ma, T. T., Z. W. Wu, Z. H. Jiang, 2012. How Does Coldwave Frequency in China Respond to a Warming Climate? *Climate Dynamics*, 39.

Man, W. M., T. J. Zhou, J. H. Jungclaus, 2012. Simulation of the East Asian Summer Monsoon During the Last Millennium with the MPI Earth System Model. *Journal of Climate*, 25(22).

Mei, W., S. P. Xie, 2016. Intensification of Landfalling Typhoons over the Northwest Pacific Since the Late 1970s. *Nature Geoscience*, 9.

Meng, W. G., J. H. Yan, H. B. Hu, 2007. Urban Effects and Summer Thunderstorms in a Tropical Cyclone Affected Situation over Guangzhou City. *Science in China Series D-Earth Sciences*, 50(12).

Mi, J. T., 2010. Study on the Landscape Pattern Change and Urban Heat Island in Macao. Beijing Forestry University, Doctoral Dissertation (in Chinese).

Mok, H. Y., Y. K. Leung, T. C. Lee, *et al.*, 2006. Regional Rainfall Characteristics of Hong Kong over the Past 50 Years. Conference on "Changing Geography in a Diversified World", Hong Kong, June 1-3, HKO Reprint No. 646.

Moon, I. J., S. H. Kim, J. C. L. Chan, 2019. Climate Change and Tropical Cyclone Trend. *Nature*, 570 (7759).

Mori, M., M. Watanabe, H. Shiogama, *et al.*, 2014. Robust Arctic Sea-Ice Influence on the Frequent Eurasian Cold Winters in Past Decades. *Nature Geoscience*, 7(12).

Park, T. W., C. H. Ho, S. Yang, 2011. Relationship Between the Arctic Oscillation and Cold Surges over East Asia. *Journal of Climate*, 24(1).

Peng, L., J. -P. Liu, Y. Wang, *et al.*, 2018. Wind Weakening in a Dense High-Rise City due to Over Nearly Five Decades of Urbanization. *Building and Environment*, 138.

Qian, C., W. Zhou, S. K. Fong, *et al.*, 2015. Two Approaches for Statistical Prediction of Non-Gaussian Climate Extremes: A Case Study of Macao Hot Extremes During 1912-2012. *Journal of Climate*, 28(2).

Ren, X. J., X. Q. Yang, X. G. Sun, 2013. Zonal Oscillation of Western Pacific Subtropical High and Subseasonal SST Variations During Yangtze Persistent Heavy Rainfall Events. *Journal of Climate*, 26(22).

Shao, X., P. Huang, R. H. Huang, *et al.*, 2014. A Review of the South China Sea Summer Monsoon Onset. *Advances in Earth Science*, 29(10) (in Chinese).

Shao, X., P. Huang, R. H. Huang, *et al.*, 2015. Predictability Analysis on Onset of South China Sea Summer Monsoon Based on the Sea Surface Temperature Anomaly. *Journal of the Meteorological Sciences*, 35(6) (in Chinese).

Takaya, K., H. Nakamura, 2005. Geographical Dependence of Upper-Level Blocking Formation Associated with Intraseasonal Amplification of the Siberian High. *Journal of the Atmospheric Sciences*, 62(12).

Tang, I. M., 2008. The Projection of Climate Change Tendency in Macao, Macao Meteorological and Geophysical Bureau (in Chinese).

Tang, I. M., S. K. Fung, K. H. Tam, *et al.*, 2008. Climate Change in Macao and the Tendency in the 21st Century. Workshop on Climate Change and Climate Prediction in the Pearl River Delta Region, Hong Kong, December 15-16 (in Chinese).

Tong, H. W., W. C. Lee, 2014. Observed Long-Term Trend in the Upper Air over Hong Kong. 28th Guangdong-Hong Kong-Macao Seminar on Meteorological Science and Technology, Hong Kong, January 13-15, HKO Reprint No. 1122.

Tong, H. W., C. P. Wong, S. M. Lee, 2017. Projection of Wet-Bulb Temperature for Hong Kong in the 21st Century Using CMIP5 Data. The 31st Guangdong-Hong Kong-Macao Seminar on Meteorological Science and Technology and the 22nd Guangdong-Hong Kong-Macao Meeting on Cooperation in Meteorological Operations, Hong Kong, February 27-March 1, HKO Reprint No. 1276.

Wang, J. L., B. Yang, F. C. Ljungqvist, *et al.*, 2013. The Relationship Between the Atlantic Multidecadal Oscillation and Temperature Variability in China During the Last Millennium. *Journal of Quaternary Science*, 28(7).

Wang, L., G. H. Chen, R. H. Huang, 2009. Spatiotemporal Distributive Characteristics of Tropical Cyclone Activities over the Northwest Pacific in 1976-2006. *Journal of Nanjing Institute of Meteorology*, 32(2) (in Chinese).

Wang, L., W. Chen, 2014. The East Asian Winter Monsoon: Re-Amplification in the Mid-2000s. *Chinese Science Bulletin*, 59(4).

Wang, L., G. Huang, W. Zhou, *et al.*, 2016. Historical Change and Future Scenarios of Sea Level Rise in Macau and Adjacent Waters. *Advances in Atmospheric Sciences*, 33(4).

Wang, Q. Q., Y. M. Feng, X. H. H. Huang, *et al.*, 2016. Nonpolar Organic Compounds as $PM_{2.5}$ Source Tracers: Investigation of Their Sources and Degradation in the Pearl River Delta, China. *Journal of Geophysical Research-Atmospheres*, 121(19).

Wang, Q. Q., X. H. H. Huang, F. C. V. Tam, *et al.*, 2019. Source Apportionment of Fine Particulate Matter in Macao, China with and without Organic Tracers: A Comparative Study Using Positive Matrix Factorization. *Atmospheric Environment*, 198.

Wang, R. F., L. G. Wu, C. Wang, 2011. Typhoon Track Changes Associated with Global Warming. *Journal of Climate*, 24(14).

Wang, W. W., W. Zhou, S. K. Fong, *et al.*, 2015. Extreme Rainfall and Summer Heat Waves in Macau Based on Statistical Theory of Extreme Values. *Climate Research*, 66(1).

Wang, W. W., W. Zhou, E. Y. Y. Ng, *et al.*, 2016. Urban Heat Islands in Hong Kong: Statistical Modeling and Trend Detection. *Natural Hazards*, 83(2).

Wang, Y., Q. Wan, W. Meng, *et al.*, 2011. Long-Term Impacts of Aerosols on Precipitation and Lightning over the Pearl River Delta Megacity Area in China. *Atmospheric Chemistry and Physics*, 11(23).

Wong, M. C., H. Y. Mok, T. C. Lee, 2011. Observed Changes in Extreme Weather Indices in Hong Kong. *International Journal of Climatology*, 31(15).

Wong, W. T., K. W. Li, K. H. Yeung, 2003. Long Term Sea Level Change in Hong Kong. *Hong Kong Meteorological Society Bulletin*, 13.

Wu, B. Y., J. Z. Su, R. H. Zhang, 2011. Effects of Autumn-Winter Arctic Sea Ice on Winter Siberian High. *Chinese Science Bulletin*, 56(30).

Wu, H. Y., Y. D. Du, W. J. Pan, 2011. The Change Characteristics of Sunshine Duration During the Past 48 Years in South China. *Acta ScientiarumNaturalium Universitatis Sunyatsen*i, 50(6) (in Chinese).

Wu, L. G., B. Wang, S. Q. Geng, 2005. Growing Typhoon Influence on East Asia. *Geophysical Research Letters*, 32(18).

Wu, L. G., C. Wang, 2015. Has the Western Pacific Subtropical High Extended Westward Since the Late 1970s? *Journal of Climate*, 28(13).

Wu, M. C., W. H. Leung, 2009. Effect of ENSO on the Hong Kong Winter Season. *Atmospheric Science Letters*, 10(2).

Wu, Z. W., J. P. Li, Z. H. Zhang, *et al.*, 2011. Predictable Climate Dynamics of Abnormal East Asian Winter Monsoon: Once-in-a-Century Snowstorms in 2007/2008 Winter. *Climate Dynamics*, 37(7-8).

Xie, B. 2016. Study on Short-lived Climate Pollutants (SLCPs) Effective Radiative Forcing and Their Effects on Global Climate. Lanzhou University, Doctoral Dissertation (in Chinese).

Xu, Y. F., L. X. Peng, J. Li, *et al.*, 2017. Relationship Between PDO and Interdecadal Change of Winter Temperature in China from 1951 to 2013. *Meteorological Science and Technology*, 45(4) (in Chinese).

Yang, S. L., Z. L. Ding, Y. Y. Li, *et al.*, 2015. Warming-Induced Northwestward Migration of the East Asian Monsoon Rain Belt from the Last Glacial Maximum to the Mid-Holocene. *Proceedings of the National Academy of Sciences of the United States of America*, 112(43).

Yang, Y. H., M. Ying, B. D. Chen, 2009. The Climatic Changes of Landfall Tropical Cyclones in China over the Past 58 Years. *Acta MeteorologicaSinica*, 67(5) (in Chinese).

Yokoi, S., Y. N. Takayabu, 2009. Multi-Model Projection of Global Warming Impact on Tropical Cyclone Genesis Frequency over the Western North Pacific. *Journal of the Meteorological Society of Japan*, 87(3).

Zhang, L. P., D. L. Li, X. Li, 2014. Definition of Onset and Retreat Date of the South China Sea Summer Monsoon and the Relation with Its Strength Under the Background of Climate Warming. *Journal of Tropical Meteorology*, 30(6) (in Chinese).

Zhang, Q., W. Zhang, X. Q. Lu, *et al.*, 2011. Landfalling Tropical Cyclones Activities in the South China: Intensifying or Weakening? *International Journal of Climatology*, 32(12).

Zhou, W., J. C. L. Chan, W. Chen, *et al.*, 2009. Synoptic-Scale Controls of Persistent Low Temperature and Icy Weather over Southern China in January 2008. *Monthly Weather Review*, 137(11).

Zhou, W., X. Wang, T. J. Zhou, *et al.*, 2007. Interdecadal Variability of the Relationship Between the East Asian Winter Monsoon and ENSO. *Meteorology and Atmospheric Physics*, 98.

Zhu, C. W., T. Nakazawa, J. P. Li, *et al.*, 2003. The 30-60 day Intraseasonal Oscillation over the Western North Pacific Ocean and Its Impacts on Summer Flooding in China During 1998. *Geophysical Research Letters*, 30(18).

Chapter 2 Climate Change Impact, Risk, and Adaptation

Hong Kong and Macao are important cities with dense population, rapid economic development, and high level of urbanization. Hong Kong-Macao region is located in the coasts of South China Sea, which stretches on the two sides of the Pearl River Delta and the Pearl River Estuary. This region is dominated by subtropical monsoon climate with significant seasonality of precipitation, and is frequently affected by natural hazards, such as typhoon, rainstorm, flooding, and heatwave. Due to this geographical setting combined with rapid socio-economic development, people and properties are highly exposed to natural hazards, and the impacts and risks of the damages are high and serious, threatening the public health and safety of this region. Under climate change, the hazards become more frequent and extreme, leading to greater challenges for the prediction, protection, and adaptation of natural hazards in this region. The negative impacts on environmental quality and public health generate uncertainties in the long-term regional sustainability in economic, social and environmental aspects.

The history of weather observations started fairly early and has continued for over one century. For example, Hong Kong Observatory (HKO) has conducted meteorological measurements since 1884, and accumulated long-term weather data as a crucial foundation for studying regional climate change. Observations of the past 100 years show that climate change has caused extensive impacts in the Hong Kong-Macao region, and various hydroclimatic characteristics and events have changed significantly, such as precipitation, temperature, typhoon, and sea level. Such changes cause considerable impacts on hazard management and city safety, such as (1) increasing risks of urban flooding due to more frequent and intensive precipitation extreme; (2) greater challenge in water resources management due to changes in precipitation pattern; (3) increases in the occurrence of super typhoons and storm surge and the rise of sea level; and (4) increases in high temperatures and heatwaves. Previous observations and studies have also indicated that climate change has brought about impacts on the eco-environment and public health in the Hong Kong-Macao region, such as deterioration in air quality, cross-boundary transport of air pollutants, increase in eutrophication and hypoxia in the coastal waters, increase in infectious and non-communicable diseases, etc. Because of the importance of the Hong Kong-Macao region and the extensive impacts of the hazards, many previous studies have evaluated the future potential climate change impacts and risks in this region, and analyzed the uncertainties. To adapt to the climate change risks, the Hong Kong-Macao region has formulated and developed unique adaptation and prevention strategies and measures based on its physical geographical setting and socio-economic conditions. Therefore, this chapter addresses the impacts, risks, adaptation strategies of climate change in the following five major aspects: (1) flood hazards and urban security, (2) water resources and water security, (3) typhoon hazards, storm surge and sea level rise, (4) quality and protection of eco-environment, and (5) public health, risk perception and prevention strategies.

2.1 Flood Hazards and Urban Security

2.1.1 Floods and water logging in cities

Hong Kong is located on the southern coasts of China, and adjacent to the South China Sea. Hong Kong is one of the cities with the highest rainfall in the Pacific region. According to the observations in 1884～2018 at the HKO, then mean annual rainfall is about 2 400 mm. Furthermore, rainfall in Hong Kong highly varies throughout the year. The highest daily rainfall in the historical records reaches 534 mm, and the highest hourly rainfall is about 145 mm. Hong Kong was attacked by several major floods in history. The flood event on 18 June 1972 was the one that caused the largest damage in terms of causality and economic loss. This event was triggered by consecutive rainy days, which eventually caused flash floods and disasters. To tackle floods and protect Hong Kong from the threats of floods, the Drainage Services Office was set up in 1891, and developed to become Drainage Services Department (DSD) in 1989. Since the establishment of DSD, three-pronged strategy in flood prevention, i.e., stormwater interception at upstream, flood storage at mid-stream and drainage improvement at downstream, has been consistently adopted, which has contributed to great success in flood control in Hong Kong. In 1994, the DSD developed a systematic approach to identify flooding blackspots, and implemented flood control engineering projects. Since then, the flooding blackspots in Hong Kong decreased substantially. Since 1995, 126 flooding blackspots were removed, and there were only 5 in 2019. Drainage improvement engineer has started to eliminate the remaining blackspots, and it is expected that all blackspots will be eliminated by 2025. Nowadays, comprehensive stormwater drainage systems have been built in Hong Kong, including 2 400 km of underground stormwater drains, 360 km of manmade waterways, 21 km of stormwater drainage tunnels, and 4 underground stormwater storage tanks. These facilities significantly alleviate the threats of floods caused by extreme precipitation in Hong Kong①. As we can see, Hong Kong suffered from severe impacts of flood hazards over the past decades. The Hong Kong government has spent a lot of effort on constructing flood prevention facilities. As the improvement of the urban drainage systems, Hong Kong can face extreme weather events. How to adapt to climate change, improve urban resilience, and promote sustainable development is a new challenge that Hong Kong is facing.

Because of its geographical location, Hong Kong is frequently affected by tropical cyclones (TCs) associated with heavy rainfall. For example, daily rainfall in Hong Kong reached 167.5 mm under the influence of Super Typhoon Mangkhut on 16 September 2018; daily rainfall was 165.3 mm on 27 August 2017 due to Super Typhoon Hato; daily rainfall reached 223.4 mm due to Super Typhoon Sarika on 19 October 2016. In the western Pacific region, the South China Sea and the east region to the Philippines are always the places highly affected by TCs (Jinnan *et al.*, 2009). An analysis based on observed records found that the intensities of TCs increased significantly since 1980s, and the proportions of extreme TCs become higher and higher (Elsner *et al.*, 2008; Kang and Elsner,

---

① Hong Kong Drainage Services Department, 2018. *Sustainability Report 2018-19*, https://www.dsd.gov.hk/TC/Publicity_and_Publications/Publicity/DSD_Sustainability_Reports/index.html.

2012; Walsh *et al.*, 2016; Lee *et al.*, 2020). Moreover, Kossin (2018) published in Nature pointed out that in the past 70 years, the translational speed of TCs slowed down by 10% globally, 16% in the western Pacific basin, and even 21% over land. Therefore, the possibility of consecutive heavy rainfall events may increase, because TCs may stay over land for longer time even without considering the changes in typhoon intensity (Lai *et al.*, 2020). These results demonstrated that Hong Kong may face greater threats in flood risks triggered by consecutive extreme rainfall events.

Global warming causes the increase in water holding capacity in the atmosphere, leading to a higher probability of precipitation (IPCC, 2013), which is also an important reason why annual precipitation in Hong Kong in recent years was higher than the climatology. For example, WMO confirmed that the annual precipitation of Hong Kong in 2016 was 3 027 mm, which was the highest in recent years and 26% higher than the climatology, i.e., 2 400 mm. Furthermore, global warming also raises the rain rate near the TCs (IPCC, 2013; Wuebbles *et al.*, 2017). Global warming also results in sea level rise, leading to greater threats of storm surge to Hong Kong. Four of the 10 highest records of sea level in Victoria Harbor since 1954 occurred in the recent 10 years, including 3.53 mCD caused by Typhoon Hagupit in 2008, 3.25 mCD caused by Typhoon Nelgae in 2011, 3.57 mCD caused by Typhoon Hato in 2017, and 3.88 mCD caused by Typhoon Mangkhut in 2018. Sea level rise is very likely to further increase flood risks in Hong Kong especially the low-lying coastal regions.

IPCC AR5 noted that in the face of global warming, the risk associated with increasing occurrence of heavy rainfall events in East Asia is continuously escalating. Average rainfall intensity in Hong Kong from 1885～2018 has also increased (Figure 1–2, Ginn *et al.*, 2010). All these indicate that Hong Kong will face severe challenges in flood risk prevention in the future.

2.1.2 Stormwater management and flood prevention strategies

The frequent rainstorms and high rainfall intensity in Hong Kong were very likely to bring about stormwater and urban flooding in the low-lying suburban and rural areas in the northern part of Hong Kong and parts of the old urban areas (Chan F. *et al.*, 2013; Chan *et al.*, 2014; Wong *et al.*, 2011). Owing to the hilly terrain in Hong Kong, extreme rainstorm events may enhance the combing effects alongside with other natural hazards such as landslides (CEDD, 2020), which cause severe damage to infrastructures, properties, injuries and even loss of life.

Due to the rapid urbanization in the 1980s, suburban and rural areas (including flood pains) were highly developed, and large areas of the land surface were transformed to the impervious surface, increasing runoff coefficient, shortening flood duration, and increasing flood peak. Also, because of the aging urban drainage systems in some old-built urban areas, the flood prevention capacity of these areas cannot reach adequate design standards, e.g., most of the standards are less than 20 years (Chui *et al.*, 2006). The above factors make the existed urbanized areas (old towns and districts) are frequently affected by urban flooding. To tackle the above hazards, the Hong Kong SAR government established the Drainage Services Department (DSD) in 1989. DSD is committed to improving the flood prevention capacity of Hong Kong, and the achievements since its establishment are significant. In recent years, DSD adopted several measures based on the flood prevention strategies (e.g., developed Drainage Master Plan (DMP) for urban areas), improved the flood prevention level from once in 50 years to once in 200 years (DSD, 2019). Compared to the flood prevention levels of once in 1～10 years in many Asian cities, the flood prevention level in Hong Kong is high and safe among the Chinese cities and ahead many Asian cities (Chan F. *et al.*, 2012).

DSD implemented the DMP to adopt various flood prevention measures across several catchment areas in Hong Kong, such as interceptor channels, stormwater storage facilities, and pumping stations. The flood prevention concepts can be classified into a three-pronged strategy, i.e., stormwater interception, flood storage, and drainage. An innovative flood prevention system to mitigate urban flood risk that includes: constructing the stormwater tunnel in the mid-hill to intercept stormwater from midstream and upstream through the interceptor channel, and then stormwater is discharged to the sea or river channel directly. The stormwater storage tank is built in the midstream for stormwater storage, these practices are temporally reducing flood risk in the flood prone area in downstream; drainage and waterways are constructed and effectively minimize the potential of flash (mountainous) floods that may affect the city center, also to improve the drainage and flood prevention capacity (DSD, 2019). Through the implementation of this strategy, the urban flooding problems of Hong Kong has been significantly alleviated. However, as combined effects of climate change and geographical settings, stormwater and urban flooding problems still exist and cause serious loss of life and properties in Hong Kong. For example, the flash flood events caused by rainstorms on 7 June 2008 and 22 July 2010 killed two people and caused economic losses of about 578 million Hong Kong Dollars (Greenpeace China, 2009).

In general, the reasons for frequent occurrence of urban flooding and waterlogging in Hong Kong can characterize into the following points: First, the low flood prevention level. When short-term intensive (heavy) precipitation amount exceeds the flood prevention design standards, urban flooding may occur; second, obstruction of the drainage systems. For instance, catch-pits and channels can be blocked by debris, rubbishes, tree fragments, mud and silt, reducing their flow capacities; third, the increases in sea level due to tropical cyclones can lead to storm surge (Lee *et al.*, 2010). For example, the waterlogging (surface water flooding) events in the low-lying areas of Sheung Wan, Eastern District, East Kowloon, North West New Territories of Hong Kong were caused by seawater intrusion in recent years.

The drainage systems in Hong Kong and Macao have been substantially improved compared to 20～30 years before. The flood protection standards now are as high as other major overseas cities such as Tokyo and London, which is in a leading position in southeast Asia (Chan F. *et al.*, 2018). The SAR governments and departments spare no efforts in solving urban flooding issues, and have adopted a variety of innovative improvement measures, such as stormwater tunnels in the mid-hill to intercept stormwater, and stormwater storage tanks in the midstream to store rainwater. These practices can alleviate the pressure of drainage systems downstream, and reduce urban flood risks in urban and low-lying areas (such as MongKok, Happy Valley, and Wan Chai in Hong Kong), making Hong Kong a role model city of urban flood management and prevention in China and even Asia.

With rapid development and climate change impacts in the Greater Bay Area, it is necessary to improve the stormwater drainage systems further in Hong Kong and Macao. The concept of a sponge city has already been adopted in Hong Kong for urban planning to strengthen the city's capacity to fight against floods. Underground stormwater storage schemes are a great achievement of Hong Kong in building a sponge city. Furthermore, the concept of sponge city has also been implemented in Hong Kong's efforts in river revitalization, rehabilitation, and rainwater harvesting. Hong Kong already has certain foundation for climate change adaptation, especially for TCs and urban floods which is worthy of being learnt from the Greater Bay Area. On the other hand, government departments in charge of infrastructure development should be further enhanced to adapt to various threats resulting from climate change. Macao also proactively learns, adopts and implements a variety of flood prevention measures. For example, after the severe damage caused by Typhoon Hato in 2017, the Macao Special Administrative Region government started to construct stormwater pumping stations in O Porto Interior and install movable tide gates with a design standard of floods with 200-year return period in 2018. Furthermore, the ordinance was made to request public carparks in low-lying areas to be closed one hour after the hoist of typhoon signal No.8 or black warning of storm surge, to prevent human lives and properties from the possible floods. For the coastal areas, floodwalls with about 2 km length and 1.5 m height will be constructed. The walls will be able to defend the 20-year floods and storm surges at the levels of Typhoon Hagupit in 2008. However, in Typhoon Mangkhut in September 2019, Macao failed to predict the typhoon accurately, and hoisted typhoon signal No. 8 only 10 minutes before typhoon signal No. 10 in Hong Kong. The time for flood prevention actions was missed due to this mistake, resulting in enormous losses in this event. Therefore, a reliable warning system is key to alleviating the impacts of flood hazards.

Therefore, to adapt to future climate change, the Hong Kong and Macao SAR governments should further improve stormwater drainage capacities, and strengthen flood prevention abilities. Apart from traditional hard engineering measures, the latest urban water management concepts such as "Sponge City" and revitalizing water bodies should be proactively introduced, also implementing reasonable urban and land planning should be promoted and implemented to reduce the threats of flood risk. In addition, governmental departments and institutions should collaborate further to promote the knowledge of climate change and stormwater management, and improve the public's perceptions and abilities to adapt to flooding.

2.1.3 Landslides and debris flows hazards

Slopes and natural terrain occupy more than 60% of Hong Kong's land area. Among the various types of landslide hazards that occur, rainfall-induced debris flows, typically consisting of a large mass of saturated soil that surge down slopes due to gravity, is perhaps the most unpredictable and detrimental. Due to steep mountainous gullies, debris flows can travel over very long distances (Ng *et al.*, 2013). Unlike other mountainous regions around the world, Hong Kong and Macao are densely populated and even a small landslide/debris flow can result in severe consequences.

The convective available potential energy, which causes rainfall events, is expected to increase substantially under global warming (Singh *et al.*, 2017). Correspondingly, landslide risk will increase. Rainfall-landslide correlation shows the trend that the density of natural terrain landslides in Hong Kong increases exponentially with normalized rainfall intensity.

To highlight the impact of extreme weather events on society, past events in Hong Kong and their consequences are given as follows:

The rainstorms of May and August 1982 triggered over 1 400 natural terrain landslides, which caused over 20 fatalities and caused serious damage particularly to squatter areas. Over 600 mm of rainfall was recorded within the four days from 28 to 31 May 1982 and more than 520 mm of rainfall from 15 to 19 August 1982 (Hudson, 1993; Lee, 1983; Wong *et al.*, 2006).

The rainstorm of November bombarded Lantau Island and triggered more than 800 natural terrain landslides and more than 300 man-made slope failures, which resulted in closure of roads and the evacuation of houses. Over 700 mm of rainfall was recorded in 24 hours (Wong *et al.*, 1997; Wong and Ho, 1995).

The rainstorm of July 1994 resulted in about 200 landslides, 5 fatalities, 4 injuries, closure of roads and evacuations of houses (Chan, 1996). From 22 to 24 July, over 600 mm rainfall was recorded at HKO, together with record-breaking maximum 1-hour, and 24-hour rainfall of 185.5 mm and 954 mm respectively recorded at Tai Mo Shan.

The rainstorm of August 1999 triggered over 200 man-made slope failures, which resulted in 1 fatality and permanent evacuation of 3 public housing blocks. Over 500 mm of rainfall was recorded in 24 hours over the southern part of the New Territories (Ho *et al.*, 2002).

The June 2008 extreme rainstorm in Hong Kong triggered more than 2 400 natural terrain landslides (GEO, 2016), including debris flows, affecting developments including highways. The North Lantau Expressway, which is the sole vehicular access to the airport, was blocked by debris flow deposits for 16 hours (GEO, 2012).

The above-mentioned extreme weather events were fortunately not centered in the most densely populated regions of Hong Kong. Furthermore, the intensities of the rainstorms that struck the more densely populated locations were not particularly severe. Otherwise, the number of landslides and fatalities would have been higher (Ho *et al.*, 2019).

The Geotechnical Engineering Office (GEO) of Hong Kong has implemented a comprehensive Slope Safety System to manage the risk posed by landslides. The existing system has generally proved to be effective in coping with prevailing conditions. However, Hong Kong is not yet prepared for future likely extreme rainfall scenarios. The impact of extreme rainfall poses new challenges that need to be addressed from a strategy, policy and technical perspective (Ho *et al.*, 2019).

Conventional strategies for managing the risk posed by landslide hazards are mainly based on quantitative risk assessment (QRA), which estimates the various components of risk, including hazard, vulnerability and exposure of facilities at risk. The QRA is an excellent tool for quantifying and estimating the expected loss. However, one of the challenges in estimating the risk for extreme events using QRA is that the failure mechanisms for extreme events may be entirely different compared to prevailing conditions, thereby imposing an implicit residual risk. Climate change has led engineers to the realization that past experiences may no longer be a reliable predictor of the future scale and frequency of events. Stress tests conducted by the GEO of Hong Kong have shown that extreme events will compromise the integrity and performance of the existing slope safety system. Therefore, research on the extent to which conventional engineering design is not suitable for dealing with gaps in the existing slope safety system to enhance preparedness and resiliency.

The most straightforward approach to mitigating landslide hazards is by constructing physical barriers (Ng *et al.*, 2016a; 2016b; 2018a; Choi *et al.*, 2019). To address the unique design challenges in Hong Kong and Macao, conventional, empirically-based protective measures used in other parts of the world do not apply to this densely-populated city where land is scarce and costly. For example, in Hong Kong, hundreds of barriers have been installed on hillsides in recent years. Under the current Landslip Prevention and Mitigation Program that was commissioned in 2010 by the Hong Kong Government, several new barriers are scheduled to be installed each year. Barriers not only require robust designs, but they also need to be sustainable.

For instance, installing single large reinforced concrete barriers (Lo, 2000) at the base of a slope is decreasingly viable in densely populated areas because of the challenges of land scarcity and the need to preserve and green the natural environment (Ng *et al.*, 2016c; Ni *et al.*, 2018; Ng *et al.*, 2019a; 2019b). Instead, several smaller flexible barriers (Song *et al.*, 2018) that blend in well with their natural surroundings are becoming the more sustainable and favorable option (Ng *et al.*, 2018b). Although the use of multiple barriers is a promising solution, scientific-based guidelines for optimizing the impact load on the next barrier in a channel is not yet available. To achieve this, fundamental research on overflow (Kwan *et al.*, 2015) and landing mechanisms between barriers must first be elucidated. Eventually, findings may lead to design guidelines that engineers can use to develop safe and sustainable mitigation measures against debris flows. However, given the potential scale of extreme landslide events, it is not feasible to solely rely on physical countermeasures.

Statistical learning or deep learning have been developed and successfully applied to the prediction of natural disasters. Such artificial intelligence (AI) techniques should be adopted to improve and enhance the present state-of-the-art methodologies of landslip warning. Data driven AI and machine learning models require high-quality data to make reliable predictions and Hong Kong is data-rich in terms of high-quality information on rainfall and landslides. Available historical data of rainfall and landslides are vital in developing and training deep learning models to establish correlations between landslide patterns and extreme rainfall events. However, as pointed out previously, historical data may not be able to predict extreme events. Nevertheless, we can only keep on obtaining real time data acquired from an extensive array of automatic rain gauges provided by Hong Kong and continuously improve rainfall-landslide correlation for a landslip early warning system, but at the same time be aware of the residual risk involved in such an approach.

In the case of extreme weather events, installing more barriers may not always be a viable or cost-effective solution. Existing experience from disaster management of natural hazards suggests a higher effectiveness of emergency management, system resilience and community resilience in limiting vulnerability and minimizing risks. Therefore, public awareness and knowledge transfer are crucial to enhancing community resiliency against the landslide threat posed by climate change. To achieve this, decision makers, scientists and engineers will need to reach far beyond conventional means to reach out to the general public. For instance, an immersive experience education environment in slope safety via the MOOC platform using emerging AR/VR technologies and narrative visualization-based technologies should be leveraged to enhance knowledge transfer. An account of GEO's expanded efforts on public education, information and communication under the Hong Kong Slope Safety System is given by Tam and Lui (2013).

Based on existing predictions of climate scenarios for Hong Kong and Macao, these scenarios unequivocally support the observation that precipitation is increasing and extreme events are occurring with greater frequency. Correspondingly, landslide risk is increasing and urgently needs to be managed. The GEO of Hong Kong has reviewed its existing landslide emergency management system based on predicted rainfall scenarios due to climate change. From their analysis, they revealed considerable insight on the potential scale of damage from extreme landslides, as well as gaps in the existing slope safety system. More importantly, we cannot solely rely on physical countermeasures to manage the risk of extreme rainfall because they are costly. Engineering approaches need to be supplemented by non-physical measures that can enhance emergency preparedness, response and recovery. These non-physical measures include leveraging the rich data base of rainfall and landslide data that Hong Kong has collected and to enhance public education to empower the general public to be more resilient against landslides.

2.2 Water Resources and Water Security

Hong Kong is dominated by the warm and humid subtropical monsoon climate with abundant precipitation. The long-term average of annual mean precipitation is about 2 400 mm, but the seasonal distribution of precipitation within the year is uneven (i.e., >80% of precipitation is concentrated in the rainy season of May-September), and the inter-annual variation is high, causing much rainwater has to be discharged to the sea via the drainage system. Due to topographical, geological, and urban population concentration factors, the local water resources in Hong Kong are inadequate and scarce. In terms of topography, almost 80% of the 1 100 km^2 land area of Hong Kong is hilly terrain, lacking low-lying plains and large rivers and lakes, making it difficult to capture and utilize the abundant rainwater. The geology of Hong Kong is dominated by impermeable granitic and volcanic rocks, making groundwater recharge difficult and leaving no significant groundwater storage available for use. Hong Kong is considered a severely water-deficit region because the annual mean total water resources in Hong Kong are only 1.08 billion m^3, and 140 m^3 per capita, which is far below the local water demands. After the Second World War, as the rapid population growth in Hong Kong, the water scarcity problem became increasingly severe, causing suspension of water supplies, becoming a common problem plaguing the Hong Kong community in the 1950s and 1960s. In particular, the early 1960s were hit by the extreme drought, and the Hong Kong government implemented strict water rationing measures and provided water supplies for only 4 hours every four days in June 1963 (Shen and Ji, 1997). To compensate for the insufficiency and instability of local water supplies and fulfill Hong Kong's water demands, Hong Kong has been importing Dongjiang water through Shenzhen Reservoir since 1965, and Hong Kong has enjoyed a 24-hour uninterrupted water supply since 1982 (Hong Kong Water Supplies Department, 2011). Nowadays, Dongjiang water has become a significant water source for Hong Kong, accounting for 70% to 80% of drinking water supplies and 50% of total water consumption (including seawater for toilet flushing) in Hong Kong. Up till now, Dongjiang water is an important and the most cost-effective water source in terms of resources, environment, and economy (Hong Kong Water Supplies Department, 2018). Hong Kong relies on three primary water sources in the past half-century, namely Dongjiang transfer as the primary source, local rainwater collection as a secondary source, and the maximum use of seawater for toilet flushing, to successfully ensure water supply security and support sustainable development.

The risk of salty tide upwelling has increased under climate change with rising sea levels and changing precipitation patterns. In addition, with human activities and other factors, the safety of Macao's water supply has been threatened (Xu and Hou, 2017). When the salty tide upwelling during the dry season in 2004-2005 affected the water supply of Macao and Zhuhai and other cities in the Pearl River Delta, China implemented the first basin-wide emergency water transfer and achieved water transfer to suppress the salty tide and alleviate the disaster through the joint dispatch of major water conservancy hubs in the upper reaches of the basin (Global Water Partnership China Committee, 2016). The upwelling of salty tides poses a major threat to the water supply security of cities along the Pearl River Estuary, such as Macao. Thus, relevant authorities have done a lot of work in the past to address this problem, and some satisfactory results have been achieved.

Therefore, understanding climate change impacts on the regional water cycle and water resources in Dongjiang is crucial for Hong Kong to adapt to the future uncertainties in long-term stable water supplies associated with climate change, rapid economic development, and population growth. This section is therefore divided into three parts. In section 2.2.1, the past and future changes in the Dongjiang water and the policies and measures of water transfer in the basin are reviewed. In section 2.2.2, the response of local water resources in Hong Kong to climate change and the measures adopted by Hong Kong to maintain water security are discussed. Finally, section 2.2.3 focuses on the impacts of salt tide upwelling on Macao's water supply under climate change and adopted adaptation strategies.

2.2.1 Dongjiang water and cross-boundary water transfer

The Dongjiang basin is dominated by the subtropical monsoon climate characterized by obvious seasonal precipitation patterns. The summer precipitation is about 72%～88% of the total annual precipitation, and the streamflow of the basin is mainly recharged by precipitation (Li J. *et al.*, 2016b).

Observations of the past years and model simulations have revealed the responses of the regional water cycle to climate change in Dongjiang: Annual precipitation changes insignificantly, while precipitation in spring and winter increases and precipitation decreases in summer and autumn, and the number of rainy days decreases significantly (Chen *et al.*, 2011; Liu *et al.*, 2015; Zhou *et al.*, 2016); temperature increases significantly, but observations of evaporation pan decreases significantly (Liu *et al.*, 2010); soil moisture decreases significantly in autumn, with insignificant changes in other seasons (Zhang *et al.*, 2018). Under the combined effects of climate change and human activities (e.g., water abstraction, reservoir operation and land use land cover change, etc.), the long-term observed annual streamflow in the major hydrological stations of the Dongjiang basin (e.g., Boluo station) changes insignificantly, while that in the dry season increases significantly (Chen *et al.*, 2012; Zhou *et al.*, 2016; Wu *et al.*, 2019). The analysis of Tu *et al.* (2015) found that reservoir allocation reduces the year-to-year variation of streamflow in Dongjiang by 33%, and the contributions of the operations of Xinfengjiang, Fengshuba, and Baipenzhu reservoirs to the reduction in annual variation are 21%, 10%, and 2%, respectively. During the dry season, the increase in precipitation, decrease in potential evapotranspiration and seasonal allocation of reservoirs are the primary reasons for the increases in streamflow, which alleviate the drought risks in the dry season in the Dongjiang basin (Wu *et al.*, 2018).

Global Climate Models (GCMs) outputs from CMIP5 indicate little or possibly slight future change (less than 10%) in annual precipitation in the Dongjiang basin. Temperature and evapotranspiration are generally increasing, but the differences in future projections of precipitation from different GCMs are considerable (Sun *et al.*, 2015; Yan *et al.*, 2015; Li J. *et al.*, 2016a; Wang *et al.*,

2017). Future projections of water resources in Dongjiang are mainly driven by outputs of GCMs under future climate scenarios, and the land-surface hydrological model is used to simulate hydrological processes in the basin and derive future streamflow. Based on five CMIP5 global climate model outputs driving the VIC (Variable Infiltration Capacity) hydrological model, Yan *et al.* (2015) found that streamflow in the dry season from 2079 to 2099 is lower than that from 1979 to 1999, and the decrease under RCP8.5 is more evident than under RCP4.5. Because streamflow in Dongjiang is mainly fed by precipitation, the differences in precipitation outputs from GCMs are a significant source of uncertainties in future projections of water resources (Li J. *et al.*, 2016a). So far, the projections of changes in water resources are mainly based on statistical downscaling. Li J. *et al.* (2016b) found that when the inter-daily variations of atmospheric factors change after statistical downscaling, even if monthly or annual means remain unchanged, the changes in daily hydrological processes can be accumulated and affect the long-term changes in hydrological variables such as water availability, which is one of the uncertainty sources for projections of water resources in the future. Therefore, to project future changes in water availability in Dongjiang under climate change, GCMs, hydrological models, downscaling methods, and future climate scenarios are crucial sources of the uncertainties.

The Dongjiang water supply scheme in 1960～2005 was mainly based on fixed water supply agreements (Hong Kong LegCo, 2017). Given the dry-wet changes in precipitation and fluctuations of water demand in Hong Kong, Hong Kong has adopted the "package deal lump sum" approach for the Dongjiang water supply agreement to reduce the overflow of water resources. The input of Dongjiang water is based on actual demand elasticity to a cap of 820 million m^3, which considers a return period of once in 100 years extreme drought scenario and ensures 99% water supply reliability (Hong Kong Water Supplies Department, 2018). The Dongjiang is the primary water source for cities including Hong Kong, Shenzhen, and Dongguan, supplying more than 40 million people in this region, but the water resources per capita are only 1 100 m^3/year, which is classified as a water-scarcity region according to the international standards. To relieve the contradiction between urban water supply and demand in this region, and to ensure water security during the dry season and basic ecological water requirements within the river, Guangdong province has implemented the Dongjiang River Basin Water Resources Allocation Scheme in 2008, and the quota of water supply allocated for Hong Kong is included in this scheme (Tu *et al.*, 2016; Wang *et al.*, 2017). With the basin's rapid population growth and economic development, Dongjiang water resources have been heavily developed. The estimated water demand of Shenzhen in 2020 is expected to exceed the allocated quota in the water allocation scheme, resulting in a water shortage (Hong Kong LegCo, 2015; Shenzhen Water Resources Bureau, 2016). The construction of the Pearl River Dealt Water Resource Allocation Project ("West-East Water Transfer") began in 2019 to divert water from Xijiang river to the eastern part of the Pearl River Delta to provide an emergency backup water source for the eastern cities. To address the challenges of future uncertainties of water quantity changes and increasing water demand, the Hong Kong Water Supplies Department (WSD) released the *Total Water Management Strategy* in 2008 and comprehensively reviewed the strategy in 2019 (Hong Kong Water Supplies Department, 2008; 2018; 2019). The review in 2019 confirmed that the diversified water supply structure has achieved its respective milestones and predicted the water supply and demand conditions till 2040 to prepare Hong Kong for the potential challenges on water supplies in the future. Furthermore, the Dongjiang water is considerably regulated by Xinfengjiang, Fengshuba, and Baipenzhu reservoirs. Therefore, future projections and relevant adaptation measures should consider the effects of reservoir regulations.

2.2.2 Local water resources and water security

Over the past decades, stable, adequate, clean and reliable freshwater supplies have been vital to the social prosperity and economic development of Hong Kong. Hong Kong has successfully ensured water supplies and water security by adopting revenue-raising and expenditure-cutting policies, a "walking on two legs" strategy with local and Dongjiang water, and "soft and hard" measures (Chen, 2001). These water development and utilization measures are unique in their local context and even are unique in the international arena and have attracted much attention (Yue and Tang, 2011). In terms of revenue-raising, the government-built reservoirs and designated about one-third of the land (about 300 km^2) as water catchments in the mid-20th century during the early stages of urban development. In the past 20 years, the rainwater collected annually is about 103-385 million m^3, accounting for 20%～30% of Hong Kong's freshwater resources. The water catchments in Hong Kong are located mainly in highly protected areas such as country parks. Development in the catchment areas is also regulated to ensure that water quality is not polluted. Secondly, the Government has implemented an innovative scheme to use seawater for toilet flushing since the 1950s, with the construction of the world's largest dual-pipe network urban water supply system. In recent years, the seawater flushing network has covered about 85% of the population in Hong Kong, saving about 300 million cubic meters of fresh water (about 24% of total water consumption in 2019) and a significant amount of energy (electricity consumption per unit of seawater is only 2/3 of that of fresh water) every year. Hong Kong is the only city in the world to use seawater for toilet flushing fully. Moreover, Hong Kong is also taking forward desalination and recycling (i.e., reclaimed water, water reuse and rainwater recycling for non-potable uses) as new water resources, etc. In terms of expenditure-cutting, soft measures include a tiered-pricing system of water and enhancement of public education, which encourages water saving by economic incentive and educational guidance to control the increase in water demands. Hard measures include the replacement and rehabilitation of about 3 000 km of aged water mains between 2000 and 2015, which has reduced the water leakage rate of water mains in Hong Kong from 25% to 15%.

Since Hong Kong's local water resources are scarce and highly varied within and between years, importing cross-boundary water is a significant approach to ensure water security. After Hong Kong experienced a water shortage in the extreme drought in the 1960s, Hong Kong has imported Dongjiang water since 1965. As a result, the annual water supply to Hong Kong has increased from a few tens of millions of cubic meters in the 1960s and 1970s to 600～800 million cubic meters since the 1990s. The proportion of Dongjiang water in the total fresh water consumption has also increased from less than 30% in the early years to 70-80% in the past three decades. Since the first water supply agreement between Hong Kong and the Guangdong Provincial Government in 1960 and up to 2017, the two governments have signed a total of 11 water supply agreements according to Hong Kong's water demands. The importation of Dongjiang water has eliminated the need for intermittent water supply cuts in Hong Kong since June 1982. Since 2006, the Dongjiang water supply agreement has adopted the "package deal lump sum" approach, in which Hong Kong pays a fixed amount of money to

Guangdong each year to receive an annual supply ceiling of 820 million m^3 of Dongjiang water to ensure Hong Kong can maintain a full day's water supply even in an extreme drought once in 100 years (Hong Kong LegCo, 2015).

Hong Kong owns one of the safest and most reliable water supply systems worldwide. The Water Supplies Department (WSD) is responsible for collecting different types of water resources and the treatment and distribution of drinking water and seawater supplies. The water supply system provides 99.99% of Hong Kong's population with drinking water that complies with the guidelines of the World Health Organization, covering the densely populated Hong Kong-Kowloon city areas, satellite city towns, and the less populated rural areas. Furthermore, the seawater flushing systems have been expanded to cover most urban areas and new towns. Hong Kong WSD has a history of almost 170 years. The government has built, expanded and maintained extensive water supply infrastructures throughout the years to ensure Hong Kong's water supply. Nowadays, the primary water supply facilities in Hong Kong are mainly (1) two marine reservoirs for storing Dongjiang and local raw water, including Plover Cove, the first reservoir in the sea in the world, and 15 conventional reservoirs. The total storage capacity of the 17 reservoirs in Hong Kong is about 586 million m^3 that is equivalent to about six months of water consumption in Hong Kong; (2) local rainwater collecting systems including tunnels, waterways, etc.; (3) 20 water treatment plants with the capacity of 4.68 million m^3 per day; (4) more than 190 pumping stations for raw water, fresh water and seawater with various sizes, more than 220 service reservoirs for the temporary storage of fresh water or seawater as well as for controlling water pressure of the supply, and more than 8 000 km of water mains to transfer fresh water and seawater to users (Hong Kong Water Supplies Department, 2018).

Over the past half-century or so, Hong Kong has successfully ensured water supply security and supported sustainable development by relying on three primary sources of water, namely Dongjiang water transfer as the primary source, local rainwater collection as a secondary source, and the maximum use of seawater for toilet flushing. However, with the impact of climate change on water resources, increasing water demand due to population growth and economic development, and the competition of water resources among cities of the Greater Bay Area, Hong Kong is facing increasing challenges in water resources management and water supply security. In view of this, the WSD implemented the Total Water Management Strategy in 2008 to integrate a variety of effective water development approaches and measures of revenue-raising and expenditure-cutting policies and the strategies of using local and Dongjiang water, with the aims to ensure the long-term stability and security of water supplies in Hong Kong (Hong Kong Water Supplies Department, 2008). These strategies balance water demand and water supply management, focusing on controlling water demand growth through promoting water conservation and exploiting new water resources. The WSD completed the review of this strategy in 2019 and confirmed that the water demand and supply management under the strategy had achieved their respective milestones. The review also predicted the water demand and supply of Hong Kong to 2040, considering the impact of climate change (Hong Kong Water Supplies Department, 2019). After implementing the strategy, the annual fresh water consumption in Hong Kong was controlled at the level of around 1 billion m^3 over the past ten years, and the fresh water consumption per capita has dropped from an average of 140 per person from 1999 to 2008 to 133 m^3 from 2009 to 2018.

The primary (Dongjiang) and secondary (local) fresh water sources of Hong Kong are affected by climate change. Dongjiang is the smallest sub-basin in the

Pearl River basin. Its water resources per capita are the lowest (only 1 100 m^3/year, a water-deficit area based on the international standards). Still, Dongjiang is the primary water source for Guangzhou, Shenzhen, Dongguan, Huizhou, Heyuan, etc., and also shoulder the critical task of supporting the water supplies in Hong Kong. With a total water supply population of over 40 million, the water development level of the basin is almost the 40% ceiling of the international consensus. As described above, the importance of Dongjiang water to Hong Kong is evident, and the impacts of climate change on Dongjiang water have been introduced in the previous section. The climate change impacts on Hong Kong's local water resources and management are twofold: (1) how the changes in the spatiotemporal patterns of local precipitation under climate change may affect the effectiveness and volume of water collection; (2) how the increase in reservoir surface evaporation in response to warmer temperatures may affect the water storage. Further research is needed on the impacts of climate change on Hong Kong's local water resources and management.

Hong Kong is adjacent to the Dongjiang basin, controlled by the same climatic conditions, and in dry years, both areas often suffer the pressure from water shortages. In order to adapt to climate change and increase the amount and diversity of local water sources, the total water management proposes to increase the number of water sources for Hong Kong's water supply from three to five (Hong Kong Water Supplies Department, 2008), i.e., adding desalination and recycling to the existing sources. The 2019 review indicated that these measures are being developed smoothly (Hong Kong Water Supplies Department, 2019). For desalination, the government has proposed to adopt reverse osmosis technology in the Tseung Kwan O Desalination Plant Project, the first stage of which can meet 5% of overall fresh water demand in Hong Kong and expand to

10% if necessary. The plant is expected to be completed by 2023. For reclaimed water, the government has commissioned a pilot scheme in 2006 to collect and reclaim sewage for non-potable purposes at Ngong Ping and Shek Wu Hui. The Water Supplies Department intends to start supplying reclaimed water for toilet flushing and other non-diversional uses in the North East New Territories in phases from 2023. The relevant work is in progress, and it is expected that the total reclaimed water consumption in the region will eventually reach 22 million m^3 per year, which is a water-saving and environmentally friendly measure. For grey water recycling and rainwater recycling, the government has formulated guidelines and been installing related facilities in new government buildings. There have been more than 100 of these projects, and the number is expanding. The development of the Anderson Road Quarry Site is one of these projects that receives much attention. A centralized grey water recycling system is planned to construct and is expected to be completed by 2023. The above two new water sources are essential non-conventional water sources that are not affected by climate change and are crucial for improving the climate resilience and resilience of water supply security in Hong Kong. In addition, the government has commissioned two other major projects to reduce wasting water resources and enhance the security of Hong Kong's water supply. The first is the Water Intelligent Network project in which sensing equipment is installed in the water distribution network to monitor the condition of the network, and smart technique is adopted to reduce water loss in the network. Secondly, to increase local rainwater collection and reduce the risk of flooding, the WSD and Drainage Services Department commissioned jointly to implement an Inter-reservoirs Transfer Scheme to reduce the overflow of reservoirs during the wet season and increase the efficiency of local rainwater collection (Hong Kong Environment

Bureau, 2015; 2017).

2.2.3 Salt tide upwelling and water supply stability

Macao is subject to a subtropical maritime climate with abundant precipitation and an average annual rainfall of over 2 000 mm. But its small area and lack of significant rivers have resulted in a shortage of local fresh water resources, with 96% of its water supply coming from the Xijiang River, supplied by Zhuhai (Yu *et al.*, 2014; Macao Maritime and Water Affairs Bureau, 2021). Local water storage facilities have an effective storage capacity of 1.9 million m^3, mainly used for water storage buffer and emergency salt transfer (Macao Maritime and Water Affairs Bureau, 2017). The Modaomen Channel, at the lower reaches of Xijiang River, is the primary water source for Macao and Zhuhai and is at the outlet of Xijiang River. During the winter and spring dry periods, when the water coming from the upper reaches of Xijiang is low, the salinity of raw water increases in the Modaomen Channel due to the backflow of seawater, affecting the safety and stability of Macao's water supply.

Under climate change, the precipitation pattern changes, affecting the incoming water from the upper reaches during the dry season. At the same time, sea level rises, and a combination of factors increases the risk of seawater backflow and salty tide upstream, threatening the security of Macao's water supply. In addition, as Macao's economy develops, its population and tourists increase, and the water consumption of Macao also gradually increases, posing further challenges to water supply (Yu *et al.*, 2014; Xu and Hou, 2017). The low precipitation in the Pearl River and the sea level rise in the second half of 2009 were the primary causes of the upwelling of the salty tide in the Modaomen

Channel during the dry period of 2009/2010, which occurred earlier and shifted the salty boundary upward compared with previous salty tide intrusions (Kong *et al.*, 2011). Model simulations show that the salty tide's upwelling in the Pearl River Estuary is sensitive to sea level rise. A 0.1 m increase in sea level can further deteriorate the salty tide condition (Yuan *et al.*, 2015). Therefore, the risk of salty tides will continue to increase under future climate change, with further changes in precipitation patterns and further sea level rise.

To cope with the salty tide's upwelling during the dry season, water authorities in Macao and the mainland have adopted various adaptation measures to ensure the safety of Macao's water supply (Macao Maritime and Water Affairs Bureau, 2017). The Pearl River Basin conducts the dry-period water scheduling during the dry season. Significant reservoirs, pumping stations and other water conservancy facilities in the basin are jointly scheduled to transfer water, suppressing salinity and control the salinity of raw water supplied to Macao, accumulating considerable experience in combating salinity. In addition, the West-East Water Transfer System in Zhuhai ensures the safety of water supply to Macao and Zhuhai during salty tides. At the same time, scientific research on the upwelling of salty tides has been intensified, and considerable foundation and experience have been accumulated in research on relevant regular mechanisms, early warming and forecasting, and salt suppression and prevention, providing a solid scientific research foundation for the water supply security of Macao and the cities in the Pearl River Delta (Global Water Partnership China Committee, 2016).

2.3 Typhoon Hazards, Storm Surge, and Sea Level Rise

2.3.1 Typhoon hazards

Typhoons are defined as tropical cyclones (TCs) with wind speed near the center exceeding 118 km/h developed from the Northwestern Pacific Ocean and South China Sea regions with high sea surface temperature (SST). In the past 50 years from 1961 to 2010, the annual number of TCs that affected the Hong Kong-Macao region is 6 on average, which mostly occurred in June – September. TCs usually bring about other primary hazards such as strong wind, rainstorm, and storm surge, as well as secondary hazards such as landslides and debris flows. After the formation of a typhoon, the wind speed near the center can reach 240 km/h, causing collapses of houses and trees and even causalities.

Furthermore, on the days preceding the approach of typhoons, Hong Kong may experience extremely hot weather and serious air pollution. The peripheral sinking airflow of a typhoon can trigger stuffy and hot weather (Lee *et al.*, 2015). On 22 August 2017, affected by Super Typhoon Hato, many regions of Hong Kong experienced high temperature of 38℃ or above, which broke the historical record. Typhoons also can worsen air quality such as lowing visibility and increasing concentrations of air pollutants. A study has found that preceding approach of typhoons towards Hong Kong (i.e., the issuance of Standby Signal No. 1 by the Hong Kong Observatory), the concentrations of $PM_{2.5}$, PM_{10}, SO_2, and NO_2 in Hong Kong increased by 26%, 28%, 46%, and 17%, respectively (Luo *et al.*, 2018). Especially when TCs are passing through Taiwan, the cross-boundary transportation of air pollutants from the Pearl River Delta to Hong Kong is increased, and the air in Hong Kong is descending, making visibility decrease and concentrations of air pollutants (e.g., O_3) increase significantly (Wei *et al.*, 2016; Chow *et al.*, 2018; Lam Y. *et al.*, 2018).

From a long-term perspective, there is no consensus on the changes in the number of TCs affecting the Hong Kong-Macao region, different observation data lead to contradictory results (Lee *et al.*, 2012; Walsh *et al.*, 2016; Lee *et al.*, 2020). Meanwhile, there is no obvious trend in the number of TCs landfalling at the east coast of Asia (Chan and Xu, 2009). However, the typhoons making landfall also become more intense significantly (Mei and Xie, 2016), and cause more severe economic losses (Wang *et al.*, 2016). For example, on 23 August 2017, Super Typhoon Hato resulted in direct economic losses of more than 3.5 billion USD, 22 deaths and 5 841 hm^2 agricultural lands affected (Benfield, 2018; HKO, 2019). On 16 September 2018, Super Typhoon Mangkhut attacked Hong Kong, Macao, and Guangdong. In this event, the 10-min mean wind speed reached 180 km/h in the Waglan Island, 450 people were injured, more than 60 000 trees were down, more than 800 flights were cancelled, and power supplies of thousands of households were affected (Choy *et al.*, 2020a; HKO, 2020).

The general public in Hong Kong and Macao has developed certain adaptation capacity in typhoon warning and prevention. In particular, Hong Kong government and related departments have accumulated rich experience. HKO usually issues No.3 Strong Wind Signal in advance as a warning, and the Education Bureau then makes an announcement through radio and television that all kindergartens are closed. If a No. 8 Gale or Strom Signal is likely to be issued, advanced notice will be advised to the public and evacuate residents in some regions in various means two hours before the issue of the signal. When a No. 8

Gale or Strom Signal is issued, the Home Affairs Department will open temporary shelters. For example, to fight against the possible threats from the Super Typhoon Mangkhut, the Security Bureau chaired an inter-departmental meeting consisting of representatives from 30 government bureau, departments and organizations including Home Affairs Department, Housing Department, Highways Department, Hong Kong Police Force, Information Services Department, etc., to formulate response plans of emergency, enhance coordination and information flow, and instructed the emergency rescue departments and related unities to get prepared (Choy *et al.*, 2020b). In Macao, the Chief Executive chaired a defense meeting to prepare preventive measures to minimize the impacts.

In the future, under global warming, although the total number of TCs may decrease, the frequency and intensity of super typhoons are likely to increase (Huang P. *et al.*, 2015; Knutson *et al.*, 2015; Tsuboki *et al.*, 2015; Walsh *et al.*, 2016; Lok and Chan, 2018; Yang *et al.*, 2018; Wehner *et al.*, 2018; Cha *et al.*, 2020). Due to the uncertainties of future projections of typhoons, studies of projections of typhoon risks under various socio-economic and climate change scenarios should be improved. Given the increasing number of super typhoons, the Hong Kong-Macao region should further improve typhoon warning capacities to announce typhoon warning to the general public and related departments in time to reduce the losses caused by typhoon hazards.

2.3.2 Sea level rise and storm surge

Thermal expansion of ocean and the melting of continental glaciers have caused the rise in the mean sea level around the world. With increasing mean atmospheric temperature, according to the projection of SROCC, under the

RCP8.5 scenario, the mean sea level in Hong Kong and its adjacent waters in 2091～2100 (with the effect of local vertical land displacement incorporated) will be 0.73 m～1.28 m higher than the 1986～2005 average. It is virtually certain that mean sea level will continue to rise beyond the year 2500. For Hong Kong and Macao, local tidal gauge data shows a clear rising trend of the annual mean sea level, and that the rate of increase in annual mean sea level is also increasing (Wong *et al.*, 2003, He *et al.*, 2016, Wang *et al.*, 2016b).

Historically, tropical cyclones and the associated storm surge had brought severe damage and casualties for Hong Kong (e.g., 1874, 1937, 1962). With rising sea level, the return period of these extreme storm surge events is shortening (Lee *et al.*, 2010; Yu *et al.*, 2018). For example, the return period of extreme surge events with sea level up to 3.5 m above Chart Datum based on past data was 50 years (Lee *et al.*, 2010), but with Super-typhoons Hato and Mangkhut brought the sea level above 3.5 m in consecutive years (i.e., 2017 and 2018), the return period of such events has now shortened by half (less than 25 years). This underscores the changing nature of storm surge under climate change and the rising threat from extreme flooding events.

For Hong Kong and Macao, a combination of different tidal, weather and geographic factors can combine to exacerbate coastal erosion and flooding. First, maximum seasonal tidal amplitude occurs around May-June and October-November, and the period May-June also coincides with the frequent occurrence of pre-summer mesoscale systems (Luo, 2017). The later mesoscale systems often lead to higher water level and larger river discharges at the Pearl River delta, setting up a background condition that is prone to flooding. Around the same time, the establishment of the upper-level easterly jet and the weakening of vertical wind shear starts to allow typhoon development over the South China

Sea and the western Pacific (Ding and Chan, 2005). Some of these cyclones, which can significantly increase wave heights and induce storm surge events (Yin *et al.*, 2017), can track towards Hong Kong and Macao. Finally, due to backwater effects and the funnel-like shape of the estuary, the combination of all the aforementioned factors can combine to cause water levels to reach a few meters (Tracy *et al.*, 2007). With the steady rise of mean sea level, the potential for coastal erosion and flooding, as measured by the shortening of the return period for severe flooding events, is great increased (Yu *et al.*, 2018).

The IPCC AR4 WGII report (2007) already noted that there is very high confidence that the hazard risks of Asian mega deltas, such as the Pearl River Delta, are expected to increase, due to large populations and high exposure to sea level rise, storm surge and river flooding. It is also well recognized that, because of the high population and the concentration of developed assets, the potential for damage is great over Hong Kong and Macao (Tracy *et al.*, 2007). However, there is much less quantitative research on the projected impacts of sea level rise or storm surge specifically carried out for this region.

More recently, combining flood damage data from the Global Active Archive of Large Flood Events developed and maintained by the Dartmouth Flood Observatory, and global sea level rise projection in the IPCC AR5 WGI report (2014), Yu *et al.* (2018) carried out a quantitative estimate of the human impact due to rising sea levels as a result of climate change. They estimated that, with the IPCC projection of global mean sea level increase of 40～80 cm by 2010, a currently 100-year return period event could have its return period shortened to about 1.5 years (i.e., occur twice in three years), and if no long-term adaptation is done, Hong Kong could face projected death toll of 15～20+, number of people displaced of 20 000～100 000; and economic loss of around 100～800 million

HKD per year by 2100. Such annual losses are staggering, and underscore the importance of taking long-term adaptive measures to reduce exposure and vulnerability of our cities. The impact estimate from Yu *et al.* (2018) is based on macroscopic population and economic data rather than process specific estimates.

Super Typhoons Hato and Mangkhut visited Hong Kong and Macao in 2017 and 2018, caused substantial damages and coastal erosion and gave us a preview of the serious threat posed by severe flooding and storm surge events in the future. On the other hand, the significant drop in fatalities in Macao from twelve in 2017 (Hato) to zero in 2018 (Mangkhut), despite the fact that Mangkhut is a much stronger storm, showed clearly the benefit and the importance of smart and well-targeted adaptation measures in reducing exposure, vulnerability, and the overall risks to the two cities.

Coastal flooding and storm surge events can impact us in many ways, including coastal erosion, infiltration of sea water into drinking system, blockage of our drainage system, readjustment of coast line because of land loss (Ayyub *et al.*, 2012), disrupt transport (Dawson *et al.*, 2016), flood critical urban underground infrastructure including subway system, parking lot, backup power systems (Tollefson, 2013), as well as changing system composition of ecologically sensitive areas like Mai Po (Tracy *et al.*, 2007). Much more local studies are needed to help qualify and quantify how sea level rise, coastal flooding and storm surge may impact individual sectors, so that more targeted monitoring and smart adaptive measures commensurate with the potential threats can be planned and established to further reduce our risks and protect the public.

Hong Kong and Macao are metropolitan cities that are by nature not self-sufficient but depend on the surrounding areas for subsistence (Lin, 2014; Sharifi and Yamagata, 2014). Both cities are dependent on the supply of food, water and energy from neighboring cities in Guangdong (e.g., Guangzhou, Zhuhai and Zhongshan), which are built on the natural flood plain of the Pearl River Delta (Huang *et al.*, 2004). Hence, even if Hong Kong or Macao are not directly hit by flood waters, extended period of flooding in any of these cities due to a major storm surge event will also cause serious disruption in our supply chains, and substantial economic and social impacts in Hong Kong and Macao. Better regional collaboration plans, and emergency protocols should be set up so that cities around the region can provide more coordinated support for each other so that the overall resilience of the city cluster in the region can be further improved (Leichenko, 2011).

2.4 Quality and Protection of Eco-environment

2.4.1 Urban high temperature and heatwave

Affected by summer monsoon and the subtropical high of the Northwestern Pacific Ocean, Hong Kong and Macao experience high-temperature weathers frequently. Meanwhile, as the effects of ENSO, heatwaves in South China show certain inter-annual oscillations (Qian C. *et al.*, 2018; Luo and Lau, 2019). However, in the long term, mean air temperature, number of hot days, and heatwaves increase in Hong Kong and Macao.

In addition to climate change, urban heat island effect also raises the risks of urban high temperature and heatwaves. Urban heat island indicates that the temperature in urban areas is usually higher than in rural areas (Oke, 1982; Giridharan *et al.*, 2004). Because of the highly-dense population, high energy consumption, compact and dense buildings, and low ventilation, the urban heat island effect is obvious in Hong Kong and Macao (Siu and Hart, 2013; Peng *et al.*, 2018). In general, the urban heat island effect in winter is stronger than in summer in Hong Kong, and the difference in urban and rural temperature can exceed 2℃. The most intensive urban heat island effect usually occurs at night, especially before sunset, and the intensity of the effect at daytime can be negative (Shi *et al.*, 2011; Wang W. *et al.*, 2016). The urban areas in Hong Kong were 0.87℃ warmer than the rural areas in 1970～2015 on average due to urban heat island effect (To and Yu, 2016). With the continuous urbanization, urban heat island effect in Hong Kong may become further stronger (Chen and Jeong, 2018).

Hong Kong and Macao are developed cities with dense population and buildings, making them very vulnerable to high temperatures and heatwaves, which can cause considerable economic losses and casualties in the cities. For example, hot temperatures and heatwaves can cause large impacts on human health, as these events can trigger infectious diseases, cardiovascular diseases, respiratory diseases, etc. (Gasparrini *et al.*, 2015; Yi and Chan, 2015; Qiu *et al.*, 2016; Sun *et al.*, 2016; 2018; Tian *et al.*, 2016). In 2010～2016, hospital admissions due to hot weather in Hong Kong increased considerably (Sun *et al.*, 2019). When air temperature is more than 28℃, the death risk is expected to increase by 2% for every 1℃ increase in temperature (Chan E. *et al.*, 2012). The long-lasting hot events are particularly harmful to human health, as the mortality rate increases by 7.99% for hot nights last for more than 5 days (Ho *et al.*, 2017). Females and the elderly are relatively more vulnerable to extreme drought events (Wang *et al.*, 2019). Furthermore, high temperatures and heatwaves exert significant impacts on the environment, ecosystem, and economy in Hong Kong and Macao. High temperature facilitates the formation of high concentration of O_3

in Hong Kong (Zhao *et al.*, 2019). The high temperatures may also increase labor costs. As the Wet Bulb Global Temperature (WBGT) increased by 1℃, the percentage of working hours would decrease by 0.33% (Yi and Chan, 2017). Under warming climate, the more frequent hot events may raise energy consumption, as energy consumptions in Hong Kong is estimated to increase by 4%～5% for every 1°C increase in temperature (Jovanović *et al.*, 2015; Ang *et al.*, 2017).

Currently, the local governments have adopted a serious of measures to alleviate the impacts of high temperatures and heatwaves. For example, the Urban Climatic Map in Hong Kong helps to understand the heat environment in the city, formulate wind environment standards and improve the urban heat comfort. In 2014, the Hong Kong SAR government published the Sustainable Building Design Considerations Guidelines to provide guidelines for street alignments to connect open areas and set up no-building areas to balance the effects of the wall buildings, with the aims to improve the building environment of Hong Kong by alleviating urban heat island effects and improving ventilation among buildings (Ren *et al.*, 2011; 2012). The Hong Kong and Macao meteorological departments use various means such as television, radio, and Internet to issue the hot temperature warning to help the public (especially vulnerable groups) informed of the latest weather condition and take appropriate measures for self-health protection (Lee *et al.*, 2015). Furthermore, Hong Kong and Macao actively prompt renewable energy for power generation to mitigate climate change impacts (Song *et al.*, 2017). Hong Kong government is encouraging the use of water-cooled air conditioning system to reduce energy consumption (Wang Y. *et al.*, 2018). Moreover, increasing urban tree coverage also can lower the temperature effectively by evapotranspiration (Tan *et al.*, 2016; Kong *et al.*, 2017). Some studies also indicated that using phase change material for building roof can effectively cool down a building by 6.8℃ (Yang *et al.*, 2017), and compared with intensive green roof, extensive green roof is more economically cheaper but also more effective in cooling (Peng and Jim, 2015).

2.4.2 Air quality and transboundary air pollution

Climate change can affect air quality in different ways. Climate change can affect the proportion, propagation, dispersion, and interactions among different pollutants. Regional climate change can be caused by climate and or the local driven factors such as urbanization. The changes of climate in a region can directly affect air quality. For example, regional climate change can slow down surface wind, directly leading to aggregation of air pollutants and deterioration of air quality of the region. Apart from these direct impacts, climate change also indirectly affects air pollutant emissions. Because of the strong interactions between climate and air quality, scientific studies have demonstrated that air pollutants can exert impacts on the climate through feedback mechanisms.

Air pollution is strongly influenced by weather/climate conditions; therefore, it is very sensitive to climate change. The IPCC report projected that the air quality in cities will continue to deteriorate in the future (IPCC, 2014) which can be attributed to the increases in anticyclone weather conditions. For example, Hulme *et al.* (2002) found that climate change would worsen air quality in Hong Kong. Yim *et al.* (2019a) used a statistical method to analyze the transboundary air pollution in Hong Kong in 4 ENSO events and 20 heatwave events during 2002～2016. They found that during the two El Niño events (2015～2016 and 2009～2010), precipitation increased, frequency of northerly wind below 700 hPa decreased, and wind speed increased, which caused the contributions of transboundary air pollutants decreased consistently. Comparatively, during the two La Niña events (2007～2008 and 2010～2011), precipitation decreased, frequency of northerly wind below 900 hPa decreased, and surface wind speed decreased, which is in favor of long-distance transportation and local accumulation of air pollutants. In addition to the impacts of the above short-term events, existing studies also evaluated the long-term impacts of climate change on air quality. Tong *et al.* (2018) projected that the air pollutant will decrease in June-August, but increase in other seasons. In the far future (2090～2099) under RCP8.5 scenario, the projected mean concentrations will change more substantial. Among different meteorological variables, surface temperature is the most correlated with the projected changes of the three pollutants. For example, under RCP8.5 scenario, the relative contributions of surface air temperature to all pollutants will be between 56.9% and 65.2% in different seasons. Furthermore, it is worth noting that other related meteorological factors, including vertical temperature gradient and the differences between air temperature and dew point temperature, also have positive impacts on pollutant concentrations. This study also estimated that the frequency of high pollutant concentrations in December-February of the following year and March-May will increase. The frequency of high pollutant concentrations greater than the 95th percentile is expected to increase by 6.4%～9.6%. The authors summarized that climate change would exert considerable impacts on the air quality of the Greater Bay Area in the future.

Liu *et al.* (2013) projected the impacts of climate change and emission changes on O_3 level in South China in October during 2000～2050 under the A1B scenario using numerical simulation methods. The projections of this study showed that the changes in radiation and surface temperature induced by climate change will lead to significant increases in the emissions of isoprene and monoterpenoids, and pointed out that surface temperature higher than 40°C may inhibit biological emissions. However, because the projected surface temperature is not that high, biological emissions of pollutants may increase in the future. Due to the increases in emissions, isoprene is expected to increase by 30～80 ppt. The authors predicted that due to climate change, the mean surface O_3 level in the afternoon would increase by 1.5 ppb. Because of the changes in anthropogenic emissions, even though O_3 in the southern part of the Greater Bay Area decreases, the mean O_3 concentration of the whole Greater Bay Area will increase by 6.1 ppb. Under the combined effects of changes in climate and anthropogenic emissions, surface O_3 level in the afternoon of the Greater Bay Area will increase by 11.4 ppb. The study results emphasized that although the influences of changes in anthropogenic emissions are large, climate change impacts on O_3 are also important.

Urbanization-induced changes in climatic conditions at the regional scale also affect air quality. Li M. *et al.* (2016) predicted that urbanization reduces daytime O_3 concentration by 1.3 ppb and increases daytime O_3 by 5.2 ppb in the Greater Bay Area. Such changes in O_3 are mainly due to stronger wind and increase in the height of atmospheric boundary layer that results in dilution in nitrogen oxides. This hence increases nighttime O_3 through titration and decreases daytime O_3 through weaker photochemistry. Another study evaluated the impacts of urbanization in the Greater Bay area on the secondary organic aerosol in March 2001 (Wang *et al.*, 2009), and estimated that urbanization could cause the concentrations of nitrogen dioxides and VOC decrease by 4.0 ppb and 1.5 ppb, respectively. Increases in temperature and decreases in wind speed can cause the concentrations of O_3 and nitrate increase by 2～4 ppb and 4～12 ppt, respectively.

The above results showed that urbanization has certain impacts on regional climate which hence affects O_3 concentration. In the highly urbanized PRD region, transboundary air pollution is significant. Wang *et al.* (2020) estimated that transboundary air pollution contributed to 46% of ozone-induced premature mortality. Yim *et al.* (2019b) estimated that urbanization increased O_3, which translated to a 39.6% increase in O_3 induced premature mortality (1 100 deaths). They also reported that urbanization would also change the atmospheric responses to emissions, and thus highlight the strong interactions between land use policies, urban climate adaptation strategies, and air quality policies, suggesting the need of beneficial strategies and policies. Yim *et al.* (2019b) proposed a precision environmental management concept that emphasizes the importance of considering the specific atmospheric condition and composition of a city when formulating its environmental policies.

2.4.3 Trends of eutrophication and hypoxia in the coastal waters around Pearl River Estuary

An estuary is an important transition zone where river water and seawater meet and mix. The estuarine hydrodynamic system is mainly regulated by freshwater discharge and salt water intrusion (Pritchard, 1967), resulting in unique biogeochemical processes of the estuarine ecosystem. The Pearl River Estuary (PRE) is a subtropical estuary embedded in the southern coast of China and connected with the continental shelf of the Northern South China Sea. The PRE has a triangular shape with Guangzhou at its northern apex, and Macao and Hong Kong at the southwest and southeast corners, respectively. The PRE has several very unique characteristics. It receives huge freshwater discharge with an annual average flow rate of around 10 000 m^3/s (Zhai *et al.*, 2005). The Pearl River is the 13th largest river in the world and the 2nd largest river in China in terms of freshwater discharge. The estuary encompasses a large area of about 1 900 km^2 and extends about 60 km from the river mouth to the open shelf. Its width varies from 10 km in the upper reach to 60 km in the lower reach and is wide enough that shelf circulation influences directly the lower estuary besides the gravitational circulation (Zu and Gan, 2015). The physical and biogeochemical processes in the PRE have strong seasonality because of the southwesterly/ northeasterly monsoon in summer/winter and the significant seasonal variation of river discharge. About 80% of the discharge from the Pearl River happens during the wet season from April to September (Zhai *et al.*, 2005). Meanwhile, variability induced by tidal forcing of 1 m magnitude exerts additional high frequency variation on the seasonal processes.

Hypoxia occurs when oxygen consumption in the water column cannot be replenished by supply. When the level of dissolved oxygen (DO) falls below 2～3 mg/L, hypoxia appears (Chu *et al.*, 2005; Dai *et al.*, 2006; Rabalais *et al.*, 2002). Due to the fast development of industrial and agricultural activities and urbanization in the past 30 years, the PRE receives a very high load of anthropogenic nutrients. The risk of eutrophication and hypoxia is rapidly increasing in the PRE, HKW (Hong Kong Waters) and the CTZ (Coastal Transition Zone) despite the massive sewage treatment projects in the region. Literature reports and our recent field surveys demonstrated that bottom hypoxia covers a wide range in this area (Qian W. *et al.*, 2018; Su *et al.*, 2017). Most eutrophication and hypoxia occurred in summer, because the buoyancy input and nutrient loading peaked during this period. Together with favorable wind and hydrodynamic conditions, phytoplankton bloom has frequently been observed in surface layer of the CTZ (Lu and Gan, 2015), and triggered strong and persistent oxygen deficiency in the bottom.

About the impacts of climate change, Figure 2–1a the summer wind velocity and wind friction velocity recorded at station Waglan Island in the south of HKW or east of CTZ from 1990 to 2015. We can find that the prevailing southwesterly winds generally decreased during this period. These suggest that the intensity of coastal upwelling was in a decreasing trend, and the water column stability regulated by wind force was constantly increasing during the whole period. There were also favorable physical conditions for the development of phytoplankton bloom in surface layer and for the maintenance of hypoxia in bottom layer. The decreasing trend of upwelling can also be demonstrated by the decreasing salinity and increasing temperature in bottom layer of the southern HKW (Figure 2–1b). Therefore, nutrient provided by the coastal upwelling is in a decreasing trend, and the rising NO_3^- concentration and the corresponding growing phytoplankton biomass in surface layer should mainly attributed to the increasing river input.

Water column ventilation serves as an important source of DO in bottom layer. Change of ocean ventilation often related to climate change such as global warming and can be affected by many processes, such as water column stratification and upwelling (Shepherd *et al.*, 2017). The increasing air temperature leads to increasing water temperature (Figure 2–2), and together with weakened the wind magnitudes (Figure 2–1a), it enhances water column stratification, and has significant ecological consequences.

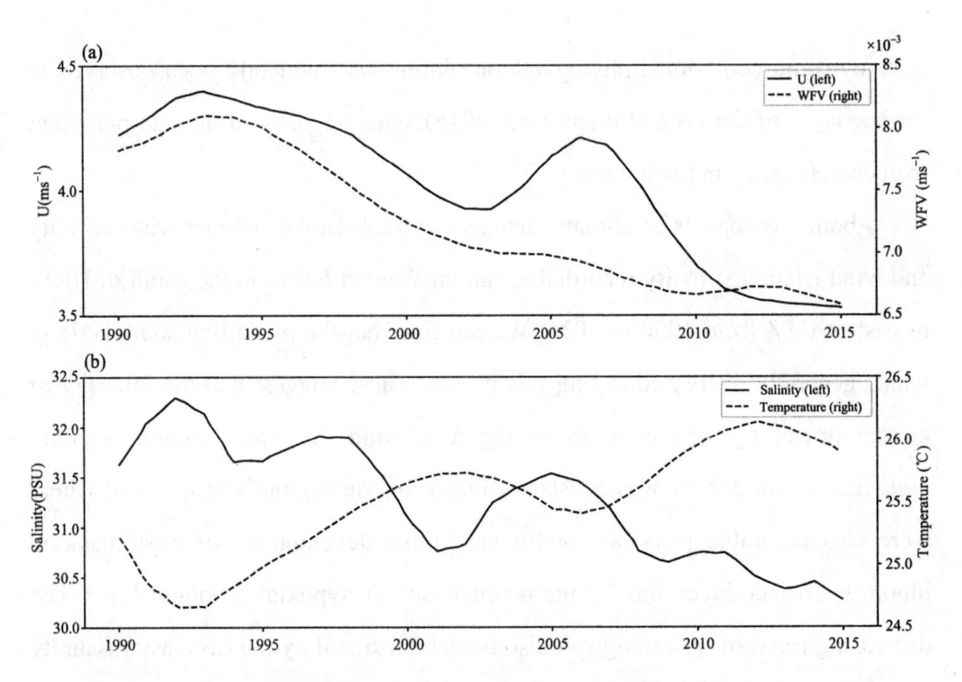

Figure 2–1 Time-series of (a) wind speed in along-shore north-eastward component (upwelling favorable) (blue solid line) and the wind friction velocity (red solid line), (b) bottom averaged seawater temperature (blue line) and salinity (red line), indicating the weakening of upwelling (Data are obtained from HKO and EPD of HK)

In summary, based on the long-term monitoring data in the ocean, atmosphere and Pearl River discharge, it is found that PRE and its adjacent coastal waters of Hong Kong have an increasing trend of eutrophication in the upper layer, hypoxia in the bottom during wet season. Both the eutrophication and hypoxia have reached an alarming level, and will lead to the acidification of the seawater in the region. Evidences show that these hazards were mainly caused by the increased nutrient loadings from Pearl River. It is also found that the changing climate, such as weakening of the wind speed and increase of air temperature may also contribute to the trends. It is also clear that the comprehensive studies of marine ecosystem and the underlying physical-biogeochemical dynamics in response to changing climate and to the rapid social-economic development in the region are urgently required for us to tackle these marine environmental issues in the region. Recent effort of OCEAN-HK project (https://ocean.ust.hk/) aims to develop science-based strategy for the eutrophication and hypoxia problem in the PRE and HKW.

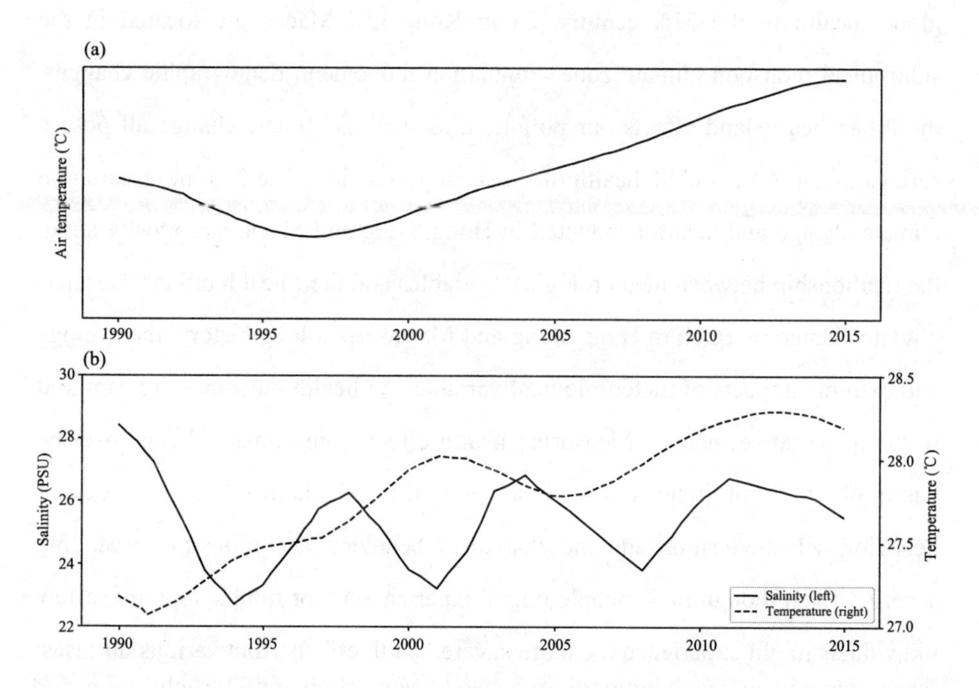

Figure 2–2 Time-series of (a) air temperature and (b) surface averaged seawater salinity (blue solid line) and temperature (red solid line) of the southern HKW (Data are obtained from HKO and EPD of HK)

2.5 Public Health, Risk Perception and Adaptation Strategies

Climate change is so far the greatest health challenge with potentially the broadest and most long-lasting effect on humanity, and is the greatest threat to global health in the 21st century. Hong Kong and Macao are located in the subtropical monsoon climate zone. Population movement, demographic changes, the urban heat island effects, air pollution, as well as climate change all pose a serious threat to the public health of the local population. The current research on climate change and health conducted in Hong Kong and Macao are mostly about the relationship between meteorological variables and their health effects. Because climate change research in Hong Kong and Macao has a long history, the average and extreme impacts of meteorological variables on health outcomes are captured in the quantitative models. Measuring health effect from climate change may be based on different indicators such as mortality, hospitalization, help seeking behavior, self-medication and the change of behavior, according to descending severity. Even though most people might experience no or mild symptoms, a few individuals might experience the more severe health effects from various diseases, including communicable disease and non-communicable disease (Chan, 2017; 2019). In recent years, meteorological variables such as temperature, humidity and rainfall have continued to be unstable, extreme weather events and natural disasters have become more frequent, and the interactions caused by anthropogenic factors, community environment and health service facilities are more complex and volatile. The research of the impact of climate change on population health is still in its infancy. This section will assess the impact of climate change on health risks among Hong Kong and Macao population and suggest public health policy to address those risks.

2.5.1 The impact of extreme temperature on the population mortality

Climate change is closely linked to the health of population (Huang Z. *et al.*, 2015) and it imposes a huge burden on the health care system. In Hong Kong, during hot season an average 1℃ increase in daily mean temperature above 28.2℃ was associated with an estimated 1.8% increase in mortality (Chan E. *et al.*, 2012). The number of cardiovascular and respiratory infection-related mortality has also increased significantly during hot days. Residents, who are female, married and with poor socioeconomic status are at high risk of adverse health outcomes. In the cold days, low temperature demonstrated a lagging effect on human health with the impact on mortality that could extend up till three weeks (Goggins *et al.*, 2013). A 1℃ drop in average daily temperature will increase 3.8% in the cumulative number of deaths of the next three weeks. When comparing the health impact of continuously hot and cold days, continuously very cold days will lead to higher accumulated mortality, especially for the elderly.

In addition to extreme temperatures, rapid urban development, high-density buildings in urban area, thermal emissions from transportation and industrial activities were also found to have caused major impacts on urban climate. A study found that urban heat island effects are directly associated with health risks. In hot days (average temperatures above 29℃), lower wind speeds were significantly associated with increased deaths, and the association was stronger in the region with high urban heat island indices (UHII) (Goggins *et al.*, 2012a).

2.5.2 The impact of extreme temperature on the population morbidity

In addition to the influence on mortality, temperature changes also have significant impacts on population morbidity. Relative humidity and air pollution level along the temperature variation were found to aggravate extreme temperature impact and led to an increase of disease related hospitalization. From 1998 to 2009, during the hot season (June-September), the number of hospitalizations increased by 4.5% for 1℃ increase in temperature above 29.0℃; during the cold season (November-March), when the temperature was between 8.2℃ and 26.9℃, the number of hospitalizations increased by 1.4% for 1℃ drop (Chan E. *et al.*, 2013).

Communicable disease

Climate change will affect the spread of infectious diseases directly or indirectly (e.g., through water, food and vector transmission), thereby affecting the risk of infection. Climate change increases water-borne intestinal disease risks through rising ambient air and sea temperature, rainfall increase, etc. In addition, climate change alters the habitats of insect vectors and may increase their reproduction rate and invasiveness, as well as shortening the extrinsic incubation period of pathogen. These all may lead to an increase in the incidence of vector-borne diseases. Urban populations are more susceptible to health impact of climate change due to its dense population, frequent traffic over borders, and urban heat island effects. Current temperature variation has already been confirmed to be associated the hospitalization due to infectious diseases in Hong Kong. During the hot days, hospitalizations due to infectious diseases may increase by 4.5% (95% CI: 2.2%~6.9%) for every 1℃ increase within the period when 10-days average temperature was 28.5℃ or higher; In addition, during the cold season, starting from 26.9 ℃ (average temperature), the number of hospitalizations increases by 1.0% (95% CI: 0.4%~1.6%) for every 1℃ drop (Chan E. *et al.*, 2013).

(1) Gastro-intestinal disease

Temperature and humidity affect the reproduction rate of intestinal infectious diseases pathogen as well as their survival time in the external environment. High temperature is beneficial to pathogen for survival and reproduction in the external environment as well as chance of disease transmission. In Hong Kong, meteorological factors have a significant impact on hand, foot and mouth disease (HFMD), salmonellosis, rotavirus and norovirus infection. Studies have suggested that there is a non-linear relationship between hand, foot and mouth disease and temperature in Hong Kong. When the average temperature is between 8℃ and 20℃ or higher than 25℃, the number of hand, foot and mouth disease admission will increase along with temperature; Meanwhile, when the relative humidity is higher than 80%, the number of hospital admission also rises sharply. In addition, the disease is also positively correlated with rainfall, strong winds, and solar radiation (Wang P. *et al.*, 2016).

(2) Salmonellosis

The incidence of salmonellosis is significantly associated with relative humidity, rainfall, and temperature. In a study about infectious diseases and temperature in Hong Kong (Wang P. *et al.*, 2018), hot temperature has a particular effect on hospitalizations of Salmonella. For every 1℃ increase in average temperature, the hospitalization rate of Salmonella would increase by 11.4% in the following 14 days. In addition to temperature, other meteorological factors are also associated with the number of general infectious disease related hospitalization. For example, average relative humidity is associated with the admission of infectious diseases with a "S" shaped statistical association. When the average relative humidity rises from 73% to 84%, the number of 10 days lagged hospitalizations has an increasing trend. Negative association was found between hospitalization and 10 days lagged average windspeed. When compared with the temperature of 13℃, there was a 6.13 (95% CI[①]: 3.52～10.67) times the risk of hospitalization at a temperature of 30.5℃. When humidity was above 60% and rainfall was between 0 and 0.14 mm, relative humidity, rainfall, and hospitalization for salmonellosis were positively associated. The extremely high level of relative humidity (96%) and micro-rain (0.02 mm) has the cumulative risk of salmonellosis admission of 2.06 (95% CI: 1.35～3.14) and 1.30 (95% CI: 1.01～1.67) in comparing to the relative humidity at 60% and zero rainfall respectively. In summary, high temperature, high relative humidity, and mild rainfall were positively associated with hospitalization for salmonellosis.

(3) Rotavirus and norovirus infection

Short-term changes in rainfall, temperature, and humidity are significantly associated with hospitalization due to rotavirus and norovirus infections among the children in Hong Kong (Wang P. *et al.*, 2018b). Rainfall was negatively associated with the rotavirus related admission due to rotavirus, but positively correlated with norovirus admission. In comparing to extremely low precipitation, extreme precipitation (99.5 mm, 99th percentile) was found to be associated with 0.40 (95% CI: 0.20～0.79) and 1.93 (95% CI: 1.21～3.09) times the risk of hospitalization due to rotavirus and norovirus infection respectively. Furthermore,

① CI: confidence interval.

the duration of association with rotavirus was notably longer than norovirus. Overall, higher temperatures were found to be associated with fewer hospitalizations for both rotavirus and norovirus infection, while higher relative humidity was generally associated with more norovirus, but fewer rotavirus, hospitalizations.

(4) Other infectious disease

International studies point out that climate change will affect the spread of other vector-borne diseases, the risk of emergence of new infectious disease such as influenza and COVID-19 etc., and exacerbate the epidemic.

Non-communicable disease

Climate change can have an impact on non-communicable diseases through both direct and indirect pathways. As shown in Table 2–1, climate change has caused an increase in respiratory and allergic, cardiovascular and mental disease among Hong Kong and Macao population.

Table 2–1 Impact of climate change on non-communicable diseases (Revised from Chan, 2019)

Effect	Climate phenomena	Route	Non-communicable disease in risk
Direct	Heat wave	Heat stress	Cardiovascular diseases Respiratory diseases
	Hot weather Drought Seasonal abnormality	Rising air pollutant concentration	Cardiovascular diseases Respiratory diseases
		Excess pollen or abnormal distribution	Respiratory and allergic diseases
	Ozone depletion	Excess ultraviolet light	Cancer Cataract
	Flooding Rainstorm	Drowning Building collapse	Harm (Physical, social and mental)

Continued

Effect	Climate phenomena	Route	Non-communicable disease in risk
Indirect	Drought Flooding	Food shortage Building damage Population migration	Poor health status Malnutrition Poor mental health
	Extreme weather event Flooding Typhoon Mountain fire	Traumatic experience Financial difficulty Loss of family members and friends	Poor mental health

(1) Respiratory system and allergic diseases

1) Chronic obstructive pulmonary disease (COPD)

During the cold season, consistent hockey-stick associations with temperature and relative humidity were found for chronic obstructive pulmonary disease admission among elderly (Lam H. *et al.*, 2018a). The minimum morbidity temperature and relative humidity were at about 21～22℃ and 82%. The lagged effects of low temperature were lagged 0～20 days.

2) Asthma

In the hot season, hospitalizations were lowest at 27℃, rose to a peak at 30℃, then a plateau was observed between 30℃ and 32℃ (Lam *et al.*, 2016). The cumulative relative risk for lags 0～3days for 30℃ vs 27℃ was 1.19 (95% CI: 1.06～1.34). In cold days, temperature is negatively associated with asthma hospitalizations. The cumulative relative risk lag (0～3 days) for 12℃ vs 25℃ was 1.33 (95% CI: 1.13～1.58). Adult asthma admissions were most sensitive to temperatures in both temperature extremes while admissions among children under 5 were least associated. Higher humidity and ozone levels in the hot temperature, and low humidity in the cold season were also associated with more asthma admissions.

3) Allergy

A study in 2017 pointed out that there are about 24% of Hong Kong residents have a history of allergies (Lam and Chan, 2018). 22.4% and 15.7% of the patients with skin and nasal reported that their symptoms worsen during hot weather respectively. In comparing to the people without a history of allergies, people with allergic symptoms described mucus secretions increase, more oral ulcers, worsen sleep quality, and depression during hot temperature.

(2) Cardio-vascular disease

Low temperatures can lead to an increase of emergency hospitalizations for cardiovascular disease. A study found that during the period 2005～2012, 7.15% of patients (37 285 patients) hospitalized for emergency treatment of the disease were attributable to hypothermia (Tian *et al.*, 2016).

1) Heart failure

Low temperature is closely related to increasing heart failure related hospital admission and mortality (Goggins and Chan, 2017). The cumulative relative risk (to 23 days) for heart failure admission was 2.63 (95% CI: 2.43～2.84) for an 11℃ vs a 25℃ day and cumulative relative risk (to 42 days) of heart failure mortality was 3.13 (95% CI: 1.90～5.16) (lag 0～42 days). The association with cold weather was stronger among older age groups and for new hospitalizations compared to recurrent ones. Both high and low relative humidity were modestly associated with more admissions.

2) Stroke

When the average daily temperature was between 8.2℃ and 31.8℃, the number of hemorrhagic stroke admission was negatively correlated with the daily mean temperature (Goggins *et al.*, 2012b). This finding indicated that lower mean temperature may result in more hemorrhagic stroke hospital admission. In addition, a 1℃ lower in 5-day average temperature was associated with a 2.7% (95% CI: 2.0%～3.4%) higher admission rate. This association was stronger among older subjects and females.

The association between ischemic stroke and temperature was weaker. Only below 22℃, with a 1℃ lower in 14-day average temperature below this threshold being associated with 1.6% higher ischemic stroke admission rate.

3) Acute myocardial infarction

Studies have found a relationship between temperature and hospital admission for patients with acute myocardial infarction in Hong Kong (Goggins *et al.*, 2013b). Patients with cardiovascular disease have an increased risk of acute myocardial infarction hospital admission in cold weather.

In the cold season, the acute myocardial infarction admission among diabetic patients was linearly and negatively associated with temperature. The cumulative relative risk of 12℃ versus 24℃ (lag 0～22 days) was 2.10 (95% CI: 1.62～2.72) (Lam H. *et al.*, 2018b). The association between myocardial infraction admission and temperature among non-diabetic patients was weaker, which the cumulative relative risk was 1.43 (95% CI: 1.21～1.69) and it only began when the temperature was below 22℃. In hot season, the acute myocardial infarction hospitalization among diabetic patients started increasing when temperature rise above 28.8℃, cumulative relative risk in lag 0～4 days (30.4℃ vs 28.8℃) is 1.14 (95% CI: 1.00～1.31). For the non-diabetic patients, the rate of infarction admission is not related to temperature.

(3) Mental health

Extreme weather events have a significant psychological impact on the affected communities, which can be directly caused by wounds, or indirectly through physical stress (such as the impact of hot weather) or environmental degradation and the destruction of important community assets. There was a linear positive correlation between temperature and the number mental disorder admission and the effect lasted about 2 days (Chan E. *et al.*, 2018). When the temperature was above 19.4℃ (25% quantile), the relative risk increased significantly. The cumulative risk of 28℃ (75% quantile) for admission to the psychological disorder (lag 0～2 days) in comparing to 19.4℃ was 1.09 (95% CI: 1.03～1.15).

In sub-disease analyses, transient mental disorders and episodic mood disorders showed a positive association with temperature while drug-related mental disorders demonstrated a positive association when temperature rose over a threshold of about 20℃. The association with transient mental disorders was strong and significant (RR 1.51 (95% CI: 1.00～2.27)), followed in strength by episodic mood disorders (RR 1.34 (95% CI: 1.05～1.71)), and then drug-related mental disorders (RR 1.13 (95% CI: 1.00～1.27)). Depressive disorder and other nonorganic psychoses showed a significantly lower risk at lower temperature. No obvious association with temperature was observed for anxiety, dissociative and somatoform disorders, schizophrenic disorders, and alcohol-related mental disorders.

2.5.3 Public health response strategy

This section summarizes the health risk perceptions of Hong Kong and Macao residents under the background of climate change and the corresponding actions taken to protect the health of urban populations.

Health protection measures against heat wave

A Hong Kong telephone survey study in 2017 showed that 45.3% of people believed that hot weather did not affect their health, among which, 28.8% were of 65 years old or older (Chan E. *et al.*, 2018b). During this period, 37.2%～97.5% of the respondents adopted at least one personal heat protection measure. The most common self-protective measure was to "drink more water" (97.5%), and the least was to "use sunscreen" (37.2%). Male participants were less likely to avoid sunlight (OR = 0.29, 95% CI: 0.13～0.65) and using sunscreen (OR = 0.41, 95% CI: 0.26～0.64). Air conditioners were less used by people with lower education levels (OR = 0.22, 95% CI: 0.05～0.91). The hot weather warning service can cover the most parts of Hong Kong, so more people take personal precautions. About 87% of the people knew that hot weather warning was issued by the HKO during the study period.

Another large telephone survey found that 90.3% of respondents had electric fans and 90% of participants had air conditioner devices (Gao *et al.*, 2020). In hot days, 91.5% of them use air conditioners, and 95.1% use fans. About half of the respondents (54.3%) reported that when indoor reached a certain temperature, cooling device would be turned on automatically. Among them, 58.1% and 95.4% of the respondents would open at 28℃ and 31℃ or higher respectively. People who are older, with lower education levels, lower household incomes, unemployment/retirement/housewives, living in public housing and with chronic diseases reported to have a lower rate of air conditioner ownership, or not using air conditioner even they feel very hot. Households with low levels of education, unemployment, living in public housing, and low-income households were the groups that less likely use air conditioner. Among the relationship, the one of household income and the use of air conditioning is the strongest. This finding is consistent with other studies (Hansen *et al.*, 2011), which explains to some extent why poor people are vulnerable to hot weather.

Although more than half of the respondents (57.7%) believe that heat waves may have an adverse effect on health, the risk perception was not significantly associated with whether they own or use air conditioner at their home. Even they think heat waves is a health risk factor, but due to variety of reasons (such as lack of coping methods), they ultimately did not take the actual action. Of course, this does not rule out that the respondent may have taken other protective measures (for example, go to places with air conditioning to avoid the heat). The above results reflect that individual financial situation appeared to be an important factor for the use of air conditioning or related device in southern Chinese metropolis, which it is consistent with other studies (Hansen *et al.*, 2011). This also explains why the poor are particularly vulnerable to hot weather.

Health protection measures against cold wave

In a telephone survey study in 2016 in Hong Kong, 45.0% of respondents were identified that they have higher underlying social and health risks (e.g., the elderly, subjects reported history of chronic diseases, living alone and receiving low-income assistance from the government), but 63.7% in that population underestimated their health risks in cold weather (Lam *et al.*, 2020). In 2016, Hong Kong citizens tended to take more protective measure against cold weather during the 2017 cold spell. People who experienced adverse health effects during the 2016 cold snap will wear more clothes during the 2017 cold spell (OR = 5.48, 95% CI: 1.26～23.83). Women (OR = 1.93, 95% CI: 1.20～3.11), those who felt cold during the study period (OR = 3.72, 95% CI 1.29～10.72) and the elderly aged 70 years or above (OR = 4.00, 95% CI: 1.12～13.19) were more likely to use heating equipment.

Typhoon preparedness

As the frequency and severity level of extreme weather events associated with climate change are increasing, understanding capacity of household preparedness is important for disaster risk management. A study in Hong Kong in 2018 examined association among disaster risk perception, family preparedness, and the immediate effects of typhoon Mangkhut (Chan *et al.*, 2019). Among 521 respondents, 93.9% and 74.3% reported they had taken routine and special emergency preparedness measures for typhoon Mangkhut respectively. 33.4% of the respondents reported that Typhoon Mangkhut had some form of influences on them. Respondents who perceive their home were at high risk during the typhoon and had routine emergency preparedness were more likely to have special emergency preparedness for the typhoon. However, there was no significant relationship between the implementation of special measures and the impact of typhoon. The study shows that current household emergency preparedness measures may not be sufficient to deal with the effects of a super typhoon. Urban context with well infrastructure and public communication remains vulnerable to extreme events and requires an effective disaster preparedness strategy to deal with extreme events related to climate change.

Health co-benefit effect in climate mitigation strategies

Many climate change mitigation measures also provide a co-beneficial effect on public health and environmental protection. A telephone survey in Hong Kong (Chan *et al.*, 2017) found that the most practiced health co-benefit behavior was using less packaging and disposable shopping bags, where 70.1% of participants claimed doing it every day. Several co-benefit behaviors had a daily practice among approximately 50% of the population, including Walk/cycle more (54.8%), Separate household waste (50.2%), Use less electricity (48.3%), and Use less AC (44.1%). Those that were practiced daily by around 30% of the population included Consume less meat (33.3%) and Shower less than five minutes every day (23.7%). Behaviors that were practiced daily by less than 10% of the population were: have one vegetarian meal a week (5.8%), buy more organic food (4.3%), and bring personal eating utensils when dining in restaurants or small eateries (4.0%). In general, female and older people are more likely to adopt co-benefit behaviors in their daily lives.

Although local government has already implemented a series of measures to promote the health co-benefit behaviors among Hong Kong's population, the research study revealed impact of these measures seemed to be diverse. For example, the Plastic Shopping Bag Levy Scheme implemented in 2015, which charges HKD 0.50 per plastic shopping bag, was demonstrated to be relatively successful because 70.1% of participants self-reported to use less packaging and fewer disposable shopping bags daily. It indicated that the plastic shopping bag charging scheme is relatively successful. Meanwhile, the Water Supplies Department organized a promotion initiative to conserve water (e.g., by taking shorter showers) through the "Let's Save 10L Water campaign", putting forward a number of suggestions. According to the domestic water consumption survey conducted by the Water Supplies Department in 2015/2016, about 98% of the respondents have adopted one or more water conservation measures, which suggested that the public was aware of the importance of water conservation. However, our findings indicated nearly half of the residents had never practiced nor considered showering less than five minutes every day. Therefore, the government should conduct more investigation and promotion on the awareness and the practice of co-benefit behaviors, especially the behaviors with low frequency in daily practice. Consideration should also be given to the role of community and health care practitioners in promoting co-benefit behavior.

References

Ang, B. W., H. Wang, X. Ma, 2017. Climatic Influence on Electricity Consumption: The Case of Singapore and Hong Kong. *Energy*, 127.

Ayyub, B. M., H. G. Braileanu, N. Qureshi, 2012. Prediction and Impact of Sea Level Rise on Properties and Infrastructure of Washington, DC. *Risk Analysis*, 32(11).

Benfield, A., 2018. *Weather, Climate & Catastrophe Insight: 2017 Annual Report*. AON Empower Results.

Campbell, S., 2005. *Typhoons Affecting Hong Kong: Case Studies APEC 21st Century COE Short-Term Fellowship Report*. Tokyo Polytechnic University.

Cha, E. J., T. R. Knutson, T. C. Lee, *et al.*, 2020. Third Assessment on Impacts of Climate Change on Tropical Cyclones in the Typhoon Committee Region – Part II: Future Projections. *Tropical Cyclone Research and Review*, 9(2).

Civil Engineering and Development Department (CEDD). 2020. *Management of Natural Terrain Landslide Risk*, Information Note 29/2020 (https://www.cedd.gov.hk/filemanager/eng/content_454/IN_2020_29E.pdf).

Chan, E. Y. Y., 2017. *Public Health Humanitarian Responses to Natural Disasters*. London: Routledge.

Chan, E. Y. Y., 2019. *Climate Change and Urban Health: the Case of Hong Kong as a Subtropical City*. London: Routledge.

Chan, E. Y. Y., W. B. Goggins, J. J. Kim, *et al.*, 2012. A Study of Intracity Variation of Temperature-Related Mortality and Socioeconomic Status Among the Chinese Population in Hong Kong. *Journal of Epidemiology and Community Health*, 66(4).

Chan, E. Y. Y., W. B. Goggins, J. S. Yue, *et al.*, 2013. Hospital Admissions as a Function of Temperature, Other Weather Phenomena and Pollution Levels in an Urban Setting in China. *Bulletin of the World Health Organization*, 91(8).

Chan, E. Y. Y., H. C. Y. Lam, S. H. W. So, *et al.*, 2018a. Association Between Ambient Temperatures and Mental Disorder Hospitalizations in a Subtropical City: A Time-Series Study of Hong Kong Special Administrative Region. *International Journal of Environmental Research and Public Health*, 15(4).

Chan, E. Y. Y., A. Y. T. Man, H. C. Y. Lam, 2018b. *Personal Heat Protective Measures During the 2017 Heatwave in Hong Kong: A Telephone Survey Study*. First Global Forum for Heat and Health, Hong Kong, 2018.

Chan, E. Y. Y., A. Y. T. Man, H. C. Y. Lam, *et al.*, 2019. Is Urban Household Emergency Preparedness Associated with Short-Term Impact Reduction After a Super Typhoon in Subtropical City? *International Journal of Environmental Research and Public Health*, 16(4).

Chan, E. Y. Y., S. S. Wang, J. Y. Ho, *et al.*, 2017. Socio-Demographic Predictors of Health and Environmental Co-Benefit Behaviours for Climate Change Mitigation in Urban China. *PLoS One*, 12(11).

Chan, F. K. S., G. Mitchell, O. Adekola, *et al.*, 2012. Flood Risk in Asia's Urban Mega-Deltas: Drivers, Impacts and Response. *Environment and Urbanization ASIA*, 3(1).

Chan, F. K. S., O. Adekola, G. Mitchell, *et al.*, 2013. Appraising Sustainable Flood Risk Management in the Pearl River Delta's Coastal Megacities: A Case Study of Hong Kong, China. *Journal of Water and Climate Change*, 4(4).

Chan, F. K. S., N. Wright, X. Cheng, *et al.,* 2014. After Sandy: Rethinking Flood Risk Management in Asian Coastal Megacities. *Natural Hazards Review*, 15.

Chan, F. K. S., C. C. Joon, A. Ziegler, *et al.*, 2018. Towards Resilient Flood Risk Management for Asian Coastal Cities: Lessons Learned from Hong Kong and Singapore. *Journal of Cleaner Production*, 187.

Chan, J. C. L., M. Xu, 2009. Inter-Annual and Inter-Decadal Variations of Landfalling Tropical Cyclones in East Asia. Part I: Time Series Analysis. *International Journal of Climatology*, 29(9).

Chan, W. L., 1996. *Hong Kong Rainfall and Landslides in 1994 (GEO Report No. 54).* Hong Kong, Geotechnical Engineering Office.

Chen, X., S. J. Jeong, 2018. Shifting the Urban Heat Island Clock in a Megacity: A Case Study of Hong Kong. *Environmental Research Letters*, 13(1).

Chen, Y. D., Q. Zhang, X. Chen, *et al.*, 2012. Multiscale Variability of Streamflow Changes in the Pearl River Basin, China. *Stochastic Environmental Research and Risk Assessment*, 26(2).

Chen, Y. D., Q. Zhang, X. Lu, *et al.*, 2011. Precipitation Variability (1956-2002) in the Dongjiang River (Zhujiang River Basin, China) and Associated Large-Scale Circulation. *Quaternary International*, 244.

Chen, Y. D., 2001. Sustainable Development and Management of Water Resources for Urban

Water Supply in Hong Kong. *Water International*, 26(1).

Choi, C. E., C. W. W. Ng, H. Liu, *et al.*, 2019. Interaction Between Dry Granular Flow and Rigid Barrier with Basal Clearance: Analytical and Physical Modelling. *Canadian Geotechnical Journal*, 57(2).

Chow, E. C. H., R. C. Y. Li, W. Zhou, 2018. Influence of Tropical Cyclones on Hong Kong Air Quality. *Advances in Atmospheric Sciences*, 35(9).

Choy, C. W., D. S. Lau, Y. He, 2020a. Super Typhoons Hato (1713) and Mangkhut (1822), Part I: Analysis of Maximum Intensity and Wind Structure. *Weather*.

Choy, C. W., D. S. Lau, Y. He, 2020b. Super Typhoons Hato (1713) and Mangkhut (1822), Part II: Challenges in Forecasting and Early Warnings. *Weather*.

Chu, P., Y. C. Chen, A. Kuninaka, 2005. Seasonal Variability of the Yellow Sea/East China Sea Surface Fluxes and Thermohaline Structure. *Advances in Atmospheric Sciences*, 22(1).

Chui, S. K., J. K. Y. Leung, C. K. Chu, 2006. The Development of a Comprehensive Flood Prevention Strategy for Hong Kong. *International Journal of River Basin Management*, 4(1).

Dai, M., X. Guo, W. Zhai, *et al.*, 2006. Oxygen Depletion in the Upper Reach of the Pearl River Estuary During a Winter Drought. *Marine Chemistry*, 102(1-2).

Dawson, D., J. Shaw, W. R. Gehrels, 2016. Sea-level Rise Impacts on Transport Infrastructure: The Notorious Case of the Coastal Railway Line at Dawlish, England. *Journal of Transport Geography*, 51.

Ding, Y., J. C. Chan. 2005. The East Asian Summer Monsoon: An Overview. *Meteorology and Atmospheric Physics*, 89(1-4).

Drainage Services Department (DSD), 2019. *Flood Prevention Strategy*. Drainage Services Department, HKSAR.

Elsner, J. B., J. P. Kossin, T. H. Jagger, 2008. The Increasing Intensity of the Strongest Tropical Cyclones. *Nature*, 455(7209).

Gao, Y., E. Y. Y. Chan, H. C. Y. Lam, *et al.*, 2020. Risk Perception of Climate Change and Utilization of Fans and Air Conditioners in a Representative Population of Hong Kong. *International Journal of Disaster Risk Science*, 11.

Gasparrini, A., Y. Guo, M. Hashizume, *et al.*, 2015. Mortality Risk Attributable to High and Low Ambient Temperature: A Multicountry Observational Study. *The Lancet*, 386(9991).

Geotechnical Engineering Office (GEO), 2012. *Geotechnical Engineering Office Report 272: Detailed Study of the 7 June 2008 Landslide on the Hillside Above North Lantau Highway and Cheung Tung Road, North Lantau*. Geotechnical Engineering Office of the Civil Engineering and Development Department, HKSAR.

Geotechnical Engineering Office (GEO), 2016. *Report on Natural Terrain Landslide Hazards in*

Hong Kong. Geotechnical Engineering Office of the Civil Engineering and Development Department, HKSAR.

Ginn, W. L., T. C. Lee, K. Y. Chan, 2010. Past and Future Changes in the Climate of Hong Kong. *Acta Meteorologica Sinica*, 24(2).

Giridharan, R., S. Ganesan, S. S. Y. Lau, 2004. Daytime Urban Heat Island Effect in High-Rise and High-Density Residential Developments in Hong Kong. *Energy and Buildings*, 36(6).

Global Water Partnership China, 2016. *Water Issues in the Pearl River Delta under Climate Change and Its Countermeasures and Control Measures*. https://www.gwp.org/globalassets/global/gwp-china_files/wacdep/_3.pdf (in Chinese).

Goggins, W. B., E. Y. Y. Chan, 2017. A Study of the Short-Term Associations Between Hospital Admissions and Mortality from Heart Failure and Meteorological Variables in Hong Kong: Weather and Heart Failure in Hong Kong. *International Journal of Cardiology*, 228.

Goggins, W. B., E. Y. Y. Chan, E. Ng, *et al.*, 2012a. Effect Modification of the Association Between Short-term Meteorological Factors and Mortality by Urban Heat Islands in Hong Kong. *PLoS One*, 7(6).

Goggins, W. B., E. Y. Y. Chan, C. Y. Yang, 2013b. Weather, Pollution, and Acute Myocardial Infarction in Hong Kong and Taiwan. *International Journal of Cardiology*, 168(1).

Goggins, W. B., E. Y. Y. Chan, C. Yang, *et al.*, 2013a. Associations Between Mortality and Meteorological and Pollutant Variables During the Cool Season in Two Asian Cities with Sub-tropical Climates: Hong Kong and Taipei. *Environmental Health*, 12(1).

Goggins, W. B., J. Woo, S. Ho, et al., 2012b. Weather, Season, and Daily Stroke Admissions in Hong Kong. *International Journal of Biometeorology*, 56(5).

Greenpeace China, 2009. *The "Climate Change Bill": Economic Costs of Heavy Rainstorm in Hong Kong*. Greenpeace China: Hong Kong.

Hansen, A., P. Bi, M. Nitschke, *et al.*, 2011. Perceptions of Heat-Susceptibility in Older Persons: Barriers to Adaptation. *International Journal of Environmental Research and Public Health*, 8(12).

He, Y. H., H. Y. Mok, E. S. Lai, 2016. Projection of Sea-Level Change in the Vicinity of Hong Kong in the 21st Century. *International Journal of Climatology*, 36(9).

Hong Kong Observatory (HKO), 2019. *Hong Kong Observatory: Tropical Cyclones in 2017*, https://www.hko.gov.hk/publica/tc/TC2017.pdf.

Hong Kong Observatory (HKO), 2020. *Hong Kong Observatory: Tropical Cyclones in 2018*, https://www.hko.gov.hk/en/publica/tc/files/TC2018.pdf.

Ho, H. C., K. K. L. Lau, C. Ren, *et al.*, 2017. Characterizing Prolonged Heat Effects on Mortality in a Sub-tropical High-density City, Hong Kong. *International Journal of*

Biometeorology, 61(11).

Ho, K., S. Lacasse, L. Picarelli, 2019. Slope Safety Preparedness for Impact of Climate Change. CRC Press.

Ho, K. K. S., H. W. Chan, T. M. Lam, 2002. *Review of 1999 Landslides (GEO Report No. 127)*. Hong Kong, Geotechnical Engineering Office.

Hong Kong Environment Bureau, 2015, *Hong Kong Climate Change Report 2015*, https://www.enb.gov.hk/sites/default/files/pdf/ClimateChangeEng.pdf.

Hong Kong Environment Bureau, 2017, *Hong Kong's Climate Action Plan 2030+*, https://www.climateready.gov.hk/files/report/en/HK_Climate_Action_Plan_2030+_booklet_En.pdf.

Hong Kong LegCo, 2015. *Water Resources of Hong Kong.*

Hong Kong LegCo, 2017. *Overview of the Supply of Dongjiang Water to Hong Kong*, https://www.legco.gov.hk/research-publications/chinese/1617fsc09-overview-of-the-supply-of-dongjiang-water-to-hong-kong-20170403-c.pdf.

Hong Kong Water Supplies Department, 2008. *Total Water Management in Hong Kong: Towards Sustainable Use of Water Resources*, https://www.wsd.gov.hk/filemanager/sc/share/pdf/TWM.pdf.

Hong Kong Water Supplies Department 2011, *Milestones of Hong Kong Water Supply*, https://www.wsd.gov.hk/filemanager/common/teaching_kit/pdf/Book2.pdf.

Hong Kong Water Supplies Department, 2018. *Annual Report 2016-2017*, https://www.wsd.gov.hk/filemanager/common/annual_report/2016_17/common/pdf/wsd_annual_report2016-2017.pdf.

Hong Kong Water Supplies Department, 2019. *Total Water Management Strategy 2019: Strategy for Sustainable Water Supply in Hong Kong*, https://www.wsd.gov.hk/filemanager/en/content_1866/twm-strategy-2019-e.pdf.

Huang, P., I. I. Lin, C. Chou, *et al.*, 2015. Change in Ocean Subsurface Environment to Suppress Tropical Cyclone Intensification Under Global Warming. *Nature Communications*, 6.

Huang, Z., H. Lin, Y. Liu, *et al.*, 2015. Individual-level and Community-level Effect Modifiers of the Temperature–mortality Relationship in 66 Chinese Communities. *BMJ Open*, 5(9).

Huang, Z., Y. Zong, W. Zhang, 2004. Coastal Inundation due to Sea Level Rise in the Pearl River Delta, China. *Natural Hazards*, 33.

Hudson, R. R., 1993. *Report on the Rainstorm of August 1982* (*GEO Report No. 26*). Hong Kong, Geotechnical Engineering Office.

Hulme, M., G. J. Jenkins, X. Lu, *et al.*, 2002. *Climate Change Scenarios for the United Kingdom: The UKCIP02 Scientific Report*, Tyndall Centre for Climate Change Research,

School of Environmental Sciences, University of East Anglia. https://artefacts.ceda.ac.uk/badc_datadocs/link/UKCIP02_tech.pdf, 2002.

IPCC, 2013. *Climate Change 2013: The Physical Science Basis. Contribution of Working Group I to the Fifth Assessment Report of the Intergovernmental Panel on Climate Change* [Stocker, T. F., D. Qin, G. K. Plattner, *et al.* (eds.)]. Cambridge University Press, Cambridge, United Kingdom and New York, NY, USA

IPCC, 2014. *Climate Change 2014: Impacts, Adaptation, and Vulnerability. Part B: Regional of the Intergovernmental Panel on Climate Change* [Barros, V. R., C. B. Field, D. J. Dokken, *et al.* (eds.)]. Cambridge University Press, Cambridge, United Kingdom and New York, NY. USA

Jovanović, S., S. Savić, M. Bojić, *et al.,* 2015. The Impact of the Mean Daily Air Temperature Change on Electricity Consumption. *Energy*, 88.

Kang, N. Y. and J. B. Elsner, 2012. Consensus on Climate Trends in Western North Pacific Tropical Cyclones. *Journal of Climate*, 25(21).

Knutson, T. R., J. J. Sirutis, M. Zhao, *et al.*, 2015. Global Projections of Intense Tropical Cyclone Activity for the Late Twenty-First Century from Dynamical Downscaling of CMIP5/RCP4. 5 Scenarios. *Journal of Climate*, 28(18).

Kong, L., X. Chen, W. Ping, *et al.*, 2011. Analysis on Severe Saltwater Intrusion in Modaomen Channel of the Pearl River Estuary in Dry Season During 2009/2010. *Journal of Natural Resources*, 26(11) (in Chinese).

Kong, L., K. K. L. Lau, C. Yuan, *et al.*, 2017. Regulation of Outdoor Thermal Comfort by Trees in Hong Kong. *Sustainable Cities and Society*, 31.

Kossin, J. P., 2018. A Global Slowdown of Tropical-Cyclone Translation Speed. *Nature*, 558.

Kwan, J. S. H., R. C. H. Koo, C. W. W. Ng, 2015. Landslide Mobility Analysis for Design of Multiple Debris-Resisting Barriers. *Canadian Geotechnical Journal*, 52(9).

Lai, Y. C., J. F. Li, X. Gu, *et al.*, 2020. Greater Flood Risks in Response to Slowdown of Tropical Cyclones over the Coast of China. *Proceedings of the National Academy of Sciences of the United States of America*, 117(26).

Lam, H. C. Y., E. Y. Y. Chan, 2018. *The Impact of High Temperature on Existing Allergic Symptoms Among an Adult Population (CCOUC Working Paper Series)*. Hong Kong: Collaborating Centre for Oxford University and CUHK for Disaster and Medical Humanitarian Response.

Lam, H. C. Y., E. Y. Y. Chan, W. B. Goggins, 2018a. Comparison of Short-Term Associations with Meteorological Variables Between COPD and Pneumonia Hospitalization Among the Elderly in Hong Kong—A Time-Series Study. *International Journal of Biometeorology*, 62.

Lam, H. C. Y., J. C. N. Chan, A. O. Y. Luk, *et al.*, 2018b. Short-term Association Between Ambient Temperature and Acute Myocardial Infarction Hospitalizations for Diabetes Mellitus Patients: A Time Series Study. *PLoS Medicine*, 15(7).

Lam, H. C., A. M. Li, E. Y. Y. Chan, *et al.*, 2016. The Short-term Association Between Asthma Hospitalisations, Ambient Temperature, Other Meteorological Factors and Air Pollutants in Hong Kong: A Time-series Study. *Thorax*, 71(12).

Lam, H. C., Z. Huang, E. Y. Y. Chan, *et al.*, 2020. Personal Cold Protection Behaviour and Its Associated Factors in 2016/17 Cold Days in Hong Kong: A Two-Year Cohort Telephone Survey Study. *International Journal of Environmental Research and Public Health*, 17(5).

Lam, Y. F., H. M. Cheung, C. C. Ying, 2018. Impact of Tropical Cyclone Track Change on Regional Air Quality. *Science of the Total Environment*, 610.

Lee, B. Y., 1983. *Hydrometeorological Aspects of the Rainstorms in May 1982*. Royal Observatory.

Lee, B. Y., W. T. Wong, W. C. Woo, 2010. *Sea-level Rise and Storm Surge-impacts of Climate Change on Hong Kong*. HKIE civil division conference.

Lee, K. L., Y. H. Chan, T. C. Lee, *et al.*, 2015. The Development of the Hong Kong Heat Index for Enhancing the Heat Stress Information Service of the Hong Kong Observatory. *International Journal of Biometeorology*, 60(7).

Lee, T. C., T. R. Knutson, H. Kamahori, *et al.*, 2012. Impacts of Climate Change on Tropical Cyclones in the Western North Pacific Basin. Part I: Past Observations. *Tropical Cyclone Research and Review*, 1(2).

Lee, T. C., T. R. Knutson, T. Nakaegawa, *et al.*, 2020. Third Assessment on Impacts of Climate Change on Tropical Cyclones in the Typhoon Committee Region. Part I: Observed Changes, Detection and Attribution. *Tropical Cyclone Research and Review*, 9 (1).

Leichenko, R., 2011. Climate Change and Urban Resilience. *Current Opinion in Environmental Sustainability*, 3(3).

Li, J., Y. D. Chen, L. Zhang, *et al.*, 2016a. Future Changes in Floods and Water Availability Across China: Linkage with Changing Climate and Uncertainties. *Journal of Hydrometeorology*, 17.

Li, J., L. Zhang, X. Shi, *et al.*, 2016b. Response of Long-term Water Availability to More Extreme Climate in the Pearl River Basin, China. International Journal of Climatology, 37(7).

Li, M., Y. Song, Z. Mao, *et al.*, 2016. Impacts of Thermal Circulations Induced by Urbanization on Ozone Formation in the Pearl River Delta Region, China, *Atmospheric Environment*, 127(2).

Lin, J., 2014. *Vulnerable and Lagging Behind: The Case of Hong Kong*. Adaptation to Climate

Change in Asia.

Liu, B., J. Chen, W. Lu, *et al.*, 2015. Spatiotemporal Characteristics of Precipitation Changes in the Pearl River Basin, China. *Theoretical and Applied Climatology*, 123(3).

Liu, D., X. Chen, Y. Lian, *et al.*, 2010. Impacts of Climate Change and Human Activities on Surface Runoff in the Dongjiang River Basin of China. *Hydrological Processes*, 24(11).

Liu, Q., K. S. Lam, F. Jiang, *et al.*, 2013. A Numerical Study of the Impact of Climate and Emission Changes on Surface Ozone over South China in Autumn Time in 2000-2050. *Atmospheric Environment*, 76.

Lu, Z., J. Gan, 2015. Controls of Seasonal Variability of Phytoplankton Blooms in the Pearl River Estuary. *Deep Sea Research Part II: Topical Studies in Oceanography*, 117.

Lo, D. O. K., 2000. *Review of Natural Terrain Landslide Debris-resisting Barrier Design*. Geotechnical Engineering Office, Civil Engineering Department, HKSAR.

Lok, C. C. F., J. C. L. Chan, 2018. Changes of Tropical Cyclone Landfalls in South China Throughout the Twenty-first Century. *Climate Dynamics*, 51(7).

Luo, M., N. C. Lau, 2019. Amplifying Effect of ENSO on Heat Waves in China. *Climate Dynamics*, 52(5-6).

Luo, M., X. Hou, Y. Gu, *et al.*, 2018. Trans-boundary Air Pollution in a City Under Various Atmospheric Conditions. *Science of the Total Environment*, 618.

Luo, Y., 2017. *Advances in Understanding the Early-summer Heavy Rainfall over South China*. The Global Monsoon System: Research and Forecast.

Macao Maritime and Water Affairs Bureau, 2017. *Water Resources and Supply in Macao 2014-2016* (in Chinese).

Macao Maritime and Water Affairs Bureau, 2021. *Water Resources in Macao*.

Mei, W., S. P. Xie, 2016. Intensification of Landfalling Typhoons over the Northwest Pacific Since the Late 1970s. *Nature Geoscience*, 9.

Ng, C. W. W., C. E. Choi, R. P. H. Law, 2013. Longitudinal Spreading of Granular Flow in Trapezoidal Channels. *Geomorphology*, 194.

Ng, C. W. W., C. E. Choi, A. Y. Su, *et al.*, 2016a. Large-scale Successive Impacts on a Rigid Barrier Shielded by Gabions. *Canadian Geotechnical Journal*, 53(10).

Ng, C. W. W., D. Song, C. E. Choi, *et al.*, 2016b. A Novel Flexible Barrier for Landslide Impact in Centrifuge. *Géotechnique Letters*, 6(3).

Ng, C. W. W., A. K. Leung, R. Yu, *et al.*, 2016c. Hydrological Effects of Live Poles on Transient Seepage in an Unsaturated Soil Slope: Centrifuge and Numerical Study. *Journal of Geotechnical and Geoenvironmental Engineering*, 143(3).

Ng, C. W. W., A. Y. Su, C. E. Choi, *et al.*, 2018a. Comparison of Cushion Mechanisms Between Cellular Glass and Gabions Subjected to Successive Boulder Impacts. *Journal of*

Geotechnical and Geoenvironmental Engineering, 144(9).

Ng, C. W. W., C. E. Choi, R. C. H. Koo, *et al.*, 2018b. Dry Granular Flow Interaction with Dual-Barrier Systems. *Géotechnique*, 68(5).

Ng, C. W. W., A. Leung, J. Ni, 2019a. *Plant-soil Slope Interaction*. CRC Press.

Ng, C. W. W., R. Chen, J. L. Coo, *et al.*, 2019b. A Novel Vegetated Three-layer Landfill Cover System Using Recycled Construction Wastes Without Geomembrane. *Canadian Geotechnical Journal*, 56(12).

Ni, J. J., A. K. Leung, C. W. Ng, 2018. Unsaturated Hydraulic Properties of Vegetated Soil Under Single and Mixed Planting Conditions. *Géotechnique*, 69(6).

Oke, T. R., 1982. The Energetic Basis of the Urban Heat Island. *Quarterly Journal of the Royal Meteorological Society*, 108(455).

Peng, L., J. P. Liu, Y. Wang, *et al.*, 2018. Wind Weakening in a Dense High-rise City due to Over Nearly Five Decades of Urbanization. *Building and Environment*, 138.

Peng, L. L. H., C. Y. Jim, 2015. Economic Evaluation of Green-roof Environmental Benefits in the Context of Climate Change: The Case of Hong Kong. *Urban Forestry & Urban Greening*, 14(3).

Pritchard, D. W., 1967. *What Is an Estuary: Physical Viewpoint, in Estuaries*. American Association for the Advancement of Science, Washington D.C.

Qian, C., W. Zhou, X. Q. Yang, *et al.*, 2018. Statistical Prediction of non-Gaussian Climate Extremes in Urban Areas Based on the First-order Difference Method. *International Journal of Climatology*, 38(6).

Qian, W., J. Gan, J. Liu, *et al.*, 2018. Current Status of Emerging Hypoxia in a Eutrophic Estuary: The Lower Reach of the Pearl River Estuary, China. *Estuarine, Coastal and Shelf Science*, 205.

Qiu, H., S. Sun, R. Tang, *et al.*, 2016. Pneumonia Hospitalization Risk in the Elderly Attributable to Cold and Hot Temperatures in Hong Kong, China. *American Journal of Epidemiology*, 184(8).

Rabalais, N. N., R. E. Turner, W. J. Wiseman Jr., 2002. Gulf of Mexico Hypoxia, Aka "The Dead Zone". *Annual Review of Ecology and Systematics*, 33(1).

Ren, C., E. Y. Ng, L. Katzschner, 2011. Urban Climatic Map Studies: A Review. *International Journal of Climatology*, 31(15).

Ren, C., E. Y. Ng, L. Katzschner, *et al.*, 2012. The Development of Urban Climatic Map and Its Current Application Situation. *Journal of Applied Meteorological Science*, 23(5) (in Chinese).

Sharifi, A., Y. Yamagata, 2014. Resilient Urban Planning: Major Principles and Criteria. *Energy Procedia*, 61.

Shen, T., B. Ji, 1997. The Study of Water Resources Features and the Water Demand and Water Supply Balance of Hong Kong. *Geographical Research*, 16(2) (in Chinese).

Shenzhen Water Resources Bureau, 2016. *The 13th Five-Year Plan for the Water Resources Development of Shenzhen*.

Shepherd, J. G., P. G. Brewer, A. Oschlies, 2017. Ocean Ventilation and Deoxygenation in a Warming World: Introduction and Overview. *Philosophical Transactions of the Royal Society A*, 375.

Shi, X., C. Lu, X. Xu, 2011. Variability and Trends of High Temperature, High Humidity, and Sultry Weather in the Warm Season in China During the Period 1961-2004. *Journal of Applied Meteorology and Climatology*, 50(1).

Singh, M. S., Z. Kuang, E. D. Maloney, *et al.*, 2017. Increasing Potential for Intense Tropical and Subtropical Thunderstorms Under Global Warming. *Proceedings of the National Academy of Sciences of the United States of America*, 114(44).

Siu, L. W., M. A. Hart, 2013. Quantifying Urban Heat Island Intensity in Hong Kong SAR, China. *Environmental Monitoring and Assessment*, 185(5).

Song, D., C. E. Choi, C. W. W. Ng, *et al.*, 2018. Geophysical Flows Impacting a Flexible Barrier: Effects of Solid-fluid Interaction. *Landslides*, 15(1).

Song, Q., J. Li, H. Duan, *et al.*, 2017. Towards to Sustainable Energy-efficient City: A Case Study of Macau. *Renewable and Sustainable Energy Reviews*, 75.

Su, J., M. Dai, B. He, *et al.*, 2017. Tracing the Origin of the Oxygen-Consuming Organic Matter in the Hypoxic Zone in a Large Eutrophic Estuary: The Lower Reach of the Pearl River Estuary, China. *Biogeosciences*, 14(18).

Sun, Q., C. Miao, Q. Duan, 2015. Projected Changes in Temperature and Precipitation in Ten River Basins over China in 21st Century. *International Journal of Climatology*, 35.

Sun, S., W. Cao, T. G. Mason, *et al.*, 2019. Increased Susceptibility to Heat for Respiratory Hospitalizations in Hong Kong. *Science of the Total Environment*, 666.

Sun, S., F. Laden, J. E. Hart, *et al.*, 2018. Seasonal Temperature Variability and Emergency Hospital Admissions for Respiratory Diseases: A Population-based Cohort Study. *Thorax*, 73(10).

Sun, S., L. Tian, H. Qiu, *et al.*, 2016. The Influence of Pre-existing Health Conditions on Short-term Mortality Risks of Temperature: Evidence from a Prospective Chinese Elderly Cohort in Hong Kong. *Environmental Research*, 148.

Tam, R. C. K., B. L. S. Lui, 2013. *Public Education and Engagement in Landslip Prevention and Hazard Mitigation*. Paper Submitted to the International Conference on Post-disaster Reconstruction – Sichuan 5.12 Hong Kong, The Hong Kong Institution of Engineers.

Tan, Z., K. K. L. Lau, E. Ng, 2016. Urban Tree Design Approaches for Mitigating Daytime

Urban Heat Island Effects in a High-density Urban Environment. *Energy and Buildings*, 114.

Tian, L., H. Qiu, S. Sun, *et al.*, 2016. Emergency Cardiovascular Hospitalization Risk Attributable to Cold Temperatures in Hong Kong. *Circulation: Cardiovascular Quality and Outcomes*, 9(2).

Tollefson, J., 2013. Natural Hazards: New York vs the Sea. *Nature*, 494(7436).

To, W. M., T. W. Yu, 2016. Characterizing the Urban Temperature Trend Using Seasonal Unit Root Analysis: Hong Kong from 1970 to 2015. *Advances in Atmospheric Sciences*, 33(12).

Tong, C. H. M., S. H. L. Yim, D. Rothenberg, *et al.,* 2018. Projecting the Impacts of Atmospheric Conditions Under Climate Change on Air Quality over the Pearl River Delta Region. *Atmospheric Environment*, 193.

Tracy, A., K. Trumbull, C. Loh, 2007. *The Impact of Climate Change in Hong Kong and the Pearl River Delta*. China Perspectives.

Tsuboki, K., M. K. Yoshioka, T. Shinoda, *et al.*, 2015. Future Increase of Supertyphoon Intensity Associated with Climate Change. *Geophysical Research Letters*, 42(2).

Tu, X., V. P. Singh, X. Chen, *et al.*, 2015. Intra-annual Distribution of Streamflow and Individual Impacts of Climate Change and Human Activities in the Dongijang River Basin, China. *Water Resources Management*, 29.

Tu, X. J., X. H. Chen, Y. Zhao, *et al.*, 2016. Responses of Hydrological Drought Properties and Water Shortage Under Changing Environments in Dongjiang River Basin. *Advances in Water Science*, 27(6) (in Chinese).

Walsh, K. J. E., J. L. McBride, P. J. Klotzbach, *et al.*, 2016. Tropical Cyclones and Climate Change. *Wiley Interdisciplinary Reviews: Climate Change*, 7(1).

Wang, D., K. K. L. Lau, C. Ren, *et al.*, 2019. The Impact of Extremely Hot Weather Events on All-cause Mortality in a Highly Urbanized and Densely Populated Subtropical City: A 10-year Time-series Study (2006–2015). *Science of The Total Environment*, 690.

Wang, L., G. Huang, W. Zhou, *et al.*, 2016. Historical Change and Future Scenarios of Sea Level Rise in Macau and Adjacent Waters. *Advances in Atmospheric Sciences,* 33(4).

Wang, M. Y., S. H. L. Yim, G. H. Dong, *et al.*, 2020. Mapping Ozone Source-receptor Relationship and Apportioning the Health Impact in the Pearl River Delta Region Using Adjoint Sensitivity Analysis. *Atmospheric Environment*, 222.

Wang, P., W. B. Goggins, E. Y. Y. Chan, 2016. Hand, Foot and Mouth Disease in Hong Kong: A Time-series Analysis on Its Relationship with Weather. *PLoS One*, 11(8).

Wang, P., W. B. Goggins, E. Y. Y. Chan, 2018a. Associations of Salmonella Hospitalizations with Ambient Temperature, Humidity and Rainfall in Hong Kong. *Environment International*, 120.

Wang, P., W. B. Goggins, E. Y. Y. Chan, 2018b. A Time-series Study of the Association of Rainfall, Relative Humidity and Ambient Temperature with Hospitalizations for Rotavirus and Norovirus Infection Among Children in Hong Kong. *Science of the Total Environment*, 643.

Wang, W., W. Zhou, E. Y. Y. Ng, *et al.*, 2016. Urban Heat Islands in Hong Kong: Statistical Modeling and Trend Detection. *Natural Hazards*, 83(2).

Wang, X., T. Yang, X. Li, *et al.*, 2017. Spatio-temporal Changes of Precipitation and Temperature over the Pearl River Basin Based on CMIP5 Multi-model Ensemble. *Stochastic Environmental Research and Risk Assessment*, 31.

Wang, X., Z. Wu, G. Liang, 2009. WRF/CHEM Modeling of Impacts of Weather Conditions Modified by Urban Expansion on Secondary Organic Aerosol Formation over Pearl River Delta. *Particuology*, 7(5).

Wang, Y., S. Wen, X. Li, *et al.*, 2016. Spatiotemporal Distributions of Influential Tropical Cyclones and Associated Economic Losses in China in 1984–2015. *Natural Hazards*, 84(3).

Wang, Y., Y. Li, S. Di Sabatino, *et al.*, 2018. Effects of Anthropogenic Heat Due to Air-conditioning Systems on an Extreme High Temperature Event in Hong Kong. *Environmental Research Letters*, 13(3).

Wehner, M. F., K. A. Reed, B. Loring, *et al.*, 2018. Changes in Tropical Cyclones Under Stabilized 1.5 and 2.0 ℃ Global Warming Scenarios as Simulated by the Community Atmospheric Model Under the HAPPI protocols. *Earth System Dynamics*, 9(1).

Wei, X., K. S. Lam, C. Cao, *et al.*, 2016. Dynamics of the Typhoon Haitang Related High Ozone Episode over Hong Kong. *Advances in Meteorology*, 2016.

Wong, C. M., H. Y. Mok, T. C. Lee, 2011. Observed Changes in Extreme Weather Indices in Hong Kong. *International Journal of Climatology*, 31(15).

Wong, H. N., Y. M. Chen, K. C. Lam, 1997. *Factual Report on the November 1993 Natural Terrain Landslides in Three Study Areas on Lantau Island (GEO Report No. 61)*. Hong Kong, Geotechnical Engineering Office.

Wong, H. N., K. K. S. Ho, 1995. *General Report on Landslips on 5 November 1993 at Man Made Features in Lantau (GEO Report No. 44)*. Hong Kong, Geotechnical Engineering Office.

Wong. H. N., F. W. Y. Ko, T. H. H. Hui, 2006. *Assessment of Landslide Risk of Natural Hillsides in Hong Kong (GEO Report No. 191)*. Hong Kong, Geotechnical Engineering Office.

Wong, W. T., K. W. Li, K. H. Yeung, 2003. Long Term Sea Level Change in Hong Kong. *Hong Kong Meteorological Society Bulletin*, 13(1-2).

Wu, C., C. Ji, B. Shi, *et al.*, 2019. The Impact of Climate Change and Human Activities on Streamflow and Sediment Load in the Pearl River Basin. *International Journal of Sediment Research*, 34(4).

Wu, J., Z. Liu, H. Yao, *et al.*, 2018. Impacts of Reservoir Operations on Multi-scale Correlations Between Hydrological Drought and Meteorological Drought. *Journal of Hydrology*, 563.

Wuebbles, D. J., D. W. Fahey, K. A. Hibbard, *et al.*, 2017. *Climate Science Special Report: Fourth National Climate Assessment (NCA4), Volume I*, U.S. Global Change Research Program, Washington, DC, USA.

Xu, S., G. Hou, 2017. Water Management During Drought Period of 2015-2016 in Pearl-River. *China Flood and Drought Management*, 27(2) (in Chinese).

Yan, D., S. E. Werners, F. Ludwig, *et al.*, 2015. Hydrological Response to Climate Change: The Pearl River, China Under Different RCP Scenarios. *Journal of Hydrology: Regional Studies*, 4.

Yang, S. H., N. Y. Kang, J. B. Elsner, *et al.*, 2018. Influence of Global Warming on Western North Pacific Tropical Cyclone Intensities During 2015. *Journal of Climate*, 31(2).

Yang, Y. K., I. S. Kang, M. H. Chung, *et al.*, 2017. Effect of PCM Cool Roof System on the Reduction in Urban Heat Island Phenomenon. *Building and Environment*, 122.

Yi, W., A. P. C. Chan, 2015. Effects of Temperature on Mortality in Hong Kong: A Time Series Analysis. *International Journal of Biometeorology*, 59(7).

Yi, W., A .P. C. Chan, 2017. Effects of Heat Stress on Construction Labor Productivity in Hong Kong: A Case Study of Rebar Workers. *International Journal of Environmental Research and Public Health*, 14(9).

Yim, S. H. L., X. Hou, J. Guo, 2019a. Contribution of Local Emissions and Transboundary Air Pollution to Air Quality in Hong Kong During El Niño-Southern Oscillation and Heatwaves. *Atmospheric Research*, 218.

Yim, S. H. L., M. Y. Wang, Y. Gu, *et al.*, 2019b. Effect of Urbanization on Ozone and Resultant Health Effects in the Pearl River Delta Region of China. *Journal of Geophysical Research: Atmospheres*, 124.

Yin, K., S. Xu, W. Huang, *et al.*, 2017. Effects of Sea Level Rise and Typhoon Intensity on Storm Surge and Waves in Pearl River Estuary. *Ocean Engineering*, 136.

Yue, D. P. T., S. L. Tang, 2011. Sustainable Strategies on Water Supply Management in Hong Kong. *Water and Environment Journal*, 25.

Yu, Q., A. K. H. Lau, K. T. Tsang, *et al.*, 2018. Human Damage Assessments of Coastal Flooding for Hong Kong and the Pearl River Delta Due to Climate Change-related Sea Level Rise in the Twenty-first Century. *Natural Hazards*, 92(2).

Yu, Y., C. Lu, Y. Li, 2014. The Current Situation and the Prospect of Water Resource in Macao.

Water and Wastewater Engineering, 40(2) (in Chinese).

Yuan, R., J. Zhu, B. Wang, 2015. Impact of Sea-level Rise on Saltwater Intrusion in the Pearl River Estuary. *Journal of Coastal Research*, 31(2).

Zhai, W., M. Dai, W. J. Cai, *et al.*, 2005. High Partial Pressure of CO_2 and Its Maintaining Mechanism in a Subtropical Estuary: The Pearl River Estuary, China. *Marine Chemistry*, 93(1).

Zhang, Q., J. Li, X. Gu, *et al.*, 2018. Is the Pearl River Basin, China, Drying or Wetting? Seasonal Variations, Causes and Implications. *Global and Planetary Change*, 166.

Zhao, W., B. Gao, M. Liu, *et al.*, 2019. Impact of Meteorological Factors on the Ozone Pollution in Hong Kong. *Environmental Science*, 40(1) (in Chinese).

Zhou, P., G. Chen, Z. Y. Liu, *et al.*, 2016. Variation Trend and Periodicity Analysis of Precipitation and Runoff in Dongjiang Watershed. *Ecological Science*, 35(2) (in Chinese).

Zu, T., J. Gan, 2015. A Numerical Study of Coupled Estuary-shelf Circulation Around the Pearl River Estuary During Summer: Responses to Variable Winds, Tides and River Discharge. *Deep Sea Research Part II: Tropical Studies in Oceanography*, 117.

Chapter 3　Climate Change Mitigation

Given the science (Chapter 1) and impacts (Chapter 2) of climate change especially with its close relevance to Hong Kong and Macao, the two SARs – being the wealthiest regions of China – also bear significant responsibilities for mitigating greenhouse gases emissions.

A variety of accounting rules to categorize and quantify greenhouse gas emissions have been developed by different communities (Table 3–1). 'Direct' emissions are commensurate with terms like 'territorial', 'production-based', or 'Scope 1' emissions, while 'indirect' emissions are commensurate with 'embodied', 'consumption-based', 'Scope 2', 'Scope 3', 'upstream' or 'downstream' emissions.

Table 3–1　Scope 1, 2 and 3 emissions under the CSG Protocol (Fong *et al.*, 2014)

Accounting Method	Scope	Definition
Production-based (PBA)	Scope 1	GHG emissions from sources located within the boundary
Consumption-based (CBA)	Scope 2	GHG emissions occurring outside of the boundary due to the use of grid-supplied electricity, heat, steam and/or cooling within the boundary
	Scope 3	All other GHG emissions that occur outside the boundary as a result of activities taking place within the boundary

Scope 1 emissions are the most widely applied in regulating a country's climate commitments such as in the Kyoto Protocol and Paris Agreement, while Mainland China adopts Scope 2 emissions in evaluating individual provinces' CO_2 emissions (State Council, 2016). For the entire globe, the three scopes are of no difference because all production and consumption take place within the boundary of the Earth. However, their gaps tend to widen when energy services (Scope 2) and goods/services (Scope 3) are sourced outside a geographic region of interest. From the perspective of GDP per capita, Hong Kong and Macao Special Administrative Regions are the most developed regions of China. Due to their small geographic areas and intensive trade with Mainland China and overseas countries, significant differences are shown across the three scopes of greenhouse gas emissions as well as corresponding mitigation options.

The chapter examines Hong Kong and Macao individually. Section 3.1 reviews the two SARs' Scope 1 greenhouse gas emissions and sources. Section 3.2 focuses on Scope 2 and Scope 3 emissions. Corresponding to Scope 1 emissions, mitigation scenarios will be constructed and explored in Section 3.3.

3.1 Greenhouse Gas Emissions and Sources

3.1.1 Hong Kong

The Environmental Protection Department of the HKSAR Government (HKEPD) has been maintaining a greenhouse gas emissions (GHG) inventory since the 1990s. In 1998, HKEPD commissioned a 'Greenhouse Gas Emission Control Study' (HKEPD 2000 study) to develop a basis for future decision-making (HKEPD, 2000). The study delivered the first Hong Kong greenhouse gas emissions inventory in accordance with the Revised 1996 Intergovernmental Panel on Climate Change (IPCC) Guidelines for National Greenhouse Gas Inventories. In 2008, HKEPD commissioned a ‘Climate Change in Hong Kong–Feasibility Study’ (HKEPD 2010 study) to review and update the local inventories of greenhouse gas emissions and removals (HKEPD, 2010), replacing the methods in the revised 1996 IPCC Guidelines as far as possible to align with the methods in the more recent 2006 IPCC Guidelines (IPCC, 2006b). More detailed information in relation to this study was also published on the HKEPD website in the form of Technical Appendices (HKEPD, 2010).

The Hong Kong greenhouse gas emissions inventory is prepared with a mix of activity data sources, predominantly derived from official statistical agencies and HKEPD. Six categories of greenhouse gases as specified in the Kyoto Protocol are included, carbon dioxide (CO_2), methane (CH_4), nitrous oxide (N_2O), hydrofluorocarbons (HFCs), perfluorocarbons (PFCs), and sulphur hexafluoride (SF_6). Additional greenhouse gases identified in the IPCC Third Assessment Report (IPCC, 2001), such as nitrogen trifluoride (NF_3), trifluoromethyl sulphur pentafluoride (SF_5CF_3) and halogenated ethers, were included in the HKEPD 2010 study. However, these three new GHGs are not required to be reported as part of the national inventory and so are not included in Hong Kong’s published annual greenhouse gas emissions inventory. Even if they were added, it is presumed that this would have a minor impact since Hong Kong does not have a significant amount of these related industrial processes. In the HKEPD 2010 study, the Global Warming Potentials (GWPs) as defined in the Revised 1996 IPCC Guidelines were used, since under the Kyoto Protocol, the Conference of the Parties decided that the GWPs in the later findings under the 2006 IPCC Guidelines should not be applied practically until the end of 2012. It is assumed that HKEPD has since applied the new GWPs as defined in the 2006 IPCC Guidelines for their annual GHG emissions after 2012 (HKEPD, 2010).

In Hong Kong, CO_2 emissions are the most dominant out of the six greenhouse gases, contributing over 90% in 2006, while CH_4 and N_2O account for about 5% and 1%, respectively, and the remaining gases are insignificant (Figure 3–1). The main reason for the dominance of CO_2 is the fact that fuel consumption for power generation, transport and other uses (which results mostly in CO_2 emissions), is the main greenhouse gas emitting activity in Hong Kong. Hong Kong does not have any significant agricultural activity (which results mostly in CH_4 and N_2O emissions), nor does it have many industrial processes operating within its borders (which can produce other greenhouse gases).

Overall greenhouse gas emissions

In terms of Hong Kong's total greenhouse gas emissions, there was a general increase from 1990 at 35.2 million tonnes(Mt) CO_{2eq} to 44.9 Mt CO_{2eq} in 2014, when it seemed to have peaked and been slowly declining to 40.7 Mt CO_{2eq} in 2017 (Figure 3–2). Although total greenhouse gas emissions in 2017 were within the same range as in the early 1990s, Hong Kong's total greenhouse gas emissions registered a dip between 1994 and 2004, due mainly to the fuel mix shift of adding nuclear power and natural gas to what was predominantly coal-fired generation in the early 1990s. However, over time the carbon reduction benefits of the fuel mix shift were overtaken by the growing amount of electricity consumed in Hong Kong, which increased by over 83% from 85 801 Terajoules (TJ) in 1990 to 157 604 TJ in 2017 (HK Census and Statistics Department, 2019a).

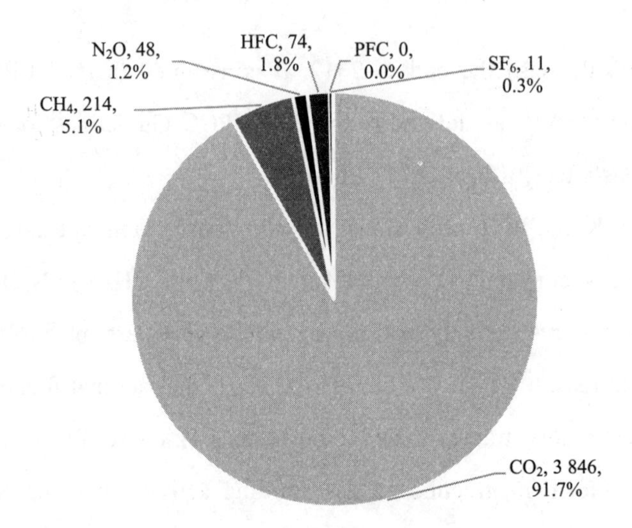

Figure 3–1 Hong Kong emissions of major greenhouse gases in 2006 (Unit: 10 000 tonnes CO_{2eq}) (HKEPD, 2010)

Hong Kong's per capita emissions between 1990 and 2017 declined by about 11% from 6.2 tCO_{2eq}/capita to 5.5 tCO_{2eq}/capita, despite over 30% reduction in the carbon intensity of the electricity generated during that period (Figure 3–2). This is due to a combination of the increase in the overall electricity demand, in part fueled by an almost 30% increase in population from around 5 704 500 people in 1990 to 7 413 100 in 2017, in addition to an over 35% rise in energy use per capita from about 0.016 TJ/capita in 1990 to 0.022 TJ/capita in 2017.

Energy consumption is closely related to GDP and economic structure. Hong Kong's carbon intensity has reduced significantly by over 58% from 3.6tonnes CO_{2eq}/HK\$100 000 (in 2017 HK\$) GDP (or 36 gCO_{2eq}/HK\$ GDP) in 1990 to 1.5 tonnes CO_{2eq}/HK\$100 000 GDP (or 15 gCO_{2eq}/HK\$ GDP) in 2017 (Figure 3–2). This largely reflects the shift in fuel mix during this period from mainly coal to including natural gas and importing nuclear power from the Mainland. The shift from an industry-based economy to a service-based one, where GDP growth occurs with less energy and thus less carbon emissions per dollar of GDP, had already occurred mostly in the 1980s. From 1984 to 1997, the share of the industrial sector dropped from 33.2% to 14.6%, while that of the commercial sector rose from 20.0% to 29.9% (Chow, 2001b). This trend corresponded to the relocation of industries to the Pearl River Delta in Mainland China and shift towards a dominantly service-oriented economy (Chow, 2001b).

Of the five major greenhouse gas emitting sectors categorized by the 2006 IPCC Guidelines, Hong Kong has reports from four of them, namely energy, waste, industrial processes and product use (IPPU), and agriculture, forestry and other land use (AFOLU). The relative contributions between the different sectors within the Hong Kong greenhouse emissions inventory, have not changed significantly over the last two and a half decades, reflecting the dominance of a more service-based economy as opposed to an industrial-based one over this period of time. This in turn has led to the electricity sector now playing a dominant role in influencing Hong Kong's total greenhouse emissions.

In Hong Kong, the entire energy sector is the largest source of greenhouse gas emissions contributing about 89% of Hong Kong's total greenhouse gas emissions (electricity generation having the largest portion at around 65%, followed by fuel consumed in the transport sector contributing about 18% and other end use of fuel contributing about 6%).

Greenhouse gas emissions from the waste sector contribute about 7%, while greenhouse gas emissions from IPPU contribute about 4% and AFOLU contribute less than 1% of Hong Kong's total greenhouse gas emissions.

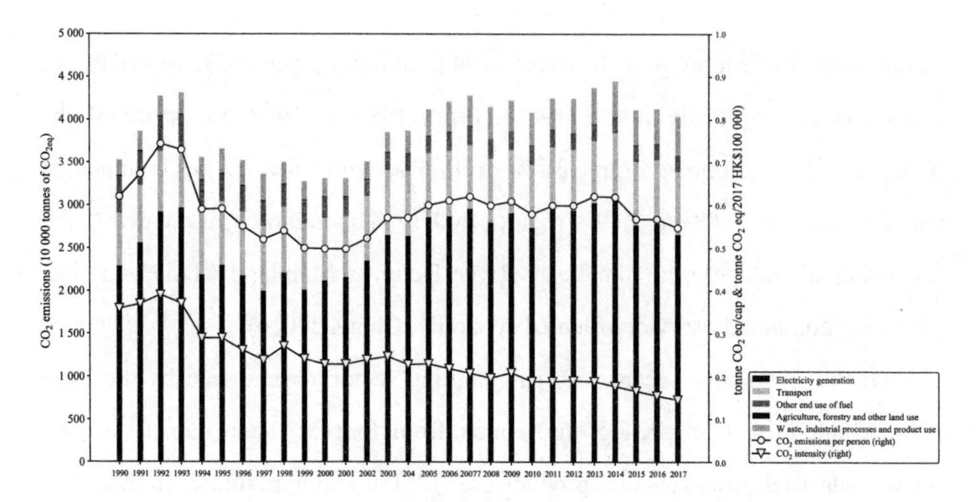

Figure 3–2 Greenhouse Gas Emissions in Hong Kong (HK Census and Statistics Department, 2019a)

Electricity generation

Out of Hong Kong's annual total of 40.7 Mt CO_{2eq} in 2017, energy-related consumption, including electricity generation, transport and other end use of fuel, contributed 36.1 Mt CO_{2eq}. The largest source was from electricity generation, accounting for around 65.4% (or 26.6 Mt CO_{2eq}), while the production of town gas accounts for only around 0.75% (or around 300 kt CO_{2eq}) of Hong Kong's total greenhouse gas emissions in 2017 (HKEPD, 2019a).

GDP and electricity consumption are closely linked, particularly in Hong Kong's context (Ho and Siu, 2007). Together with economic development and structural transformation, electricity consumption in Hong Kong has experienced significant changes in the past decades in terms of the relative importance of various sectors. In 1975, the shares of the industrial and commercial sector use were similar at around 38% each, but by 2017 the industrial sector use had dropped to 7% while commercial use had risen to 64% (Chow, 2001b). This trend corresponded to the relocation of industries to the Pearl River Delta in Mainland China and the shift towards a dominantly service-oriented economy (Chow, 2001b). With hot and humid long summers and mild winters, Hong Kong's electricity is increasingly consumed for air conditioning when the society gets wealthier. Strong seasonal cycles are highly visible for electricity consumption in commercial and domestic sectors (Fung *et al.*, 2006). Accordingly, a warming climate, in addition to strong heat island effects, tends to increase electricity consumption in Hong Kong. An increase of ambient temperature by 1℃ tends to increase electricity consumption in domestic, commercial and industrial sectors would increase by 9.2%, 3.0%, and 2.4%, respectively (Fung *et al.*, 2006).

Electricity consumption in Hong Kong has a weak response to price changes. A 40% spike of electricity tariff was projected to reduce electricity consumption only by 0.81%, while the substitution effect would increase town gas consumption by 5.12%, offsetting the resulting less CO_2 emissions from less electricity consumption (Woo *et al.*, 2018). Accordingly, price-based policies may not be effective for Hong Kong's decarbonization (Woo *et al.*, 2018).

Given the electricity consumption level, fuel mix is a crucial factor in determining Hong Kong's CO_2 emissions. Hong Kong's electric sector relies heavily on fossil fuels and has experienced several major energy transitions for electricity generation. Before 1982, Hong Kong's local electricity generation completely relied on oil (Chow, 2001a). Since 1982, coal has been playing an increasingly important role from 0% in 1981 to 23.5% in 1982 and 95.2% in 1987, while the share of oil collapsed (Chow, 2001a). Hong Kong has no local production of fossil fuels and all must be imported either from Mainland China or overseas. With growing electricity consumption and the energy transition toward coal in 1982, the significant import of steam coal was initiated in 1981 with 52 thousand tonnes and then immediately jumped to 1 451 thousand tonnes in 1982 before peaking at 11 828 thousand tonnes in 1993 (Chow, 2001a). The import of nuclear electricity into Hong Kong from the Daya Bay Nuclear Power Station resulted in a significant decrease of locally generated electricity and substantially reduced the demand for coal. Coal import peaked in 1993 at 11 828 thousand tonnes and declined to 8 450 thousand tonnes in 1994 (Chow, 2001a). Correspondingly, CO_2 emissions from Hong Kong also dipped significantly in 1994. In 1980s, most coal import came from South Africa and Australia. The share of South Africa rapidly shrank in 1990s and by 1998, most imports had been from Indonesia and Australia (Chow, 2001a).

In 1996, natural gas came into Hong Kong's fuel mix for local electricity generation to account for 24.0% and the share increased to 33.1% in 1998 (Chow, 2001a). Natural gas is generally transported through a pipeline or a liquefied natural gas tanker (LNG). Due to this technological characteristic, Hong Kong is relying on Mainland China for 100% of its natural gas supply (HK Census and Statistics Department, 2018). Hong Kong relied on the Yacheng 13-1 gas field in South China Sea after the field was developed in 1995 (Wilburn and Roberts, 1995). In 2008, the National Energy Administration of China signed a memorandum with the HKSAR government to support Hong Kong's natural gas supply (NEA, HKSAR Government, 2008), whereby in addition to the existing sources, natural gas will be supplied to Hong Kong via the West-to-East natural gas pipeline and an LNG terminal in the Mainland.

However, between 2002 and 2015, coal still remained as the dominant source of fuel for electricity generation accounting for 74.3% on average of greenhouse gas emissions in the electric sector, while the shares of natural gas and oil were

25.1% and 0.6%, respectively (To and Lee, 2017c). Hence in order to meet the Government's 2020 fuel mix targets of shifting from the 2012 existing mix of 53% coal, 22% gas and 23% nuclear to 50% gas, 25% nuclear and 25% coal plus renewable energy by 2020, this would entail reducing coal-fired electricity generation and substituting with more generation from natural gas, a floating LNG terminal was proposed to directly import LNG into Hong Kong from the international market (Environmental Resources Management, 2018). In 2019, the HKSAR Government approved Castle Peak Power Company Limited and The Hongkong Electric Company Limited to construct and operate an offshore Liquefied Natural Gas (LNG) terminal in the waters to the east of the Soko Islands. The terminal will enable a Floating Storage and Regasification Unit vessel (FSRU vessel) and an LNG carrier to be moored at a double berth jetty. The regasified LNG from the FSRU vessel will be supplied to the gas receiving stations at the Black Point Power Station (BPPS) and the Lamma Power Station (LPS) via two separate subsea gas pipelines (HK Lands Department, 2019).

The transport sectors

The transport sector was the second largest source of greenhouse gas emissions, contributing around 18% (or 7.2 Mt CO_{2eq}) of total in 2017 (Figure 3–2). This sector includes fuel combustion arising from domestic transport only within Hong Kong, in the road, rail, marine and civil aviation sectors. Although the currently published HKSAR greenhouse gas emissions inventory does not provide a breakdown of the sources, the HKEPD 2010 study did provide such details (HKEPD, 2010). In 2006 as the most recent year, road was the largest contributor that emitted 81.9% of all greenhouse gases from the transport sector (Figure 3–3). Shipping came next with a share of 11.2%, while aviation was insignificant, responsible for only 0.3%. Although Hong Kong has a major seaport and a key international airport, only a small fraction of their related CO_2 emissions in transboundary transport are counted as Hong Kong's Scope 1 emissions due to HKSAR's small territory.

The transport sector consumed 89 819 TJ of energy in 2016, while freight and passenger sub-sectors contributed 31% and 69% respectively (HK Electrical and Mechanical Services Department, 2019). For the period 2006～2016, the energy consumption of the transport sector decreased by 6.8%, which is mainly because of the 29.3% decrease in freight transport, while passenger transport energy use increased by 8.5 %. For passenger transport in Hong Kong, the car and motorcycle segment consumed most of the energy, 23 344 TJ (26%), followed by the bus of 18 918 TJ (21%), the taxi of 12 288 TJ (13.7%), and the rail of 2 951 TJ (3.3%).

In regard of land transport of passengers in Hong Kong, electrified trains (subways and light-rail) carried an average of 4.7 million passengers per day in 2016 (Lang, 2018). The 13 000 buses, burning gasoline or diesel, carry another several million passenger daily. Food and consumer goods are distributed around the territory by about 113 000 light trucks, also burning fossil fuels, which constantly crowd the streets and alleys of the city. The total number of motor vehicles licensed increased by 26.1% from 607 796 in 2010 to 766 200 in 2017, including motorcycle, private car, buses, taxis, goods vehicles, light buses and government cars, while the average daily vehicle-kilometer was 37.4 million vehicle-kilometers in 2017 (HK Census and Statistics Department, 2019b). Despite longer average commuting distances, Hong Kong's transport passenger tasks are mostly carried by energy efficient rail transport (Leung *et al.*, 2018).

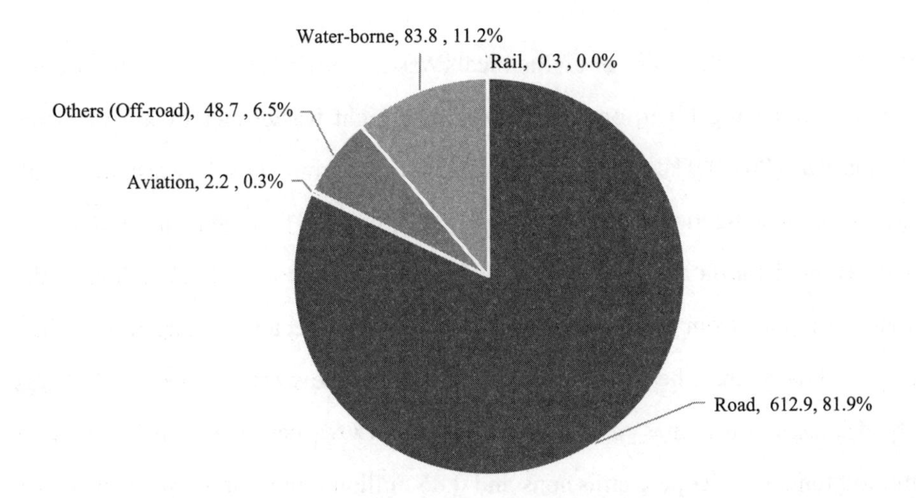

Figure 3–3 Hong Kong greenhouse gas emissions from the transport sector in 2006 (Unit: 10 000 tonnes CO_{2eq}) (HKEPD, 2010)

Other sectors

(1) Other end use of fuel

The 5.6% (or 2.3 Mt CO_{2eq}) from 'other end use of fuel' in 2017 includes fuel combustion on industrial (e.g., manufacturing and construction), commercial and domestic premises (HKEPD, 2019a). Greenhouse gas emissions between 1990 to 2006 from manufacturing and construction industries were declining over time, while greenhouse gas emissions from commercial and domestic premises were slowly increasing (HKEPD, 2019a). If we apply the 2006 ratios to the 2017 data for total GHG emissions, we can obtain an estimate of around 5% (or 2.0 Mt CO_{2eq}) coming from commercial and domestic premises, while only 0.7% (or 274 kt CO_{2eq}) was coming from the manufacturing and construction industries.

(2) Waste

The waste facilities serving Hong Kong include three landfill sites, seven

Refuse Transfer Stations, one Chemical Waste Treatment Centre, one Animal Waste Composting Plant, one Incinerator, and eight wastewater treatment plants (Dong *et al.*, 2017). Carbon emissions mainly come from (1) solid waste disposal, (2) wastewater treatment and discharge, (3) biological treatment of solid waste, and (4) incineration and open burning of waste. Dong *et al.* (2017) estimated the GHG emissions from these major sources in the waste sector of Hong Kong. Their results showed that the total GHG emission from the waste sector in 2010 was about 2.5 million tonnes of CO_{2eq} and 3.57 kg CO_{2eq} per capita, including 1.68 million tonnes of Scope 1 emissions and 0.86 million tonnes of Scope 2 emissions from purchased electricity. Among the Scope 1 emissions, landfill contributed 1.73 million tonnes, while sewage and others contributed negatively due to harvested CH_4 from sewage. The study also showed projected results of GHG emissions in 2020 that the total emissions would increase by 13% at the 2010 levels. GHG emissions from landfills decrease, while total GHG emissions from the entire waste sector increase, mainly due to the combustion of petroleum products in incinerators and the operation of biological treatment. Sensitivity analysis reveals that the increased emission can be offset when solid waste disposal is reduced by 40%. Therefore, it is strongly recommended to improve the recycling and reuse rate of waste in Hong Kong.

As among different phases of a product's life cycle disposal phase usually assumes greater significance for carbon emissions, Muthu *et al.* (2011) have conducted a study of the carbon emissions for different disposal options of different types of shopping bags in Hong Kong. They found that reuse of shopping bags resulted in lower carbon emissions, compared to that of recycling or deposing to landfill. They found that a 5% increase in reuse options of shopping bags might result in around 20% of carbon emission reduction.

In 2017, the waste sector was the third largest greenhouse gas emissions source, contributing 6.9% (or 2.8 Mt CO_{2eq}) of the total 40.7 Mt CO_{2eq} of Hong Kong greenhouse gas emissions (HKEPD, 2019a). CH_4 is the predominant greenhouse gas, contributing around 90% of all the greenhouse gas emissions generated from the waste sector, while N_2O and a small amount of CO_2 make up for the remaining 10%. Solid waste disposal constitutes the largest portion of the waste-related greenhouse gas emissions, accounting for around 90% of the total greenhouse gas emissions arising from the waste sector in 2006 (HKEPD, 2010). CH_4 is the only greenhouse gas emitted from this source. If we apply the 90% ratio to the 2017 data, then the greenhouse gas emission arising from landfills in 2017 would be around 2.5 Mt CO_{2eq}. It should be noted that since 2006, more landfill gas has been utilized to generate energy (HKEPD, 2017).

Greenhouse gas emissions from wastewater treatment and discharge constitute the second largest emission source category within the waste sector, contributing around 9% of the total greenhouse *gases* generated from the waste sector in 2006 (HKEPD, 2010). Applying this ratio to the 2017 data of 2.8 Mt CO_{2eq}, this would translate into approximately 250 kt CO_{2eq}. Greenhouse gas emissions from this sub-sector include only CH_4 from both domestic and industrial treatment, and N_2O from only domestic wastewater treatment. Historical data from the HKEPD 2010 study has shown that N_2O emissions dominate comprising roughly 80% of total greenhouse gas emissions from wastewater treatment and discharge (HKEPD, 2010). Direct CO_2 emissions result only from waste incineration activities. Hong Kong has no MSW incineration since 1997 but only a Chemical Waste Treatment Centre (CWTC). Greenhouse gas emissions from waste incineration declined between 1997 and 1998 (HKEPD, 2019a).

(3) Industrial Processes & Product Use (IPPU)

The IPPU sector is the second smallest contributor to Hong Kong's total greenhouse gas emissions (after the AFOLU sector), accounting for 4.3% or 1.7 Mt CO_{2eq} in 2017 (HKEPD, 2019a). The greenhouse gas emissions from the IPPU sector are: mainly CO_2 arising from cement production; HFCs and PFCs arising from the use of substitutes to replace Ozone Depleting Substances (ODSs); and SF_6 used in electrical equipment such as power transformers.

Cement production started as the main contributor to Hong Kong's greenhouse gas emissions in the IPPU sector but ended up being the second largest by 2006. A temporary halt in clinker production (a part of the cement manufacturing process that produces CO_2) between 2002 and 2005 corresponded the significant reduction of greenhouse gas emissions over the period, even though cement production continued. The increase in 2006 reflected the resumption of the clinker process. If we apply the approximate 2006 ratio of 39% to the 2017 IPPU data of 1.7 Mt CO_{2eq}, then the greenhouse gas emission arising in 2017 from cement production would be around 663 kt CO_{2eq}.

The ODSs substitutes seem to have come up from behind to become the largest contributor by 2006, accounting for 53% of the greenhouse gas emissions from the IPPU sector. Substitutes for ODSs are the sole emitters of HFCs and PFCs within the Hong Kong greenhouse gas emissions inventory. If we apply the 2006 ratio of 53% to the 2017 IPPU data, then the greenhouse gas emission arising in 2017 from cement production would be around 901 kt CO_{2eq}. In 2006, SF_6 accounted for around 8% of the greenhouse gas emissions from the IPPU sector. This category is the sole emitter of SF_6 within the Hong Kong greenhouse gas emissions inventory. If we apply the 2006 ratio of 8% to the 2017 IPPU data, then the greenhouse gas emission arising in 2017 from cement production would be around 136 kt CO_{2eq}.

(4) Agriculture, Forestry and Other Land Use (AFOLU)

The AFOLU sector is the smallest contributor to Hong Kong's total greenhouse gas emissions, accounting for 0.1% or 30 kt CO_{2eq} in 2017 (HKEPD, 2019a). It is also the only sector that has the capacity to contribute to carbon removal due mainly to the potential gain in carbon stock from forestland. Consequently, it is not possible to apply ratios to estimate the various contributions for the sub-sectors within this category given that the land category can result in negative greenhouse gas emissions.

The greenhouse gas emissions from the AFOLU sector are mainly: N_2O from manure management and soil management activities; CH_4 from enteric fermentation and biomass burning; and CO_2 from liming and biomass burning.

The livestock sub-sector is further divided into enteric fermentation and manure management. CH_4 is the only greenhouse gas emitted from animals' enteric fermentation, while both CH_4 and N_2O are emitted from manure management. In 2006, livestock constituted about 56% of the all Greenhouse gases arising from the AFOLU sector, with manure management being the major component accounting for 47% of the AFOLU sector emissions.

Aggregate sources and non-CO_2 emissions sources on land, which includes biomass burning; liming; and soil management and indirect N_2O emissions from manure management. In 2006, this category constituted about 44% of the all Greenhouse gases arising from the AFOLU sector, with soil and indirect N_2O manure management emissions being the major component accounting for 37% of the AFOLU sector emissions.

GHG emissions were calculated from forest land, cropland and wetlands as these are the land types present in Hong Kong. Land has the potential capability of storing over 460 kt CO_{2eq}/year, which in the past could cover 100% of the greenhouse gases arising from the AFOLU sector. The 30 kt CO_{2eq} net emissions in 2017 signals that the emissions have surpassed the storage capacity, which could be due to increased AFOLU emissions, or decreased land carbon storage capacity, or a mixture of both.

The peaking of Scope 1 emissions

Hong Kong's Scope 1 emissions should have peaked in 2014 as they were 9.4% less in 2017 (Figure 3–2). They increased by only 15.6% in 2017 from 1990 (Figure 3–2), while its population and GDP per capita grew by 29.6% and 110.6% respectively (Figure 3–2). It reflects that the greenhouse gases emission per capita and per unit of GDP had declined significantly by 11% and 58% respectively (Figure 3–2), largely due to deindustrialization (Figure 3–4) and efficiency improvements. Electricity generation accounts for about two thirds of Scope 1 emissions. In 1970s, the industrial sector was the largest electricity consumer. However, the steady deindustrialization of Hong Kong economy has been driving down its importance to 3.08 TWh in 2018, even significantly less than the level in 1980. Over the same period, commercial and residential sectors have grown to be responsible for most of Hong Kong's electricity consumption, being 29.36 TWh and 11.66 TWh respectively. The overall level of electricity consumption has remained very stable since 2014, fluctuating narrowly between 43.78 TWh and 44.20 TWh, although population and GDP per capita have been rising consistently (Figure 3–5).

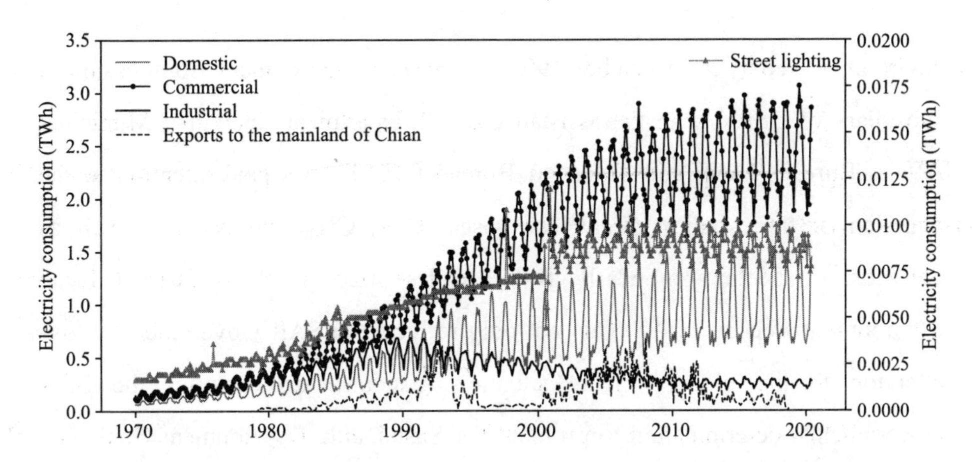

Figure 3–4 Monthly electricity consumption in Hong Kong by sector (HK Census and Statistics Department, 2019c)

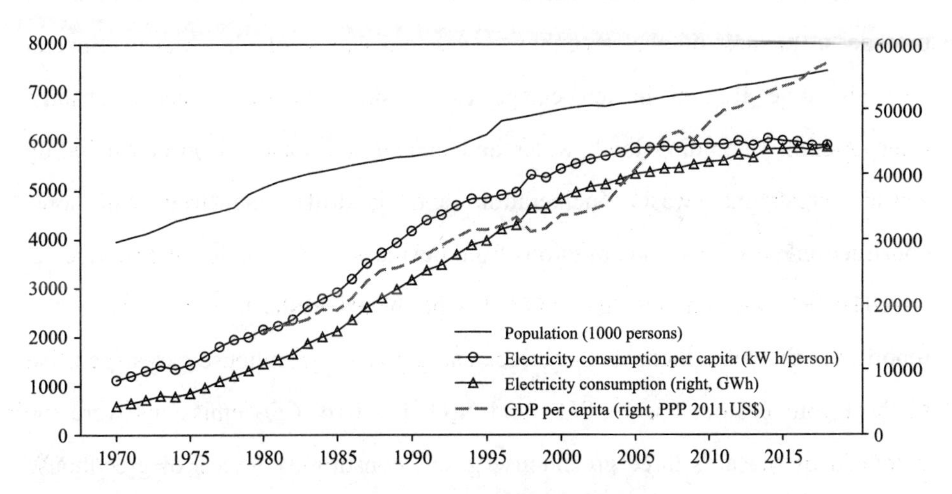

Figure 3–5 Electricity consumption and related indicators in Hong Kong (HK Census and Statistics Department, 2019c; IMF, 2019)

3.1.2 Macao

Overall greenhouse gas emissions

Since Macao's environmental protection bureau DSPA issued Macao's first environment report in December 1999, the Macao Environment Commission, a subsidiary of DSPA, with the assistance of 22 departments including Municipal Affairs Bureau (IAM) and Transport Bureau (DSAT), has paid attention to the emission of three major greenhouse gases (CO_2, CH_4, and N_2O) as well as emission by sector for nearly two decades from 1999 to 2018 (Macao Environment Report, 1999). This not only reflects the SAR Government's close attention to the global climate situation, but also demonstrates the SAR Government's determination to promote the Sustainable Development Goals and build a low-carbon society.

Based on the State of the Environment Report 1999-2017, this section will provide a picture of the greenhouse gas emissions of the Macao Special Administrative Region in ten categories of six industries: local electricity, transportation (including land, water and air), urban waste disposal (including sewage treatment, waste incineration and landfill), industrial emissions, construction emissions, and emissions from businesses, households and services.

The list of greenhouse gas emissions published in the Macao Environment report over the years includes only three categories of greenhouse gases specified in the Kyoto Protocol: CO_2, CH_4, and N_2O. In 2016, CO_2 emissions were the dominant of Macao's three greenhouse gases, constituting 94.8% of greenhouse gas emissions in 2016, while CH_4 and N_2O accounted for 1.9% and 3.2%, respectively (Figure 3–6). The main reason for the dominance of carbon dioxide is the consumption of fossil fuels in the power industry, transportation and other uses. Macao's CH_4 emissions come mainly from land transportation, while N_2O comes from municipal sewage treatment.

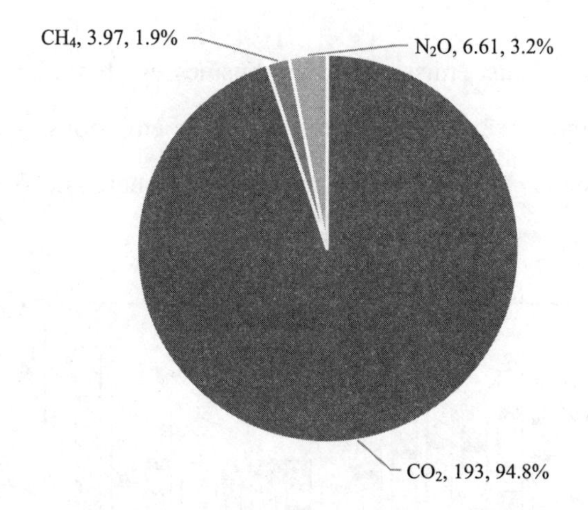

Figure 3–6 Major greenhouse gas emissions (in 10 000t CO_{2eq}) of Macao in 2016 (Macao DSPA, 2018)

Total greenhouse gas emissions in Macao (Figure 3–7) increased from 1.733 million tonnes CO_{2eq} in 2000 to 2.036 million tonnes CO_{2eq} in 2016, and peaked at 2.146 million tonnes of CO_{2eq} in 2005, due to a sharp increase in the greenhouse gas emission of local electricity generation. Greenhouse gas emissions then began to fall and reached an all-time low of 1.461 million tonnes of CO_{2eq} in 2012, largely due to an increase in the proportion of natural gas-fired electricity in local electricity. Between 2000 and 2016, Macao's per capita greenhouse gas emissions has shown an overall decreasing trend, declining from 4 tonnes of CO_{2eq} in 2000 to 2.5 tonnes in 2014, before rebounding to 3.2 tonnes in 2016.

Based on the average industrial emissions levels from 2000 to 2016, local electricity remains Macao's largest source of greenhouse gas emissions, constituting 38.6% of total emissions. This was followed by the transport sector accounting for 29.0%. Within the transport sector, the emission of land transport, sea transport and air transport accounted for 15.1%, 10.2% and 3.8%, respectively.

The share of the waste treatment sector in total emissions is approximately

14.6%. Greenhouse gas emissions from businesses, households and services accounted for about 12.0% of greenhouse gas emissions. Construction and industrial emissions accounted for the lowest share at 3.1% and 2.7%, respectively.

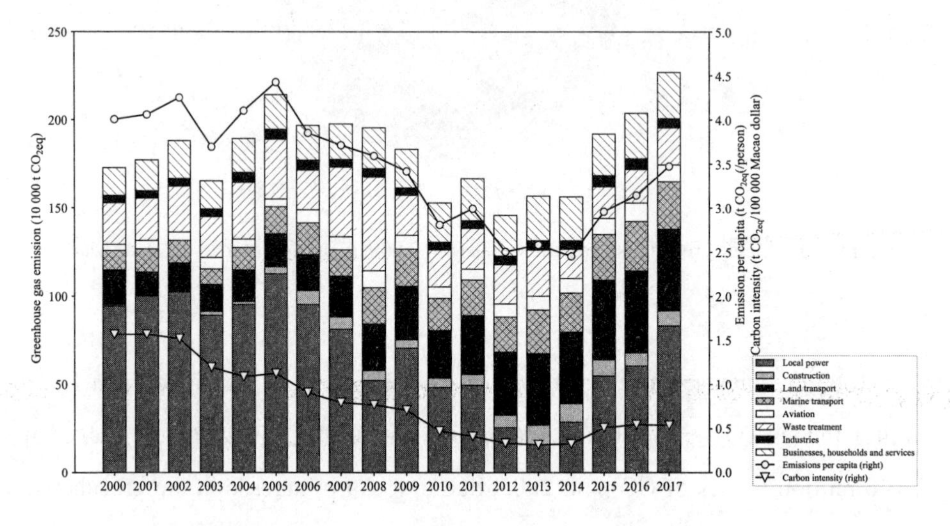

Figure 3–7 Macao's greenhouse gas emissions by sector (Macao DSPA, 2018)

Power generation

Within Macao's annual emissions of 2.036 MtCO_{2eq} in 2016, about 0.605 Mt came from local power generation in Macao, the second largest emission source accounting for 29.7% of the total annual carbon emissions. Macao's electricity is mainly imported from the China Southern Power Grid and is supplemented by local heavy oil and natural gas-fired power generation. At present, Macao's own power facilities include two power plants, a waste incineration center and a number of small photovoltaic power generation projects, which combine to form the local power supply system in Macao. In 2016, Macao's local electricity generation accounted for 15.0% of total power, the waste incineration center accounted for 3.1%, and the rest were from the China Southern Power Grid in southern China (CEM, 2016). About 67.5% of electricity is used for commercial purposes, 22.1% for residential, 2.5% for industrial and 7.9% for government and public lighting (CEM, 2016).

The two existing power plants in Macao are Coloane Power Station A (CCA) and Coloane Power Station B (CCB). Among them, CCA is mainly fueled by heavy oil and diesel fuel, while CCB is fueled by diesel and natural gas, with a total installed capacity of 407.8 MW, and both owned by Macao's electricity company CEM (CEM, 2017). The Macao Waste Incineration Centre uses the heat generated by its combustion to generate electricity from municipal waste. At full capacity, approximately 28.7 MW of electricity can be generated by burning municipal waste, of which 7 MW (about 24%) is used for its own operation and the remaining 21.7 MW (approximately 76%) is delivered to CEM's public grid. A number of photovoltaic power generation systems have been put into use at sites such as the Coloane Power Stations, and the power generation of the photovoltaic system in Macao in 2017 was 1 448 kWh (CEM, 2017).

Song *et al.* (2018a) points out that between 2010 and 2014, CEM had an average greenhouse gas emission factor of 0.71 (kg CO_2eq/kWh) for oil-fired power generation, 0.42 (kg CO_2eq/kWh) for gas-fired power, and 0.95 (kg CO_2eq/kWh) for waste incineration power, suggesting that gas-fired power is the cleanest power source for Macao. Despite relatively high carbon emissions, waste incineration power serves as an effective way of waste disposal, thus deserving promotion as clean energy.

Data for total greenhouse gas emissions from Macao's local power generation (excluding waste incineration) is published in the environmental report by Macao

Environmental Protection Bureau (DSPA) (Figure 3–7). There is an overall decreasing trend from 947 thousand tonnes of CO_2 equivalent in 2000 to 605 thousand tonnes in 2016 with the lowest level of 173 thousand tonnes in 2013. The main reason is the rapid decline in local electricity generation in Macao and the shift to electricity imports from China Southern Power Grid which reduced Scope 1 emissions.

The transport sectors

Macao has a convenient urban transport system composed of a comprehensive public transport system, a large number of private cars, and transportation hubs including Outer Harbor Ferry Terminal and Taipa Ferry Terminal among other ports, as well as Macao International Airport. In 2016, Macao's transportation industry was the largest source of greenhouse gas emissions, with annual emissions reaching 0.839 Mt CO_{2eq} or 41.2% of total emissions. Greenhouse gas emissions from the transport sector include three categories: land, marine and air transport. Among them, emissions from land transport were the highest, accounting for approximately 22.4% of total emissions in 2016 (about 0.457 Mt CO_{2eq}), while marine transport and air transport constituting 14.2% (about 0.29 Mt CO_{2eq}) and 4.6% (about 0.093 Mt CO_{2eq}), respectively.

Due to Macao's small size and intensive cross-border traffic, the results of the Macao Environment Report do not distinguish between local and cross-border emissions, both of which are included in Scope 1 emissions.

Land transport emission refers particularly to those from fuel consumption of motor vehicles. In 2016, the fuel consumption of Macao's onshore transport in 2016 was approximately 5 745 TJ, about 34.0% of total fuel consumption of the entire transport sector (Macao DSEC, 2016). By February 2019, the number of registered automobiles in Macao has reached 115.5 thousand (including 915 operating buses) and the number of registered motorcycles has reached 123.5 thousand. From 2000 to 2016, total greenhouse gas emission from land transport increased from 189 thousand tonnes of CO_2 equivalent to 457 tonnes (Figure 3–7). The upward trend in greenhouse gas emission from land transport basically conforms to the steady increase in the number of private cars and motorcycles.

Based on exhaust discharge and energy consumption of test vehicles, Song *et al.* (2018b) used life cycle assessment (LCA) to calculate greenhouse gas emissions from buses and light motor vehicles in Macao including indirect emissions from energy consumption and direct emissions from exhaust. Light motor vehicles are found to be major emitters of land transport with a share of 48% in 2014. From 2001 to 2014, greenhouse gas emissions from light motor vehicles in Macao nearly doubled from 125 thousand tonnes of CO_2 equivalent to 248 thousand tonnes (Figure 3–8), while greenhouse gas emission from buses increased from 37 thousand tonnes in 2005 to 50 thousand tonnes in 2016.

Greenhouse gas emissions of marine transport stem from ship use (including freight vessels and passenger vessels) and dock operation. In 2016, the fuel consumption of Macao's marine transport is 3 092 TJ (including 3 085 TJ of light diesel and 7 TJ of electricity). The share of energy consumption of marine transport in the total fuel consumption of the transport sector was about 21.4%, accounting for 9.9% of the energy consumption of all end-uses in Macao (Macao DSEC, 2016; DSPA, 2018). From 2000 to 2016, the amount of emissions increased from 111 thousand tonnes of CO_2 equivalent to 290 thousand tonnes (Figure 3–7) with an overall rising trend despite fluctuations in several years.

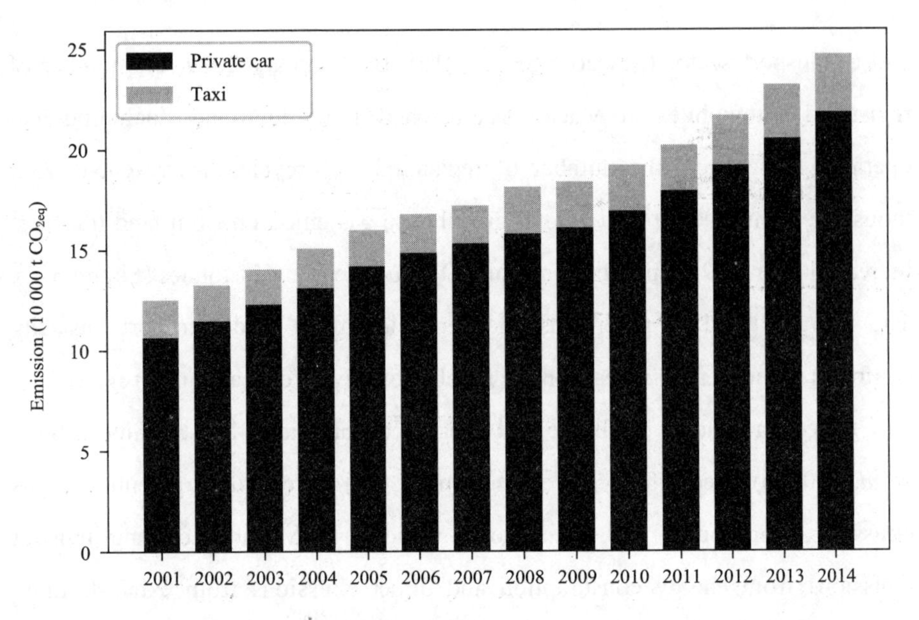

Figure 3–8 Greenhouse gas emissions from light motor vehicles in Macao (Song *et al.*, 2018c)

Greenhouse gas emissions of air transport include those from aircraft movements while exclude those from fuel combustion of flights or airport operation. In 2016, the total fuel consumption of Macao's air transport was 7 970 TJ (including 7 853 TJ of aviation kerosene and 117 TJ of electricity), accounting for about 47.2% of the total fuel consumption by the transport sector and almost all are from aviation kerosene (Macao DSEC, 2017). From 2000 to 2016, greenhouse gas emissions increased from 32 000 tonnes of CO_{2eq} to 93 000 tonnes (Figure 3–7). The annual aircraft movements of Macao International Airport (MIA) have been continually increasing (29 000 in year 2000 and 59 000 in 2017) while the emission per movement has been reducing. Significant improvement has been made in emission reduction: Based on the 2012 level (0.49 tonnes of CO_{2eq}), the target to reduce emission per movement by 20% was met in 2016 and the target of 28.7% was met in 2017. According to statistics on MIA's greenhouse gas emission by source in 2017, electricity consumption is the largest source of emission with a share of 94.9% (19.4 thousand tonnes of CO_{2eq}), followed by refrigerants which constitutes 3.7% (762 tonnes of CO_{2eq}) (MIA, 2018).

Other sectors

(1) Businesses, households, and services

Business, households and services is the third largest source of greenhouse gas emissions in 2016, accounting for approximately 12.75% of total emissions (or 0.26 Mt CO_{2eq}). This value refers to direct emissions with building as the main carrier and energy consumption as the main form (mainly LPG and electricity) (Scope 1 emission). In 2016, Macao's households along with businesses, catering and hotel services accounted for 30.8% of total energy consumption, of which the share of households was about 15.0% (Macao DSPA, 2017). Households consumed about 4 686 TJ, business, catering and hotel consumed about 4 966 TJ, while the gaming industry consumed about 7 338 TJ (Macao DSEC, 2016). Due to low carbon emission during LPG and electricity consumption, direct greenhouse gas emissions are small. Therefore, although Macao's businesses, households and services consume a large proportion of energy, its carbon emissions are not high. From 2000 to 2016, greenhouse gas emission from these sectors increased from 157.8 thousand tonnes of CO_{2eq} to 259.6 thousand tonnes (Figure 3–5).

(2) Waste treatment sector

At present, waste disposal facilities in Macao include four sewage treatment plants, a waste incineration center and a construction waste landfill. Among them, the average daily disposal capacity of the four sewage treatment plants is 211 000 m^3, the Macao Waste Incineration Centre can handle about 1 728 t of household waste per day, and the Macao Construction Waste Landfill has been in use for 13 consecutive years since it was opened in 2006. In 2017, the volume of construction waste was 2.933 million m^3 (including 1.408 million m^3 of sea mud (Macao DSPA, 2018). Greenhouse gas emissions from Macao's waste treatment sector include sewage treatment, waste incineration and landfilling, but the statistics here include emissions from the incineration of municipal solid waste in the Macao Waste Incineration Centre, emissions from waste-to-power, construction waste, furnace slag and fly ash landfilling. In 2016, greenhouse gas emission from Macao's waste treatment sector was 194 000 t CO_{2eq}, representing about 9.5% of the total annual emission. The waste treatment industry has become the fourth largest source of greenhouse gas emissions in Macao in 2016.

From 2000 to 2016, greenhouse gas emission from urban waste disposal in Macao showed a slow declining trend, falling from 237 000 tonnes of CO_{2eq} to 194 000 tonnes (Figure 3–7). A significant rebound occurred from 2008 to 2010 due to the expansion of incinerators with the incineration of huge accumulated refuse that leads to emission surge. Among three ways of waste disposal, refuse incineration is the largest greenhouse gas emitter. In recent years, in order to effectively reduce the greenhouse gas emissions from waste incineration, the Macao SAR Government issued the Macao Solid Waste Resource Management Plan (2017-2026) in 2017 and formulated Macao's solid waste policy, waste reduction target and action plan for the next ten years, striving to achieve the goal of reducing the amount of municipal solid waste by 30% by 2026 from 2016 levels.

(3) Construction sector

With rapid economic development, Macao has undertaken a large number of infrastructure projects which has gradually become a noteworthy source of greenhouse gas emissions. In 2016, Macao's construction industry accounted for about 4.11% of its annual carbon emissions (i.e., 84 000 tonnes of CO_{2eq}). Meanwhile, the output of Macao's construction industry increased from 0.11 billion USD in 2000 to 2.94 billion USD in 2015, contributing approximately 6.5% to Macao's total GDP in the same year. Data for greenhouse gas emissions from Macao's buildings refer particularly to emissions from electricity and fuel consumption during the construction process while exclude emissions from the production of steel, cement and tiles used in construction.

From 2000 to 2016, greenhouse gas emissions from buildings increased from 13 thousand tonnes of CO_{2eq} to 84 thousand tonnes with an obvious rising trend (Figure 3–7). The greatest emission occurred in 2014 at 109 thousand tonnes of CO_2 equivalent. The year-by-year increase in greenhouse gas emission from buildings is consistent with the current growth of infrastructure construction and building renovation projects in Macao.

Using the Life Cycle Assessment Methodology (Life Cycle Assessment, LCA), Zhao *et al.* (2019) estimated the greenhouse gas emissions of urban buildings in Macao between 1999 and 2016, including the construction phase, the use phase, and the disposal phase. The study points out that the total greenhouse gas emissions of buildings in Macao have increased rapidly from 1.52 million tonnes of CO_{2eq} in 1999 to 5.91 million tonnes of CO_{2eq} in 2016. In terms of building life cycle, the use phase is the main source of greenhouse gas emissions, accounting for 65.8% of total emission, followed by the construction phase (33.72%). In addition, the study compared greenhouse gas emissions from buildings in different LCA stages. The results show that the average greenhouse gas emission per unit of urban construction in Macao is 1.47 t CO_{2eq}/m^2, while use-phase residential and non-resident buildings show apparent differences: in

2016, the greenhouse gas emission per unit of residential buildings was 65.3 kg CO_{2eq}/m^2 per year, while those of non-residential buildings was as high as 3 316.4kg CO_{2eq}/m^2 per year, far higher than the emissions of commercial construction units in China's mainland. It is noteworthy that the disposal phase (end-of-life, Eol) of buildings demonstrates "negative emission", which indicates the environmental benefits from the recycling of materials.

(4) Industrial sector

Greenhouse gas emission of Macao's industrial production includes those from textile manufacturing dominated by export processing, production of construction cement, as well as the production and supply of water, electricity and gas while excludes those from sewage and waste treatment industry. Industry is the smallest contributor to total greenhouse gas emissions in Macao, accounting for about 2.7% of the annual carbon emissions of industrial production in Macao in 2016, or 55 000 tonnes of CO_{2eq}. Total industrial greenhouse gas emissions were 47 000 tonnes of CO_{2eq} in 2000 and 55 000 tonnes of CO_{2eq} in 2016. This is in line with the contracting trend of in Macao's industries (Figure 3–7).

Macao Cement Factory is one of the few large private enterprises in Macao, participating in a series of landmark processes in the process of Macao's economic development, such as the Macao International Airport Project, the new bridge and so on. In addition, the Macao Cement Plant has actively expanded its operations in mainland China and Hong Kong through its subsidiaries. Chen *et al.* (2017a) examined greenhouse gas emission from Macao's cement production in various years, but their results are significantly different from those in Macao's environmental report. Besides opposite trends about the same period, there are large discrepancies in data results that greenhouse gas emission from cement production alone exceeds total industrial greenhouse gas emission in the environmental report. The reason is that Macao's environmental report only counted greenhouse gas emission from local cement industry, while Chen *et al.* (2017a) included emission from imported cement consumed in Macao. For example, Macao's industrial greenhouse gas emission in 2013 was 53.6 thousand tonnes while emission from consumed cement was 103 thousand tonnes.

The peaking of Scope 1 emissions

Macao's Scope 1 emissions grew rapidly between 2013 and 2017, mainly due to a significant increase in local power generation, which has not yet peaked. For Macao, the key to whether Scope 1 emissions can peak as soon as possible lies in its power development policy. If the new demand is met by imported electricity and local power generation remains low, other policies such as energy conservation and waste reduction may bring Macao's Scope 1 emissions to peak as soon as possible, otherwise it will continue to increase.

3.2 Greenhouse Gases Emissions in Alternative Accounting Rules

This section mainly focuses on Scope 2 and Scope 3 emissions corresponding to their accounting rules (Table 3–1). Their gaps with Scope 1 emissions are discussed and the unique features of Hong Kong and Macao are distinguished.

3.2.1 Hong Kong

Scope 2 emissions

Hong Kong and Mainland China buy and sell electricity from each other. Prior to 1994, Hong Kong was an exporter of electricity to Mainland China with the amount increasing significantly in the early 1990s to 5 776 GWh in 1993 (Chow, 2001a). In 1993, electricity from the Daya Bay Nuclear Power Station began entering the electric grid of Hong Kong, entirely within the coverage of China Light and Power (CLP) Hong Kong Limited. Only after the plant was fully commissioned in 1994 did Hong Kong become a net importer with import reaching 6 430 GWh while export was only 1 714 GWh (Chow, 2001a). According to CLP Group's Annual Report 2018, this has increased to around 12 426 GWh of electricity in 2017 and up to around 12 501 GWh in 2018.

Nuclear power is generally considered to emit little (Scope 1) direct carbon emissions during its operational phase. The fossil fuel-generated electricity for export emits GHG emissions that are accounted into Hong Kong's Scope 1 emissions. When accounting for Hong Kong's Scope 2 emissions, this part of GHG emissions that arise from electricity export should be deducted and Hong Kong's Scope 2 emissions due to electricity consumption should be less than the Scope 1 emissions. However, the declining trend in electricity export from a peak of 4 528 GWh in 2006 to 556 GWh in 2018 and zero in the first ten months of 2019 (HK Census and Statistics Department, 2019c), indicates that the opportunity to make such deductions may cease over time.

Electricity generation and consumption, especially their geographical separation due to transmission, are critical in accounting CO_2 emissions. In evaluating the performance of Mainland provincial governments in achieving CO_2 mitigation goals, the Chinese Central Government clarified the accounting rules with the following formula (National Development and Reform Commission, 2014): CO_2 emissions = CO_2 emissions from fossil fuel (coal, oil and natural gas) consumption + CO_2 emissions from imported electricity – CO_2 emissions from exported electricity. The emission factor of electricity is the average for each province.

Hong Kong is not regulated under this framework. However, if this accounting method was applied, given that Daya Bay Nuclear Power Station is connected with Hong Kong through a dedicated transmission line (thus the emission factor can be accounted just for nuclear electricity but not Guangdong province's average), Hong Kong's CO_2 emissions should be lower than the currently reported Scope 1 emissions, even after 1994, when Daya Bay was fully commissioned.

From a lifecycle analysis (LCA) perspective, there are carbon emissions associated with nuclear power during its construction phase, the production of the equipment and fuel it uses, its decommissioning phase and the disposal of its waste. We would need to refer to publicly available emission factors to estimate Daya Bay's LCA carbon emissions. Much of the research that is publicly available has shown a significantly wide range of carbon emission estimates due to the wide range of assumptions applied for all the different value chain components, with a resultant range between 3 to 220g CO_{2eq}/kWh (Sovacool, 2008; Warner and Health, 2012).

The Daya Bay nuclear power plant itself does not publish its annual carbon emissions, but CGN Power Company Limited which owns 75% of the Daya Bay nuclear power plant, does publish aggregated Group greenhouse gas emissions arising mainly from the Group's use of energy during construction, production and operation of the Group's assets (which in 2018 were around 258 000 tonnes of CO_{2eq}) (CGN Power Co., Ltd, 2019), but not calculated on an LCA basis. From the perspective of life cycle emissions, in 2010, locally generated electricity in Hong Kong emitted 778.8 g CO_{2eq}/kWh, while nuclear electricity from Daya Bay Nuclear Power Station had an emission factor of 36.7 g/kWh to result in an overall emission factor of 722.1 g CO_{2eq}/kWh (To *et al.*, 2012). In 2015, the overall emission factor was estimated to be 700.3 g CO_{2eq}/kWh (To and Lee, 2017c). The most recent IPCC Fifth Assessment Report adopted harmonized estimates from the Warner and Heath review (Warner and Heath, 2012a) which used consistent gross system boundaries and values for some of the significant system parameters, yielding a range of 4 to 110g CO_{2eq}/kWh, with a median of 12g CO_{2eq}/kWh (IPCC, 2014). Furthermore, studies on the different carbon emissions arising from different nuclear technologies gave an average greenhouse gas emission factor for pressurized water reactor (PWR), which is the technology utilized at Daya Bay, as approximately 11.87g CO_{2eq}/kWh (Kadiyala *et al.*, 2016).

Scope 3 emissions

(1) Consumption-based accounting

Some recent research on the question of carbon leakage has concluded that from a global technical accounting perspective, there does not seem to be a large amount of carbon leakage as "the difference in accounting seems relatively small for most large emitters such as China (–16%) or the USA (+11%)" (Franzen and Mader, 2018), implying that PBA remains a valid accounting approach despite the perception of its inadequacies in disincentivizing carbon transfer behavior.

There was also some research exploring the relevance and possible future of

CBA concluding that "the established PBA model is unlikely to yield its place any time soon to its more controversial CBA alternative" (Afionis *et al.*, 2017). Part of the reason is that the CBA approach does have its challenges, including being based on relatively complicated input-output matrices which involve more assumptions than the PBA approach, thus introducing more uncertainties and potential inaccuracies for the CBA approach compared to the PBA approach.

Despite the fact that PBA is more established and forms the critical backbone of any country's greenhouse gas mitigation strategy, there is no dispute that CBA can help better inform climate change-related policy making and deter the transfer of greenhouse gas emissions to countries without emissions targets (in a way that can undermine global mitigation efforts). Undoubtedly "the institutional framework to enable countries to function in such a system needs developing" (Fong *et al.*, 2014), but given CBA' sability to enable anyone or any party with enough consumer or procurement power to influence the design and development of low carbon products and manufacturing processes and research findings pointing out the growing importance of CBA emissions (Hertwich and Wood, 2018), efforts in analyzing CBA results (Stockholm Environment Institute, 2017; Liddle, 2018) and to improve the accuracy of CBA methods continue to grow (Chen *et al.*, 2018).

Studies have been focusing on the embodied carbon in buildings. Ng and Chau (2015) found that the average embodied energy of a standard high-rise commercial building in Hong Kong was 94 GJ/m^2. Despite the distinction in terms of the amount of embodied energy between high-rise buildings and low-rise counterparts, no clear correlation is found between building height and embodied energy intensity. Ng and Chau (2015) studied the energy saving potential of high-rise commercial buildings in Hong Kong with a focus on the End-of-Life of building materials. They found that proper recycling of construction wastes could help to save 53% of the embodied energy of materials used in the next new built building, while reusing could save 6.2%. Based on the LCA results of different building materials, they draw the conclusions that recycling strategy should be implemented for the building elements containing large amount of concrete (e.g., upper floor construction), while reusing instead of recycling should be adopted for the building parts with high aluminum content (e.g., windows). Moreover, energy use for material manufacturing and transport should be prioritized in reducing the embodied energy, of which the transport energy needs to be taken into extra consideration for the case of Hong Kong (Wang *et al.*, 2018). Gan *et al.* (2017) estimated the embodied carbon in buildings with a case study on a 60-story composite core-outrigger building in Hong Kong. The results showed that structural steel and rebar from traditional blast furnace accounted for 80% of the embodied carbon in the core-outrigger building, while ready mixed concrete contributes only 20%. Hence, they pointed out the possible ways to reduce the embodied carbon by using more efficient construction materials. If steel is produced from electric arc furnace with 100% recycled steel scrap as the feedstock, the embodied carbon of the building can be reduced by over 60%. As for ready-mixed concrete, 10%～20% embodied carbon reduction in buildings can be achieved by utilizing cement substitutes (35% fly ash or 75% slag). More studies also pointed out that the embodied carbon mainly came from the use of the different materials during the life of the building (Chen and Ng, 2016; Dong *et al.*, 2015). The embodied energy use ratio could be as much as 45% in Hong Kong's Zero Carbon Building (Ayaz and Yang, 2009). The existing green building assessment scheme in Hong Kong, BEAM-Plus, has been criticized for ignoring the embodied GHG emissions in buildings. The comparisons of embodied carbon in new and refurbished residential buildings have also been conducted. Langston *et al.*'s (2018) study showed that the mean embodied carbon for refurbished buildings in Hong Kong is 33%～39% lower than new-build projects. Embodied carbon ranges from 645～1 059 $kgCO_{2eq}/m^2$ for new-build and 294～655 $kgCO_{2eq}/m^2$ for refurbished projects.

Hong Kong's economy has been transformed from manufacturing to services to result in less CO_2-intensive economic activities. However, significant amounts of greenhouse gas emissions are embedded in the goods and services that are consumed but not produced in Hong Kong. Accordingly, consumption-based accounting methods often accounted much higher levels of greenhouse gas emissions for Hong Kong as well as other consumption and trading centers (Yang *et al.*, 2015). In 2004, Hong Kong was ranked No. 2 in all global economies in terms of consumption-based emissions through net imports, being 9.2 tonnes of CO_{2eq}/person/year or 64 million tonnes of CO_{2eq}/year (Davis and Caldeira, 2010). This accounting method also ranked Hong Kong to be the No. 7 highest emitter in the world with 14.8 tonnes of CO_{2eq}/person/year (Davis and Caldeira, 2010). Net imports accounted for 62% of Hong Kong's greenhouse gas emissions in 2004, while Mainland China had 1 147 million tonnes of CO_{2eq} from net exports, accounting for more than one fifth of total emissions (Davis and Caldeira, 2010). Another study gave an even higher number for Hong Kong's consumption-based emission at a level of 33 tonnes of CO_{2eq}/person/year in 2009, or 241 million tonnes/year (Chen *et al.*, 2016). Mainland China is the largest recipient and exporter of Hong Kong's consumption-based emissions (Guo *et al.*, 2015; Liu and Ma, 2011). More studies are needed to better understand consumption-based CO_2 emissions in Hong Kong.

(2) Transboundary transport

According to To and Lee (2017a), the total cargo throughput of Hong Kong in 2014 was 326.32 million tonnes and the total GHG emissions in the logistics sector were 37.32 Mt CO_{2eq}, which accounted for 41.1% of the total GHG emissions associated with the consumption of fossil fuels in Hong Kong. Air freight, sea freight and land freight cargo accounted for 65.5%, 33.4% and 1.1% of the total emissions, respectively. To (2015) estimated the carbon intensity of the logistic sector in Hong Kong and showed that 534 tonnes of CO_{2eq} were added per million HKD in 2012. Some heavy goods and foods arrive in Hong Kong in ships burning bunker oil (Lang, 2018). Lang's study has indicated that eventually these arrivals of goods by ship will dwindle, as the fuel for powering these ocean-traveling cargo ships becomes increasingly costly or unavailable. About 5 million tonnes of cargo throughput by air every year, including goods and luxury foods from around the world, as well as baggage and mail (Lang, 2018). Air freight is the largest contributor to the total GHG emissions in the sector.

In regard to air travel, Hong Kong's international airport handles about 1 000 flights per day, landing or departing more than 72 million passengers ("passenger throughput") during 2017 (Lang, 2018). The airport employs about 73 000 staff, and these arrivals and related revenue for the city support many other jobs and services outside the airport.

Fuel trade statistics provide direct insights on CO_2 emissions that either take place or are resulted from the refilling of oil products in Hong Kong. Counting all greenhouse gas emissions from imported fossil fuels, whereby Hong Kong would have emitted 10.7 million tonnes of CO_{2eq} in 1970 and this number would rise to 99.1 million tonnes of CO_{2eq} in 2015, or 13.6 tonnes per capita (To and Lee, 2017b).

Hong Kong imports a significant amount of oil products, including aviation gasoline and kerosene (7 787 395 kiloliters in 2017), unleaded motor gasoline (640 911 kiloliters), gas oil, diesel oil and naphtha (7 398 379 kiloliters), fuel oil (8 248 682 kiloliters) and LPG (378 110 tonnes)(HK Census and Statistics Department, 2018). There is no local refinery facility in Hong Kong to refine crude oil and thus no such import was made. 99.9% of aviation gasoline and kerosene were used by aircrafts in 2017, while ships consumed 98.4% of fuel oil and 46.4% of gas oil, diesel oil and naphtha (HK Census and Statistics Department, 2018). Air flights of the Hong Kong International Airport are either to/from the mainland and overseas and ships are largely cross-boundary as well. Accordingly, most of their associated emissions are not counted in Scope 1, but in Scope 3.

This chapter estimates greenhouse gas emissions from Hong Kong's net import of fossil fuels, including CO_2, CH_4 and N_2O. CO_2 dominates the overall greenhouse gas emissions, being responsible for 99.3%～99.4% annually over the past three decades. CH_4 and N_2O account for about 0.3% each. The ratio between energy-related Scope 1 emissions and the implied emissions from net import of fossil fuels has been declining steadily in the past three decades, from 82.6% in 1990 to 36.9% in 2017 (Figure 3–9). The net import of oil products, especially for cross-boundary shipping and aviation, should account for most of the differences from energy-related Scope 1 emissions. In 2017, the net import of coal, natural gas and oil products implied 25.8, 6.6 and 65.5 million tonnes of CO_2 equivalent emissions (Figure 3–9), while energy consumption led to 36.1 million tonnes of CO_2 equivalent Scope 1 emissions (HKEPD, 2019a).

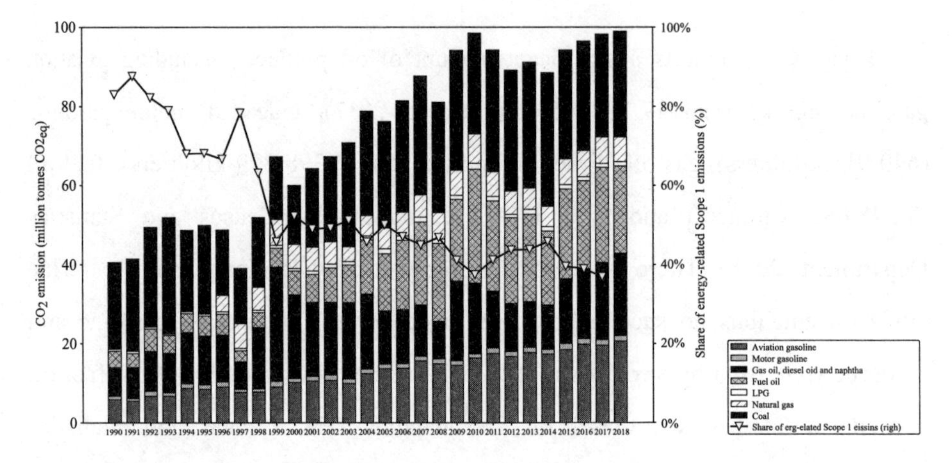

Figure 3–9　Implied greenhouse gas emissions (including CO_2, CH_4 and N_2O) from Hong Kong's net import of fossil fuels (Data on the net import of fossil fuels are from Hong Kong Energy Statistics Annual Report (HK Census and Statistics Department, 2019a). Emission factors are from IPCC (IPCC, 2006b). Global Warming Potential values for CO_2, CH_4 and N_2O are from IPCC AR5 (IPCC, 2013))

(3) The peaking of Scope 2 and 3 emissions

Scope 1 emissions cover all locally generated electricity, regardless whether it is for local consumption or for export. We adopt the estimation, 36.7 g CO_{2eq}/kWh (To *et al.*, 2012), that is directly for the Daya Bay Nuclear Power Station to estimate Hong Kong's Scope 2 emissions. The results are shown in Figure 3–8. Hong Kong's Scope 2 emissions have been consistently less than Scope 1 emissions because it exports coal-fired electricity to Mainland China and imports a significant amount of low-carbon nuclear electricity, although in 2019 the situation might be reversed as no electricity was exported to Mainland China from January to October 2019 and the nuclear electricity import remained stable (HK Census and Statistics Department, 2020). Nevertheless, their differences have been minimal and should be kept small in the future because of nuclear electricity's low emission factor.

From 1990 to 2017, Hong Kong's Scope 2 emissions increased only insignificantly by 19.0% from 35.2 to 40.7 million tonnes of CO_{2eq} (Figure 3–10). From the 2014 level, a 9.3% decrease had taken place in 2017 to indicate that the Scope 2 emissions should also have peaked in 2014.

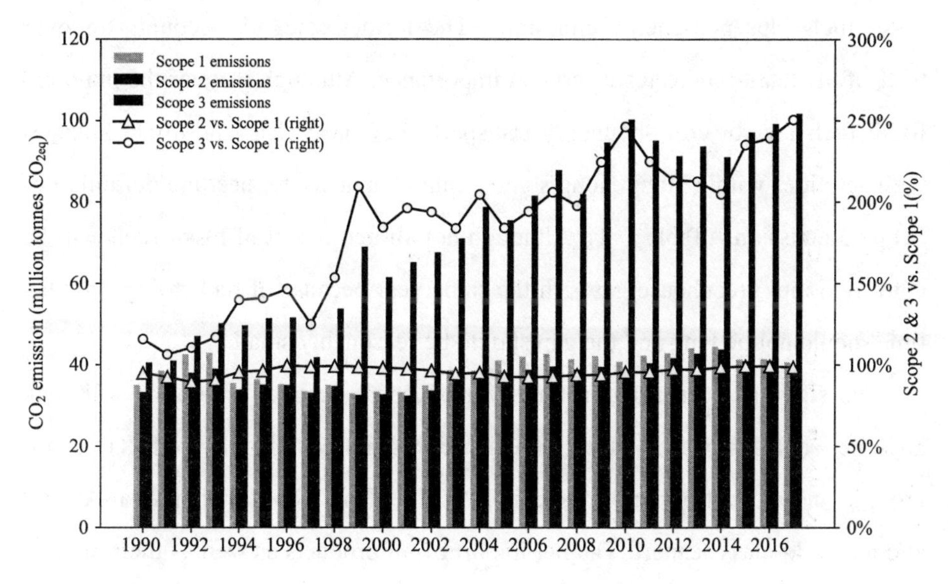

Figure 3–10 Hong Kong's Scope 1, 2 and 3 greenhouse gas emissions[①]

---

①Data for Scope 1 emissions are from official sources (HKEPD, 2019. Greenhouse Gas Emissions in Hong Kong). Scope 2 emissions are calculated as: (1) Scope 1 emissions – (2) emissions embedded in exported electricity to Mainland China + (3) emissions embedded in imported electricity from Mainland China. Emissions in category (2) = Scope 1 emissions from electricity generation * Electricity export to Mainland China / Total local electricity generation. The emissions in category (3) = Nuclear electricity import from Mainland China * Emission factor of nuclear electricity. Considering statistical data availability and for achieving a clear boundary, Scope 3 emissions count those from the net import of fossil fuels (primarily for transboundary transport; see Figure 3–10 for details) and other Scope 2 emissions (including emission from net electricity trade and non-energy emissions). It is noted that the accounting of Scope 3 emissions heavily depends on its defined scope.

Scope 3 emissions have not been regularly examined for Hong Kong partly due to its blurred scope without a clear-cut consensus. As in Figure 3–10 and due to the convenient availability of statistical data, we define Hong Kong's Scope 3 emissions as the sum of implied greenhouse gas emissions from net import of fossil fuels plus non-energy emissions. The former category accounts for over 95% of the totalto indicates its crucial importance. Although most of the imported fossil fuels are for cross-boundary transport, they have clear and direct linkages with activities within Hong Kong's geographic boundary to meet the definition of Scope 3 emissions (Table 3–1). Although not all net import of fossil fuels will be converted into greenhouse gases in the same year because of fuel storage, the lag time should not be long and the gap should be reasonably small.

No sign has been apparent that Hong Kong's Scope 3 emissions will peak anytime soon. Most of Hong Kong's consumption of oil products is for cross-boundary transportation, especially at the Hong Kong International Airport and its world-class seaport. The net import of oil products as well as their implied CO_2 emissions have kept rising up quickly with no sign of levelling off (Figure 3–10). In 2017 with latest available official data, Hong Kong's Scope 1 emissions were 40.7 million tonnes of CO_2 equivalent and the transport sector was responsible for 7.23 million tonnes with slight decline in the past two decades (Figure 3–2). However, the net import of oil products led to 72.43 million tonnes of CO_2 emissions in 2017, being doubled in two decades. Over the period of 1990～2017, Scope 3 emissions increased by 142.3% (Figure 3–10). They were 20.2% higher than Scope 1 emissions in 1990 but the difference widened substantially to 151.9% in 2017 (Figure 3–8). From the 2014 level, although Scope 1 and Scope 2 emissions dropped by 9.4% and 9.3% in 2017 respectively, Scope 3 emissions increased by 10.7% (Figure 3–10).

3.2.2 Macao

Scope 2 emissions

More than 70% of Macao's electricity is imported from Mainland China, and this portion of greenhouse gas emissions should be attributed to Macao's Scope 2 emissions (Song *et al.*, 2018a). Macao's electricity deals with Mainland China date back to the 1980s. Since 1985, Macao's electric power company CEM is the service franchise for Macao's electricity supply. Since 2005, CEM has increased its imports of electricity from China Southern Power Grid, and imported electricity has gradually become a major source of electricity consumption in Macao. In 2013, Macao imported electricity accounted for 92% of the total electricity demand (approximately 4.06 TWh) (Chen *et al.*, 2017a). In 2017, Macao's total electricity supply increased to 5.42 TWh with 1.47 TWh of local generated power (including 170 GWh from waste incineration), while electricity imports from China Southern Power Grid accounted for approximately 73.0% (3.95 TWh) (Macao DSEC, 2019). The share of imported electricity rebounded to 88.2% in 2018. Mainland China has published average CO_2 emission factors for regional power grids since 2010, so this report will use the average CO_2 emission factor (566 grams CO_2/kWh) for China Southern Power Grid from 2010 to 2012 as the emission factor of Macao's imported electricity to calculate the greenhouse gas emission, and the results are shown in Figure 3–11. In 2018, greenhouse gas emission from Macao's imported electricity was 2.779 Mt CO_{2eq}, about 25 times of the levels in 2000 (0.11 Mt CO_{2eq}).

Similar to Hong Kong, Macao's Scope 2 greenhouse gas emissions are defined as Scope 1 emissions plus the emissions implied by electricity trade.

Unlike Hong Kong, Macao's Scope 2 emissions were much higher than Scope 1, significantly increasing by 144.4% from 1.843 million tonnes in 2000 to 4.505 million tonnes in 2017, while the share of imported electricity is 79.9% (Figure 3–11).

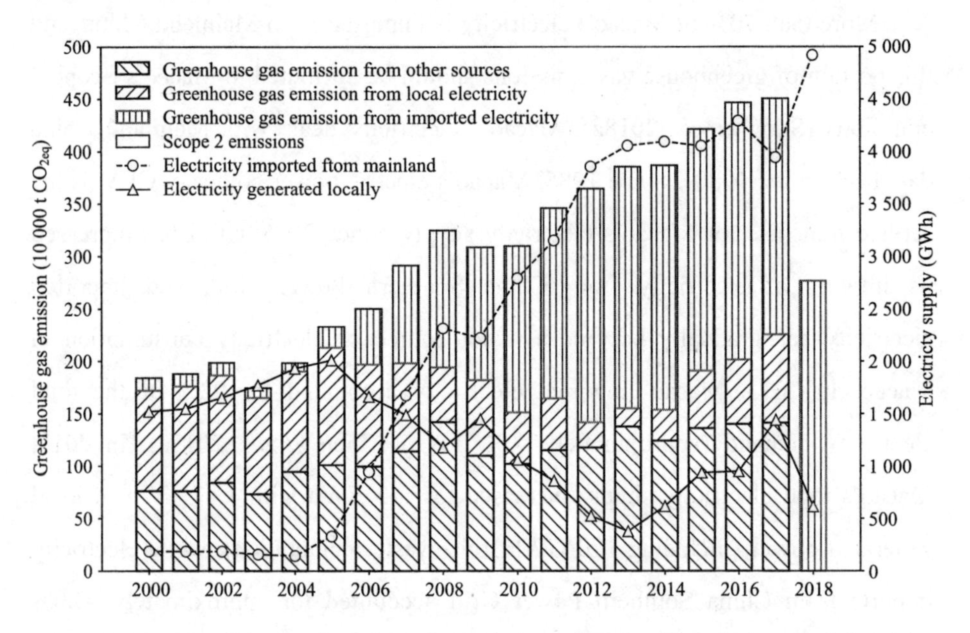

Figure 3–11 Electricity supply and Scope 2 emissions of Macao (Data for imported electricity come from Macao Statistical Yearbooks; data for emission factors come from China's national climate strategic center (National Center for Climate Change Strategy and International Cooperation (NCSC), 2013, 2014); Scope 1 emissions from local electricity and other sources come from Macao DSPA, 2018)

Scope 3 emissions

(1) Consumption-based accounting

Macao has heavily relied on trade to alleviate its resource shortage. With rapid economic growth, Macao's reliance on external trade is increasing. Chen *et al.* (2017b) investigated Macao's trade-embodied carbon inflow and outflow from 2000 to 2013, including commodity trade and service trade. Macao is a typical import-dependent city (almost all of the local goods are imported) but also a world-famous leisure tourism destination, so commodity trade refers to carbon inflow, and service trade refers to carbon outflow dominated by tourism and gaming. As shown in Figure 3–12, both imported carbon and exported carbon kept a growing trend. Imported carbon was greater than exported carbon but the gap is narrowing.

From 2010 to 2013, the carbon balance gap of Macao's trade narrowed from 2.96 million tonnes of CO_{2eq} to 0.294 million tonnes. In terms of Macao's carbon imports, the total amount increased from 5.03 million tonnes of CO_{2eq} in 2000 to 17.4 million tonnes in 2013, the major contributor changed from textile and apparel to energy products including electricity, natural gas and tap water, and its major trading partners include Mainland China, European Union and Japan. Further analysis of Macao's imported carbon emission reveals that the role of Mainland China as Macao's major source of imports had strengthened with increased share in Macao's imported carbon from 73.4% in 2000 to 83.6% in 2013 (Figure 3–13).

In terms of carbon exports, Macao's export services include gaming, hotels, shopping, postal services and tourism services. In addition, Macao still has a small number of exports of goods, mainly textiles and apparel. Figure 3–12 shows that the carbon outflow of Macao's services trade has been growing since 2000. In 2013, Macao's total carbon exports amounted to 17.1 million tonnes of CO_{2eq}, 2.6 times of the level in 2000. Since 2007, the vigorous development of the tertiary service industry has led to a significant increase in Macao's trade carbon exports and has further narrowed the carbon balance gap in Macao's trade. From 2007 to

2013, carbon balance of Macao's trade carbon balance reduced from 4 million tonnes of CO_{2eq} to 0.294 million tonnes.

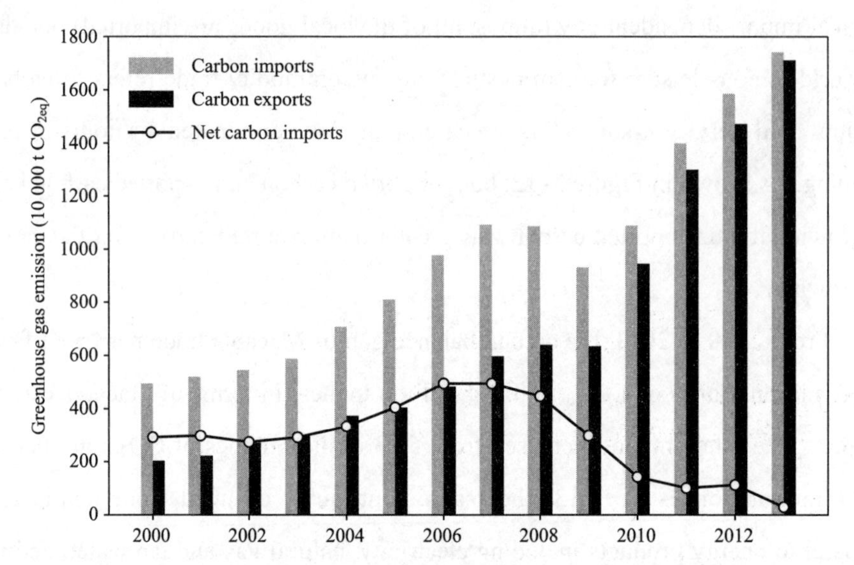

Figure 3–12 Macao's trade-embodied carbon flow (Chen *et al.*, 2017a)

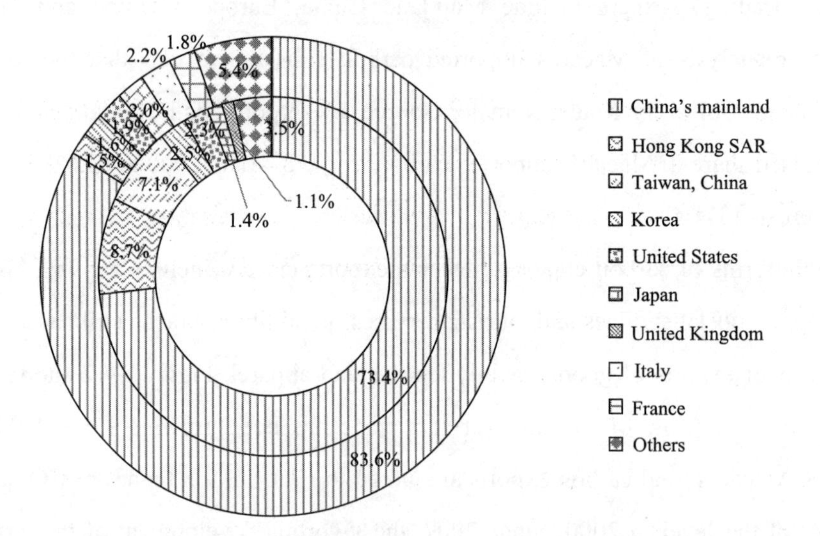

Figure 3–13 Macao's imported carbon (Inner ring represents 2000, outer ring represents 2013) (Chen *et al.*, 2017b)

Although Macao's construction industry accounted for a relatively low proportion of total annual emissions at 4% (i.e., 88 000 t CO_{2eq}) in 2017, this data does not consider the implied carbon of building materials such as cement and steel, or carbon emissions at the use phase of buildings. In 2016, carbon emissions from cement accounted for 34.6% of total emission form raw materials (about 0.433 Mt CO_{2eq}) of new construction in Macao, and carbon emissions from steel accounted for 54.4% of total emissions from raw materials (about 1.552 Mt CO_{2eq}) (Zhao *et al.*, 2019). From the perspective of the building life cycle, the implied carbon ratio of buildings in Macao is high, and the reuse of construction waste can effectively reduce the implied carbon emissions. In 2016, for example, recycling of construction waste could reduce emission by about 86 000 tonnes of CO_{2eq}.

(2) Transboundary transport

According to the Macao Statistics and Census Bureau's Macao Statistics Database, in 2017 a total of about 4.819 million cross-border vehicles in Macao traveled to and from Mainland China via four border stations, including the Border Gate station and the Cotai Border Station, which was 2.3 times the amount of cross-border vehicle traffic in 2000 (about 2.053 million vehicles). Among the border stations, the Border Gate station is the first choice for cross-border car traffic to and from Mainland China, accounting for 72% of the cross-border vehicle traffic (Macao DSEC, 2017).

In terms of air travel, as income levels increases, more and more residents are opting for safer and more convenient air travel modes, including helicopter flights to and from the Pearl River Delta and ordinary commercial flights. Statistics from Macao International Airport show that the number of flights taking off and landing at Macao Airport increased from 29 000 in 2000 to 59 000 in 2017. Airport passenger throughput increased from 3.239 million in 2000 to 7.166 million in

2018, but cargo throughput decreased from 68 100 tonnes in 2000 and 227 200 tonnes in 2005 to 37 500 tonnes in 2017 (MIA, 2020).

There are currently three passenger terminals in Macao, namely the Outer Harbor Terminal (commonly known as the Macao Hong Kong and Macao Terminal), part of the Inner Harbor Terminal Area and Taipa Ferry Terminal. The Inner Harbor Terminal Area retains only a portion of passenger ships to and from the Mainland of China, and the Outer Harbor Terminal and Taipa passenger transport mainly provide passenger services to and from Hong Kong and Mainland China. In 2017, Macao passenger vessels reached 139 000 shifts, about 1.8 times of the level (approximately 75 000) in 2000 (Macao DSEC, 2017).

The Macao Cargo Terminal includes the Inner Harbor and the Ká-Hó Harbor. Inner Harbor is located on the western side of the Macao Peninsula and consists of 34 terminals, mainly freight, inland and fisheries, and the Ká-Hó Harbor has four terminals, including the Coloane Power Company Terminal, the Ká-Hó Harbor Container Terminal, the Ká-Hó Harbor Macao Cement Terminal and the Ká-Hó Harbor Fuel Terminal. Among them, the Coloane Power Company Terminal is limited to the transportation of raw materials and equipment related to power generation, cement terminal is limited to the transport of cement production-related raw materials and equipment, fuel terminal is limited to the transport of oil, natural gas and other dangerous goods.

In terms of greenhouse gas emissions (Figure 3–14), the total amount of greenhouse gas emissions from cross-border transportation increased from 0.351 Mt CO_{2eq} in 2000 to 0.903 Mt CO_{2eq} in 2018, cross-border land transport is not included because data is not available. Overall, greenhouse gas emissions in cross-border transportation fluctuated considerably, with a significant upward trend between 2000 and 2007 from 0.351 Mt CO_{2eq} to 0.83 Mt CO_{2eq}. Starting in

2008, total greenhouse gas emissions from cross-border transport began to decline and rebounded after 2012. Greenhouse gas emissions from water transport is more stable, with average emissions of 171 000 tonnes of CO_{2eq} from 2000 to 2018, accounting for an average of 26.6% of total greenhouse gas emissions from cross-border traffic. The trend of cross-border air transport is consistent with the trend of total greenhouse gas emissions from cross-border traffic. In 2000, the greenhouse gas emissions from cross-border transportation were 244 000 tonnes of CO_{2eq}, and increased to 2.9 times of this level in 2018.

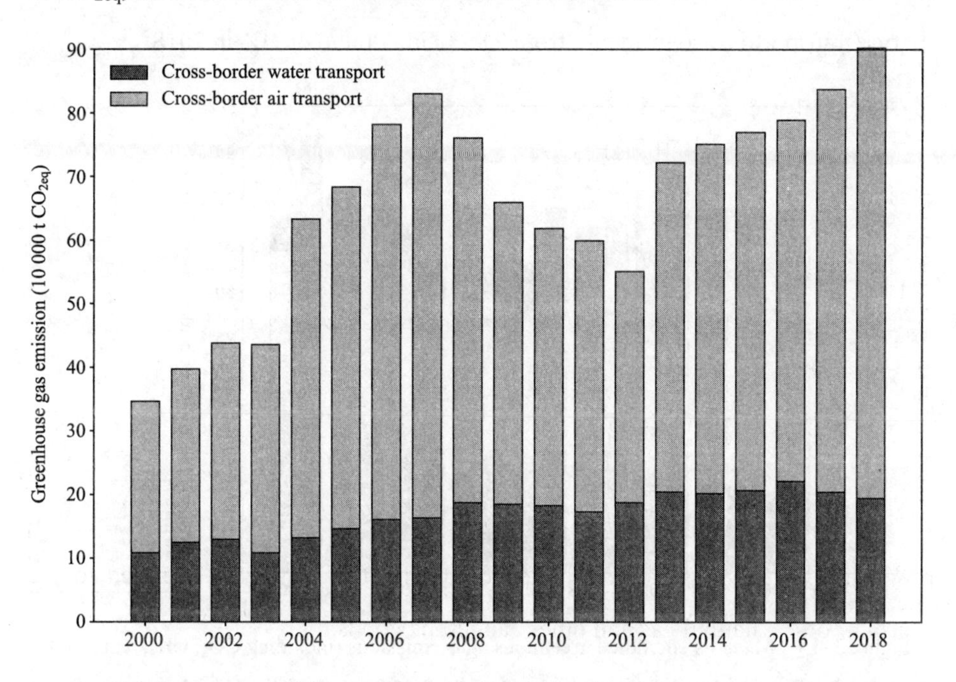

Figure 3–14 Greenhouse gas emissions from cross-border transport in Macao (Data for fossil fuel activity levels from the Macao Statistical Yearbook; data for emission factors of light diesel and aviation kerosene from IPCC (2006a))

Based on the Macao Statistical Yearbook and IPCC data, the greenhouse gas emissions from Macao's net imports of fossil fuels (including CO_2, CH_4, and N_2O)

has been estimated. Overall, Macao's imports of fossil fuels led to an increase of greenhouse gas emissions from 1.647 million tonnes of CO_{2eq} in 2000 to 1.917 million tonnes in 2018. During this period, greenhouse gas emissions were high between 2004 and 2007, averaging at 2.348 million tonnes of CO_{2eq} per year, and then gradually declined to 1.446 million tonnes in 2012. In 2018, the share of gasoline, kerosene, aviation kerosene, light diesel, heavy oil, LPG and natural gas were 13.3%, 0.3%, 36.9%, 26.4%, 6.1%, 6.2%, and 10.6%, respectively. The share of greenhouse gas emissions from energy consumed in cross-border aviation and marine transport is also growing, from 21.2% in 2000 to 46.9% in 2018.

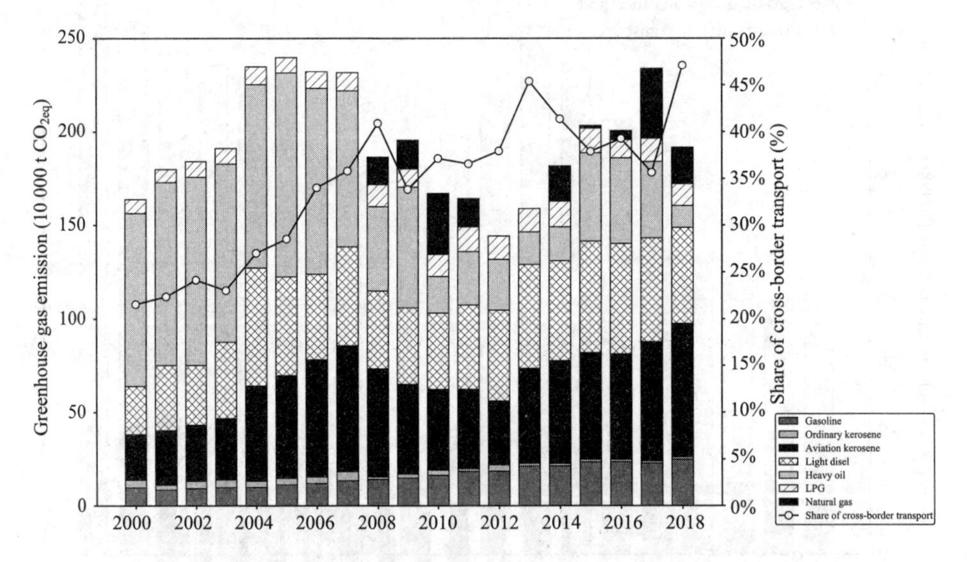

Figure 3–15 Macao's embodied greenhouse gas emissions (including CO_2, CH_4, and N_2O) (Data for (Data for fossil fuel activity levels from the Macao Statistical Yearbook; data for emission factors from IPCC (2006a); GWP values for CO_2, CH_4, and N_2O from IPCC AR5 (IPCC, 2013))

(3) The peaking of Scope 2 and 3 emissions

As Macao substantially increased its electricity imports from China Southern

Power Grid, its ratio of Scope 2 to Scope 1 emissions increased significantly, from 106.4% in 2000 to 248.5% in 2014, and then decreased to 198.6% with the increase of local power generation and Scope 1 emissions (Figure 3–16). Macao's Scope 2 greenhouse gas emissions have not yet shown signs of peaking (Figure 3–16), but with Mainland China's emission reduction efforts, especially the continual decline of the emission factor for China Southern Power Grid, emissions from imported transport are expected to decline significantly in the future, thus driving the peaking of Scope 2 emission. Scope 3 emissions are equivalent to Scope 2 emissions plus the implied greenhouse gas emissions from energy consumed in cross-border water transport and aviation. As the dominant effect of electricity import and the fact that Macao is not an important shipping hub, the trends in Scope 3 emissions are similar to those of Scope 2 without signs of peaking (Figure 3–16).

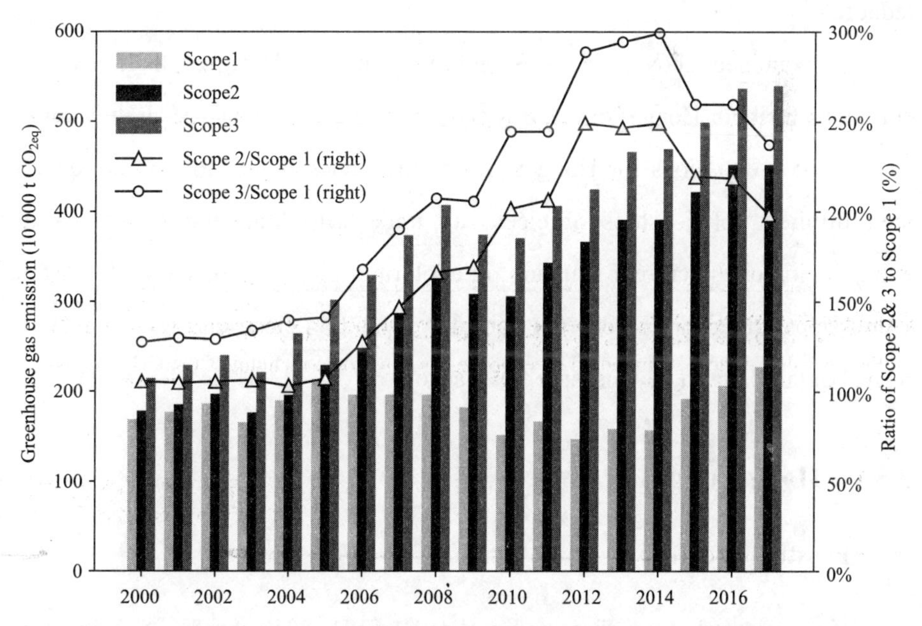

Figure 3–16 Comparison of Macao's greenhouse gas emissions in Scope 1, 2 and 3

3.3 Mitigation Scenarios

This section's discussion on mitigation focuses only on Scope 1 emissions. For Hong Kong and especially Macao, one convenient strategy for mitigating Scope 1 emissions is to move them into the categories of Scope 2 and Scope 3. Electricity generation could be outsourced beyond their individual boundaries, specifically to Mainland China, to replace locally generated electricity. Thus Scope 1 emissions can be proportionally avoided, while Scope 2 emissions may be increased, depending on the CO_2 intensities of local and imported electricity. When importing more goods and services and producing less locally, the embedded Scope 3 emissions would increase but Scope 1 emissions would be reduced.

As examined above, Scope 3 emissions are much higher than Scope 1 emissions for both Hong Kong and Macao. Scope 2 emissions are slightly lower than Scope 1 emissions for Hong Kong but are much higher for Macao, due to their different relationships of electricity trade with Mainland China. As an extrapolation of the current situation, in exploring mitigation scenarios for Scope 1 emissions, this chapter does not consider expanding electricity trade for Hong Kong but takes it as a key measure for Macao.

3.3.1 Hong Kong

Existing scenarios

It is acknowledged that under the HKEPD 2010 Study, a Mitigation

Assessment to compare potential scenarios out to 2030 with 2005 as a base year, was also included (HKEPD). In the Mitigation Assessment, there was a Base Case and three Scenarios with differing ambition, ranging from: Scenario 1 "AQO Scenario" which assumes measures to meet new AQO (air quality objectives) standards; Scenario 2 "Accelerated Scenario" which builds upon Scenario 1, but includes additional efforts on measures to increase energy efficiency and reduce energy demand, particularly in the building and transport sectors; and Scenario 3 "Aggressive Scenario" which builds upon Scenario 2, but accelerates the integration of the power system in Hong Kong with its neighboring areas. Details of the assumptions included can be found in Table 3.1 in Appendix B of the HKEPD 2010 Study. The main conclusions from the assessment included: total carbon emissions for all three scenarios falling below the 2005 level of 42 Mt CO_{2eq} only by 2030, by 6 to 36%; the carbon intensity per unit of GDP for all three scenarios continue to fall well below the 2005 level of 0.03 kg CO_{2eq}/HK$ GDP by 37 to 57% in 2020 and 56 to 70% in 2030; and the carbon emission per capita for all three scenarios also continue to fall below the 2005 level of 6.16 t CO_{2eq}/capita by 10 to 38% in 2020 and 23 to 48% by 2030. It should be noted that a predominant driver of the drop in both measures of carbon intensity particularly between 2020 and 2025 is the assumption that most of the existing coal-fired power plant units are scheduled to be decommissioned during this period.

This assessment reinforced the fact that although Hong Kong is on its way to reducing its carbon emissions, the more significant carbon reductions required cannot be achieved without: further increasing the share of nuclear import to at least 35% (Scenario 2) or even up to 50% (Scenario 3); at least 15% energy savings in all existing commercial buildings; at least 50% reduction in cooling demand and 50% energy savings in all new buildings; new vehicles being 20% more efficient than in 2005, with all petrol being 10% ethanol and all diesel being 10% biodiesel; fully utilizing all landfill gas and all gas generated from wastewater treatment; and having enough Integrated Waste Management Facilities (IWMF) to manage all of Hong Kong's Municipal Solid Waste (MSW), as well as at least two Organic Waste Treatment Facilities (OWTF). This assessment included the assumption of only using technologies that were considered proven at the time.

Further to the conclusions of the HKEPD 2010 Study, in 2017 the HKSAR Government announced carbon reduction targets of 26 to 36% reduction by 2030 compared to the 2005 level. For more details, please refer to Chapter 4. It is also noted that in 2019 the HKSAR Government commenced a Public Engagement exercise on the Long Term Decarbonisation Strategy for Hong Kong (Council for Sustainable Development, 2019).

Future Scenarios 2040

Existing scenarios, as introduced above, are not ambitious enough. This report designs mitigation scenarios with reference to the recent IPCC report on 1.5℃ warming (IPCC, 2018), which requires the world to achieve net zero CO_2 emissions in as early as 2040 following a straight mitigation slope. Due to Hong Kong and Macao's constraints in providing negative emissions such as through biofuel or ecosystem uptake, we explore necessary measures for reducing their Scope 1 emissions by 90% from the 2017 level with the most recent available greenhouse gas emissions official data. Existing studies are scarce on Hong Kong and Macao's mitigation scenarios, while it is not this chapter's intention to conduct a thorough original research. Accordingly, in estimating necessary measures for achieving the mitigation scenarios, we adopt the concept of "stabilizing wedges" to identify big strokes for Hong Kong and Macao's future mitigation (Pacala and Socolow, 2004). In the scenario analysis, we only examine necessary technical measures for providing major mitigation opportunities, without analyzing the economic, political and social feasibilities. In order to reach more details and greater depth, more modelling studies should be conducted in the future for Hong Kong and Macao.

In 2017, energy consumption and waste management were responsible for 36.1 and 2.8 million tonnes of CO_{2eq}, respectively, with the remaining 1.8 million tonnes from industrial processes and others (HKEPD, 2019a). We demonstrate one possible pathway that brings down Hong Kong's greenhouse gases emissions by 90% in 2040, or 4.1 million tonnes from 40.7 million tonnes in 2017. Six major technical measures are designed for achieving the 90% reduction goal, including suspension of fossil-fuel-fired electricity export, full electrification of energy consumption, energy conservation, lower emission factor due to fuel mix change, CCS (CO_2 capture and storage) and waste to energy. The scenario and mitigation measures are summarized in Figure 3–17 and described below in details.

(1) Suspension of fossil-fuel-fired electricity export

The export of fossil-fuel-fired electricity from Hong Kong to Mainland China (4 828 TJ or 1 341 GWh in 2017) is assumed to be zero in 2040, while the import of nuclear electricity from Mainland China (45 274 TJ or 12 576 GWh) remains unchanged (HK Census and Statistics Department, 2019a). This will reduce Hong Kong's emissions by 1.0 million tonnes to 39.7 million tonnes.

(2) Electrification of energy consumption

As analyzed below, full electrification of Hong Kong's energy consumption would lead to more greenhouse gases emissions, or 45.7 million tonnes. The assumptions are made with differentiations between (i) the residential, commercial and industrial sectors and (ii) the transport sector.

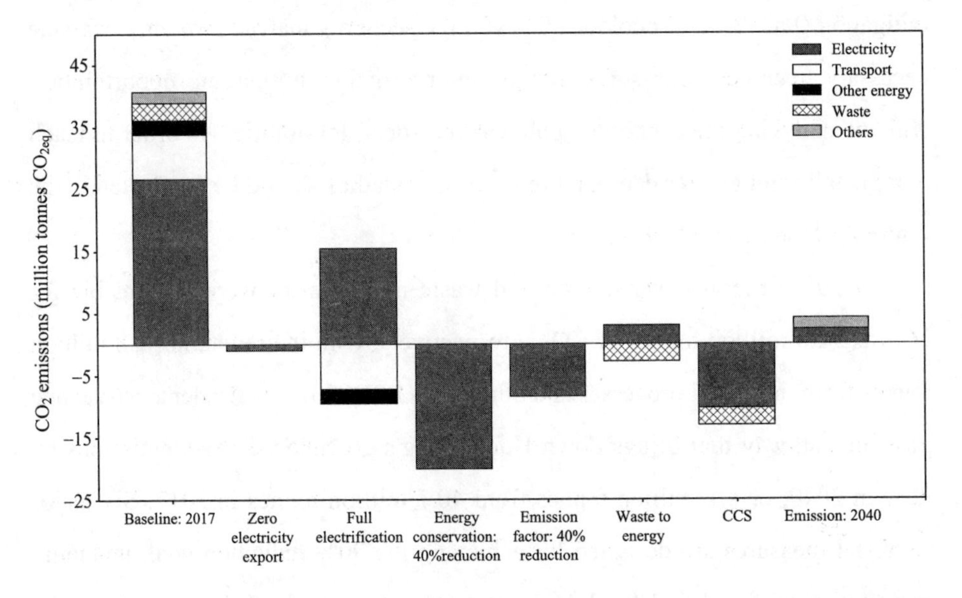

Figure 3–17 Hong Kong's mitigation scenario and measures for Scope 1 emissions (Hong Kong's emissions in 2017 is taken as the baseline. After suspending fossil-fuel-electricity export to Mainland China, five mitigation measures are planned, one after another, for achieving the 90% reduction goal.)

1) Electrification of the residential, commercial and industrial sectors

Residential and commercial sectors, generally in buildings, accounted for a great majority of Hong Kong's end-use energy consumption. Residential sector contributes to around 21% of total energy consumption in Hong Kong, which was 59 992 TJ in 2017 (HK Electrical and Mechanical Services Department, 2019). The average per capita energy consumption in the Residential Sector increased by about 2.6% between 2007 and 2017, while the total energy consumption of the Residential Sector increased by 9.6%, which is mainly because of the increase in both the population and household number in Hong Kong. The life cycle GHG emission of a standard public residential building was estimated to be 4 980 kg CO_{2eq}/m^2, of which 85.8% was from operating energy consumption, 12.7% for materials, 1.1% for renovation, 0.3% for end-of-life of the building, and 0.1% for other factors (Yim *et al.*, 2018). Compared to public residential buildings, private residential units may have a relatively higher emission rate. Humphrey (2004) found that a standard private sector block in Hong Kong had 6% more energy consumption per square meter. Although the operating energy consumption of private building contributes less emissions than that of the public housing block, the estimated carbon emissions of its construction, repair and maintenance, and demolition are significantly higher. In 2017, private housing energy consumption accounted for 50% of the total energy consumption in the Residential Sector, while public housing units only used about 28%. Among the operating energy consumption, the largest proportion of energy was used for air conditioning (27%), cooking (26%) and providing hot water (24%) (HK Electrical and Mechanical Services Department, 2019). Therefore, various carbon reduction measures should be attempted and evaluated, such as the use of energy efficient equipment, renewable energy, recycled/recyclable materials, and eco-design by utilizing natural lighting and ventilation (Yim *et al.*, 2018).

Commercial sector takes up a more significant portion of total energy consumption, which was 125 158 TJ and about 44% of the total energy consumption in 2017. In the Commercial Sector, air conditioning was also the largest end user, consuming 25% of the total energy, followed by cooking (13%), lighting (12%) and hot water & refrigeration (12%) (HK Electrical and Mechanical Services Department, 2019). Wang *et al.* (2018) have conducted a life cycle analysis on the energy use of ten real-life high-rise office buildings in Hong

Kong. Their results showed that the average life cycle energy consumption of these office buildings was 51.78～73.64 GJ/m^2 over a 50-year study period. Building operation consumes 78%～89% and the rest is taken by the embodied energy. As air conditioning and lighting are the two major energy consumers in the commercial sector, the results derived from studies analyzing the decision schemes on energy conservation in Hong Kong's shopping malls showed that shopping mall managers in Hong Kong decided to set the indoor air temperature a bit lower than thermal neutral with a rational explanation that tolerance to hotness is lower than that to coldness, which affects the duration of customer staying in the malls (Kwok *et al.*, 2017a; 2017b). The results on lighting in shopping malls are also more or less the same. The prevailing weather conditions in Hong Kong from 2009～2100 may lead to an increase for annual cooling load and energy use in office buildings by 10.7% and 4.9% compared with 1979～2008 levels (Lam *et al.*, 2010).

Electricity was the dominant energy type with 42 127 TJ and 104 758 TJ or 11 702 GWh and 29 099 GWh respectively for residential and commercial sectors in 2017 (HK Electrical and Mechanical Services Department, 2019). Industrial sector consumed 12 705TJ of energy, including 7 963 TJ of electricity (2 212 GWh). A significant proportion of the non-electricity energy consumption in these sectors is heat, such as for cooking and hot water. Considering efficiency and behavioral differences when using gas and electric ovens, their net energy consumption is largely equivalent to each other as found in a European Commission study (BIO Intelligence Service, 2011). Accordingly, as an approximate estimation, the non-electricity energy consumption is converted into electricity by equivalent heating values (1 kWh = 3.6 MJ). For all energy consumption in these sectors to be fully electrified, 54 960 GWh, or 11 946 GWh more, of electricity will be needed.

2) Electrification of the transport sector

Considering low-carbon transport options, Lang (2018) pointed out that in some cities, bicycles can carry a substantial part of the load for commuting and distribution of goods, but unlike in many cities in Mainland China, there is virtually no bicycle-commuting in Hong Kong. At present, with the streets congested with busses, trucks, cars, and taxis, it would be impossible at any time soon to make room for safe bicycle traffic on most of these roads, and there is virtually no public support or advocacy for increased use of bicycles, except for recreational uses in some areas along the waterfront. In any case, many residential buildings are in hilly areas which would make bicycle transport difficult except for the hardiest cyclists. Hong Kong's transport policy generally discourages cycling in core urban areas and sees cycling more of a recreation or local commuting mode in the New Territories. This is finally changing with a plan for the expansion of cycling tracks (but largely in the New Territories) and the introduction of new, dockless bike-sharing schemes in various new towns in the New Territories (Leung *et al.*, 2018).

Increasing vehicle ownership was the largestcontributor to the increase of passenger transport energy use in Hong Kong (Boey and Su, 2014). Private car ownership increased by 33.2% between 2010 and 2017 (HK Transport Department, 2018). The attitude of Hong Kong's inhabitants toward private vehicles was revealed by a survey among young people indicating that "over 70% of the respondents showed some intention of buying a car in the future" (Cullinane, 2003). The problem raised by another survey among car owners is that once a car is purchased, it rapidly becomes a necessity and the feeling is reinforced over time (Cullinane and Cullinane, 2003). Consequently, an average

Hong Kong car owner manages to drive over 5 000 km a year (over 13 km per day) in an area approximately 20 by 20 km large (Poudenx, 2008). Wang *et al.*'s (2012) study employed econometric models to estimate the direct rebound effect for private passenger transport in Hong Kong when vehicles became more efficient. They found that the magnitudes of the direct rebound effect are respectively 45% and 35% for 1993～2009 and 2002～2009, which indicated that there was a declining trend in the direct rebound effect for passenger transport over time. It also indicated that the direct rebound effect needed to be taken into consideration in assessing the impact of energy efficiency measures on reducing energy use in Hong Kong. Chow (2016) investigated the impacts of different spatial-modal strategies on reducing commuting emissions in Hong Kong. The findings illustrated that if Hong Kong were reconstructed to be a city with multiple CBDs, a dual-centric strategy is desirable because both minimum and maximum commuting are shorter than that of a tri-centric strategy. Moreover, the modal strategy to actively promote rail usage shows more impacts on emissions reduction and car usage should be maintained at the current level.

We envision an electric future for Hong Kong's transport sector when all vehicles are converted with electricity as the only fuel. Hong Kong's transport sector consumed 88 414 TJ of energy in 2017 (31% of total), in which 3 129 TJ (869 GWh) was in the form of electricity (HK Electrical and Mechanical Services Department, 2019). According to the U.S. Department of Energy, electric vehicles have efficiencies of about 59%～62%, while those gasoline-based ones are of 17%～21%. Because diesel-based vehicles tend to be more efficient than gasoline-based ones, we use a conversion rate of 3.18 to convert the non-electricity energy consumption in the transport sector into a future one that relies entirely on electricity. According, 8 309 GWh of electricity will be needed.

Additional measures should be put into place to control the rising demand for transportation in order to keep the energy demand flat.

As a result, full electrification will elevate electricity consumption in these end-use sectors to be 63 269 GWh, or 19 386 GWh higher than the 2017 level. Furthermore, in 2017, the ratio between Hong Kong's electricity consumption and the power industry's own use and loss was 10.0%. It will then indicate that Hong Kong should supply 69 589 GWh of electricity for satisfying the demand. With 45 274 TJ or 12 576 GWh of imported electricity from Mainland China and no further export, as assumed earlier from the 2017 situation, Hong Kong should generate 57 013 GWh of electricity in local power plants. When applying the same emission factor in 2017 of local electricity generation (26.6 million tonnes/36 917 GWh = 721 grams CO_{2eq} per kWh) (HK Census and Statistics Department, 2019a; HKEPD, 2019a), it indicates that the 41.1 million tonnes of CO_{2eq} will be emitted from the energy system. In addition to the 2.8 million tonnes from waste management and 1.8 million tonnes from other sources, 45.7 million tonnes will be emitted.

3) Energy conservation

As a crucial method of CO_2 mitigation worldwide, energy efficiency and conservation should be technologically enhanced and energy conservation behaviors should be improved. We assume a 40% reduction of electricity demand in 2040 to reach 41 753 GWh, which requires 29 177 GWh of local electricity generation with 21.0 million tonnes of CO_{2eq} emissions. The overall emission level will drop to 25.6 million tonnes.

4) Emission factor and fuel mix

Furthermore, the emission factor of local electricity generation is also assumed to be reduced by 40% to reach 432 grams CO_{2eq} per kWh, similar to the level for natural gas-fired electricity. This will bring down Hong Kong's greenhouse gases emissions from the energy system to 12.6 million tonnes of CO_{2eq} and overall, to 17.2 million tonnes of CO_{2eq}. Hong Kong's fuel mix has been stable in the past two decades with coal accounting for about half of electricity supply, natural gas one quarter and the imported nuclear electricity another quarter (HK Environment Bureau, 2014). One official plan proposes to increase the share of natural gas to 60% in Hong Kong's electricity supply and reduce the share of coal below 20% (HK Environment Bureau, 2014). A survey on Hong Kong consumers' willingness to pay found stronger support to this final plan, being 48%～51% relative to the current electricity bills, while that for the increase of nuclear electricity import was 32%～42% (Cheng *et al.*, 2017). This natural gas plan was finally adopted. The fuel mix is about to have major changes over the next decades with the replacement of retiring coal-fired power plants with natural gas power plants (HK Environment Bureau, 2017).

Energy transition toward renewables is crucial for the global mitigation of greenhouse gas emissions. However, in the densely populated city of Hong Kong, it may face three major challenges, including (i) education and engagement with the public about a sustainable future, (ii) alignment of stakeholders' economic interests, and (iii) absorption capacity of emerging technologies (Ng and Nathwani, 2010). Renewables are not expected to play any sizable role in Hong Kong's future fuel mix in the official plans. Hong Kong proposed an offshore wind farm in southeastern waters, but the plan has not been materialized. Building-integrated photovoltaic (BIPV) systems have been evaluated in Hong Kong specific solar contexts and building situations, while locations and orientations are key factors to influence the PV systems' performance (Lu and Yang, 2010; Yang *et al.*, 2004). Together with the renegotiation of the Scheme of

Control, Hong Kong started feed-in-tariff to encourage the local generation of renewable electricity. The rates per kWh were set for systems of three different sizes: (a) \$5 for $\leqslant$10 kW; (b) \$4 for >10 kW to $\leqslant$200 kW; and (c) \$3 for >200 kW to $\leqslant$1MW (HK Offshore Wind Limited, 2006).

5) CO_2 Capture and Storage (CCS)

CO_2 Capture and Storage (CCS) technology is utilized for realizing the ecosystem wedge and negative emissions (National Academies of Sciences Engineering and Medicine, 2019). Xu and Liu (2015) examined an alternative plan that was proposed and assessed for Hong Kong to continue using coal but utilize CO_2 capture and storage (CCS) technologies for reducing CO_2 emissions. An 800-km natural gas pipeline was used to supply natural gas to Hong Kong from a field in South China Sea. The plan proposed to capture CO_2 in Hong Kong's coal-fired power plants and then transport it in a reversed direction with the pipeline to conduct enhanced gas recovery and CO_2 storage in the depleted natural gas Yacheng 13-1 field (Xu and Liu, 2015). Adequate CO_2 storage capacity exists in this gas field to support this plan (Zhang *et al.*, 2014). This plan could also be applied to natural-gas fired power plants.

We assume that the CCS technology could avoid 85% of CO_2 from Hong Kong's power plants (IPCC, 2005). Accordingly, 1.9 million tonnes of CO_{2eq} will still be emitted into the atmosphere from the energy system after CCS, or 6.5 million tonnes in total.

6) Waste to energy

In 2017, Hong Kong dumped 3.9 million tonnes of municipal solid waste into landfills (HKEPD, 2019b), while waste management incurred 2.8 million tonnes of CO_{2eq}. We assume that in 2040, these wastes will be incinerated for electricity generation with CCS. When incinerated, one tonne of municipal solid waste could lead to 0.7～1.2 tonnes of CO_2 emissions and generate about 300～600 kWh of electricity (Johnke, 2001). Our estimation adopts their average values and the emission factor is then 2.1 kg CO_{2eq}/kWh, much higher than that of even coal-fired electricity. Accordingly, the incineration of all 3.9 million tonnes of municipal solid waste will incur 3.7 million tonnes of CO_{2eq} and 1 763 GWh of electricity. Because landfills account for nearly all greenhouse gas emission in Hong Kong's waste management (Dong *et al.*, 2017), we assume that the original 2.8 million tonnes of CO_{2eq} in 2017 from waste management will be entirely eliminated. The electricity from waste replaces local fossil fuel-fired electricity. Without CCS, waste to energy could increase greenhouse gases emission by net 0.15 million tonnes. After CCS, electricity generation from fossil fuels and municipal solid waste incineration will emit 2.3 million tonnes of CO_{2eq} to bring the overall emissions at 4.1 million tonnes of CO_{2eq}.

3.3.2 Macao

The mitigation of Scope 1 emissions in Macao can be achieved by full electrification of energy use, energy conservation and waste reduction, and increased purchase of electricity. As mentioned earlier, Macao's electricity development policy is the key to the peaking of Scope 1 emissions and reducing emissions. Full electrification allows for imported electricity to gradually replace other local emissions; energy conservation can reduce the demand for imported electricity, and waste reduction can reduce greenhouse gas emissions from local waste treatment. This strategy is equivalent to shifting Scope 1 emissions to Scope 2, whereas the implied greenhouse gas emissions of outsourced electricity depend on the external electricity and emission factors. In contrast to Hong Kong, the emissions and emission reduction trends in Scope 1 and 2 emissions of Macao are closely related to Mainland China, particularly the China Southern Power Grid.

Full electrification, energy saving and waste reduction

(1) Transport

The transportation sector is an important source of greenhouse gas emissions in Macao, especially land transportation and water transportation. The two main types of fuels are gasoline and light diesel, with accounting for 22% and 14% of greenhouse gas emissions in 2016. Macao currently promotes various mitigation policies in terms of imported new cars, in-use cars, fuel oil and environment-friendly vehicles. To further implement low-carbon transport, the government set standards in unleaded gasoline and light diesel (Macao SAR Government, 2016) and launched preferential taxation of environment-friendly vehicles (Macao SAR Government, 2018) to change residents' preference. Based on successful experience in other places, the SAR government launched "Elimination of Heavy and Light Two-stroke Motorcycle Funding Scheme" which has eliminated 52% of motorcycles in Macao. With the implementation of various policies, there has been extremely small room for traditional energy vehicles to reduce greenhouse gas emission. New energy vehicles have become the core of low-carbon transportation strategies, especially electric vehicles. Electric vehicles will theoretically achieve zero direct emissions during the use phase, actually transferring responsibility for emission reduction to power generation.

(2) Waste management

Paper/cardboard, plastics and organic matter are the main physical components of municipal solid waste in Macao, with a ratio of 24.4%, 21.0% and 38.5% respectively in 2016 (Macao DSPA, 2017). On one hand, the growing waste is approaching the maximum capacity of the Macao Waste Incineration Centre, and on the other hand, it is constantly adding to greenhouse gas emissions from the waste treatment sector. In addition, the burial of fly ash and construction waste is also threatening Macao's stretched available land area.

In 2017, Government of the Macao Special Administrative Region issued the Macao Solid Waste Resource Management Plan (2017-2026) which sets the target for solid waste reduction in Macao for the next decade, and plans to reduce the per capita solid waste disposal by 30% (0.63kg per capita) by 2026. By recycling resource waste such as paper/cardboard and plastic, the amount of solid waste treated by the Macao Waste Incineration Centre can be reduced, thereby reducing greenhouse gas emissions. For organic components in solid waste such as kitchen waste, it can be considered for conversion into bioenergy.

(3) Industries, construction, businesses, and households

With the improvement in living standards and technological advancement, electrical and electronic appliances have become an integral part of household and commercial activities. "Report on the Energy Efficiency Survey of the Macao Special Administrative Region" (Macao GDSE, 2019) pointed out that local residents currently exhibit a relatively high level of energy-saving behavior in the early-stage use of electronic products, but there is still room for improvement. In line with its development positioning of "World-class Tourism and Leisure Center", the SAR Government has been committed to promoting low-carbon life initiatives to residents. Each year, the Macao International Environmental Cooperation Forum (MIECF) is held to strengthen international environmental cooperation and deepen residents' awareness of environmental protection (Macao International Environmental Cooperation Forum, 2019). Targeted at Macao's booming hotel industry, the Macao Environmental Protection Bureau (DSPA) has established the "Macao Green Hotel Award" since 2007 to actively encourage the green development of the local hotel industry. For residents, companies and schools, the Macao Environmental Protection Bureau organizes annual activities such as "Environment Fun", "Low Carbon Festival" and "Green Enterprise Partnership Program" to encourage "green travel" and implement the low-carbon life initiative in all aspects of residents' life. Additionally, the Macao Environmental Protection Bureau (DSPA) issued the "Macao Environmental Protection Plan (2010-2020)" and the public sector environmental management plan to highlight the role of the public sector and improve social environmental performance.

For Macao, electrical and electronic appliances, especially air conditioners, have become the chief source of energy consumption. Therefore, improving their energy efficiency has become an important measure of urban energy conservation and emission reduction. For this purpose, Macao Government has established "Environmental Protection and Energy Conservation Fund" to subsidize energy efficient products (Macao GDSE, 2019). In terms of air conditioning, only non-government offices and hotels turned down their setting temperature in 2017, while households and restaurants turned up their setting temperature. In summer, a 1℃ increase in the setting temperature of air conditioners leads to 7%～10% increase in energy efficiency. The larger the difference between indoor and outdoor temperature, the greater the improvement in energy efficiency, which suggests huge energy saving potential of air conditioning in Macao (over 20% referring to 26℃ in Mainland China) (Song *et al.*, 2017). In addition, the promotion of energy labels helps residents choose energy efficient electrical and electronic appliances. Taking the 268 L domestic refrigerator as an example, one with energy efficiency Level 1 saves about 0.7 kWh/day of electricity compared to one with Level 5.

Power sector developments

The local power industry has long been an important source of greenhouse gas emissions in Macao. In the future emission reduction scenario, the proportion of imported electricity need to continue to rise. Macao can switch to natural gas instead of heavy oil as the main fuel for power generation. Song *et al.* (2018a)'s points out that the average emission factor for natural gas and oil-fired power is 0.42 (kg CO_{2eq}/kWh) and 0.71 (kg CO_{2eq}/kWh), respectively, so that greenhouse gas emissions can be reduced by 41% per kWh. The widespread use of renewable energy such as solar energy can also bring about emission reduction benefits, but its share and potential are small, with only 1 448 kWh of electricity generated in the pilot photovoltaic system in 2017 (CEM, 2017).

References

Afionis, S., M. Sakai, K. Scott, *et al.*, 2017. Consumption-based Carbon Accounting: Does it Have a Future? *Wiley Interdisciplinary Reviews: Climate Change*, 8.

Ayaz, E., F. Yang, 2009. Zero Carbon Isn't Really Zero: Why Embodied Carbon in Materials Can't Be Ignored, https://www.di.net/articles/zero_carbon/.

BIO Intelligence Service, 2011. *Domestic and Commercial Ovens (Electric, Gas, Microwave), Including Incorporated in Cookers: Task 3 Consumer Behaviour and Local Infrastructure*, https://www.eceee.org/static/media/uploads/site-2/ecodesign/products/lot22-23-kitchen/lot22-task3-final.pdf.

Boey, A., B. Su, 2014. Low-carbon Transport Sectoral Development and Policy in Hong Kong and Singapore. *Energy Procedia*, 61.

CGN Power Co., Ltd., 2019. *Environmental, Social & Governance Report 2018*.

Chen, B., Q. Yang, J. Li, *et al.*, 2017a. Decoupling Analysis on Energy Consumption, Embodied GHG Emissions and Economic Growth—the Case Study of Macao. *Renewable and*

Sustainable Energy Reviews, 67.

Chen, B., Q. Yang, S. Zhou, *et al.*, 2017b. Urban Economy's Carbon Flow Through External Trade: Spatial–temporal Evolution for Macao. *Energy Policy*, 110.

Chen, G. W., T. Wiedmann, Y. F. Wang, *et al.*, 2016. Transnational City Carbon Footprint Networks–Exploring Carbon Links Between Australian and Chinese Cities. *Applied Energy*, 184.

Chen, Y., S. T. Ng, 2016. Factoring in Embodied GHG Emissions When Assessing the Environmental Performance of Building. *Sustainable Cities and Society*, 27.

Chen, Z. M., S. Ohshita, M. Lenzen, *et al.*, 2018. Consumption-based Greenhouse Gas Emissions Accounting with Capital Stock Change Highlights Dynamics of Fast-developing Countries. *Nature Communications*, 9.

Cheng, Y. S., K. H. Cao, C. K. Woo, *et al.*, 2017. Residential Willingness to Pay for Deep Decarbonization of Electricity Supply: Contingent Valuation Evidence from Hong Kong. *Energy Policy*, 109.

Chow, L. C. H., 2001a. Changes in Fuel Input of Electricity Sector in Hong Kong Since 1982 and Their Implications. *Energy Policy*, 29.

Chow, L. C. H., 2001b. A Study of Sectoral Energy Consumption in Hong Kong (1984-97) with Special Emphasis on the Household Sector. *Energy Policy*, 29.

Council for Sustainable Development, 2019. *Support Group on Long-term Decarbonisation Strategy*.

Cullinane, S., 2003. Hong Kong's Low Car Dependence: Lessons and Prospects. *Journal of Transport Geography*, 11.

Cullinane, S., K. Cullinane, 2003. Car Dependence in a Public Transport Dominated City: Evidence from Hong Kong. *Transportation Research Part D: Transport and Environment*, 8(2).

Davis, S. J., K. Caldeira, 2010. Consumption-based Accounting of CO_2 Emissions. *Proceedings of the National Academy of Sciences of the United States of America*, 107.

Dong, Y. H., A. K. An, Y. S. Yan, *et al.*, 2017. Hong Kong's Greenhouse Gas Emissions from the Waste Sector and Its Projected Changes by Integrated Waste Management Facilities. *Journal of Cleaner Production*, 149.

Dong, Y. H., S. T. Ng, 2015. A Life Cycle Assessment Model for Evaluating the Environmental Impacts of Building Construction in Hong Kong. *Building and Environment,* 89.

Environmental Resources Management, 2018. *Hong Kong Offshore LNG Terminal*.

Franzen, A., S. Mader, 2018. Consumption-based Versus Production-based Accounting of CO_2 Emissions: Is There Evidence for Carbon Leakage? *Environmental Science & Policy*, 84.

Fung, W. Y., K. S. Lam, W. T. Hung, *et al.*, 2006. Impact of Urban Temperature on Energy Consumption of Hong Kong. *Energy*, 31.

Fong, W. K., M. Sotos, M. Doust, *et al.*, 2014. *Greenhouse Gas Protocol: Global Protocol for*

Community-Scale Greenhouse Gas Emission Inventories: An Accounting and Reporting Standard for Cities.

Gan, V. J. L., J. C. P. Cheng, I. M. C. Lo, *et al.*, 2017. Developing a CO_2-e Accounting Method for Quantification and Analysis of Embodied Carbon in High-Rise Buildings. *Journal of Cleaner Production*, 141.

Guo, S., G. Shen, J. Yang, *et al.*, 2015. Embodied Energy of Service Trading in Hong Kong. *Smart and Sustainable Built Environment*, 4.

Hertwich, E. G., R. Wood, 2018. The Growing Importance of Scope 3 Greenhouse Gas Emissions from Industry. *Environmental Research Letters*, 13.

HK Census and Statistics Department, 2018. *Hong Kong Energy Statistics Annual Report*.

HK Census and Statistics Department, 2019a. *Hong Kong Energy Statistics Annual Report*.

HK Census and Statistics Department, 2019b. *Hong Kong Annual Digest of Statistics*.

HK Census and Statistics Department, 2019c. *Table 127: Electricity Consumption*.

HK Census and Statistics Department, 2020. *Hong Kong Energy Statistics Annual Report*.

HK Electrical and Mechanical Services Department, 2019. *Hong Kong Energy End-Use Data 2019*, https://www.emsd.gov.hk/en/energy_efficiency/energy_end_use_data_and_consumption_indicators/hong_kong_energy_end_use_data/.

HK Environment Bureau, 2014. *Planning Ahead for a Better Fuel Mix: Future Fuel Mix for Electricity Generation Consultation Document*.

HK Environment Bureau, 2017. *Hong Kong Climate Action Plan 2030+*.

HKEPD, 2000. *Greenhouse Gas Emission Control Study*.

HKEPD, 2010. *A Study of Climate Change in Hong Kong-Feasibility Study*.

HKEPD, 2017. *Landfill Gas Utilization*.

HKEPD, 2019a. *Greenhouse Gas Emissions in Hong Kong*.

HKEPD, 2019b. *Hong Kong Waste Treatment and Disposal Statistics*.

HK Lands Department, 2019. *Proposed Construction of Hong Kong Offshore Liquefied Natural Gas (LNG) Terminal*.

HK Offshore Wind Limited, 2006. *Hong Kong Offshore Wind Farm in Southeastern Waters*.

HK Transport Department, 2018. *The Annual Traffic Census 2017*.

Ho, C. Y., K. W. Siu, 2007. A Dynamic Equilibrium of Electricity Consumption and GDP in Hong Kong: An Empirical Investigation. *Energy Policy*, 35.

Humphrey, S., 2004. *Whole Life Comparison of High-rise Residential Blocks in Hong Kong*.

IMF, 2019. *World Economic Outlook Databases-October 2019 Edition*.

IPCC, 2001. Climate Change 2001: *The Scientific Basis. Contributing of Working Group I to the Third Assessment Report of the Intergovernmental Panel on Climate Change*. Cambridge University Press, Cambridge, United Kingdom and New York, NY. USA.

IPCC, 2005. *IPCC Special Report on Carbon Dioxide Capture and Storage.* Prepared by Working Group III of the Intergovernmental Panel on Climate Change. Cambridge University Press, Cambridge, United Kingdom and New York, NY. USA.

IPCC, 2006a. *2006 IPCC Guidelines for National Greenhouse Gas Inventories, Volume 2 Energy*. Prepared by the National Greenhouse Gas Inventories Programme, Eggleston, H.S., L. Buendia, K. Miwa, *et al.* (eds). Published: IGES, Japan.

IPCC, 2006b. *Guidelines for National Greenhouse Gas Inventories*, Prepared by the National Greenhouse Gas Inventories Programme, Eggleston, H. S., L. Buendia, K. Miwa, *et al.* (eds) Published: IGES, Japan.

IPCC, 2013. *Climate Change 2013: The Physical Science Basis. Contribution of Working Group I to the Fifth Assessment Report of the Intergovernmental Panel on Climate Change.* Cambridge University Press, Cambridge, United Kingdom and New York, NY, USA.

IPCC, 2014. *Climate Change 2014: Mitigation of Climate Change. Contribution of Working Group III to the Fifth Assessment Report of the Intergovernmental Panel on Climate Change.* Cambridge University Press, Cambridge, United Kingdom and New York, NY, USA.

IPCC, 2018. *Global Warming of 1.5 ℃. An IPCC Special Report on the Impacts of Global Warming of 1.5 ℃ Above Pre-industrial Levels and Related Global Greenhouse Gas Emission Pathways, in the Context of Strengthening the Global Response to the Threat of Climate Change, Sustainable Development, and Efforts to Eradicate Poverty*. World Meteorological Organization, Geneva, Switzerland.

Johnke, B., 2001. *Chapter 5.3: Emissions from Waste Incineration*, in: IPCC Good Practice Guidance and Uncertainty Management in National Greenhouse Gas Inventories.

Kadiyala, A., R. Kommalapati, Z. Huque, 2016. Quantification of the Lifecycle Greenhouse Gas Emissions from Nuclear Power Generation Systems. *Energies*, 9.

Kwok, T. F., Y. Xu, P. T. Wong, 2017a. Complying with Voluntary Energy Conservation Agreements (I): Air Conditioning in Hong Kong's Shopping Malls. *Resources, Conservation and Recycling*, 117.

Kwok, T. F., Y. Xu, P. T. Wong, 2017b. Complying with Voluntary Energy Conservation Agreements (II): Lighting in Hong Kong's Shopping Malls. Resources. *Conservation and Recycling*, 117.

Lam, J. C., K. K. Wan, T. N. Lam, *et al.*, 2010. An Analysis of Future Building Energy Use in Subtropical Hong Kong. *Energy*, 35(3).

Lang, G., 2018. Urban Energy Futures: A Comparative Analysis. *European Journal of Futures Research*, 6.

Langston, C., E. Chan, E. Yung, 2018. Hybrid Input-output Analysis of Embodied Carbon and

Construction Cost Differences Between New-build and Refurbished Projects. *Sustainability*, 10(9).

Leung, A., M. Burke, J. Cui, 2018. The Tale of Two (Very Different) Cities–Mapping the Urban Transport Oil Vulnerability of Brisbane and Hong Kong. *Transportation Research Part D: Transport and Environment*, 65.

Liddle, B., 2018. Consumption-based Accounting and the Trade-carbon Emissions Nexus. *Energy Economics*, 69.

Liu, L., X. M. Ma, 2011. CO_2 Embodied in China's Foreign Trade 2007 with Discussion for Global Climate Policy. *Procedia Environmental Sciences*, 5.

Lu, L., H. X. Yang, 2010. Environmental Payback Time Analysis of a Roof-mounted Building-integrated Photovoltaic (BIPV) System in Hong Kong. *Applied Energy*, 87(12).

Macao DSEC, 2016. *Yearbook of Statistics*.

Macao DSEC, 2017. *Yearbook of Statistics*.

Macao DSEC, 2019. *Yearbook of Statistics*.

Macao DSPA, 2018. *Report on the State of Environment of Macao 2000-2017*.

Macao International Airport (MIA), 2020. *Airport Data*.

Muthu, S. S., Y. Li, J. Y. Hu, *et al.*, 2011. Carbon Footprint of Shopping (grocery) Bags in China, Hong Kong and India. *Atmospheric Environment*, 45.

National Academies of Sciences Engineering and Medicine, 2019. *Negative Emissions Technologies and Reliable Sequestration: A Research Agenda*, The National Academies Press, Washington, DC.

National Development and Reform Commission, 2014. *Assessment Method on the Mitigation of CO_2 Emissions per Unit of GDP*, NDRC Climate [2014] No. 1828.

NEA, HKSAR Government, 2008. *Memo on the Supply of Natural Gas and Electricity*.

Ng, A. W., J. Nathwani, 2010. Sustainable Energy Policy for Asia: Mitigating Systemic Hurdles in a Highly Dense City. *Renewable & Sustainable Energy Reviews*, 14.

Ng, W. Y., C. K. Chau, 2015. New Life of the Building Materials—Recycle, Reuse and Recovery. *Energy Procedia*, 75.

Pacala, S. W., R. H. Socolow, 2004. Stabilization Wedges: Solving the Climate Problem for the Next 50 Years with Current Technologies. *Science*, 305(5686).

Poudenx, P., 2008. The Effect of Transportation Policies on Energy Consumption and Greenhouse Gas Emission from Urban Passenger Transportation. *Transportation Research Part A: Policy and Practice*, 42.

Song, Q., J. Li, H. Duan, *et al.*, 2017. Towards to Sustainable Energy-efficient City: A Case Study of Macau. *Renewable and Sustainable Energy Reviews*, 75.

Song, Q., Z. Wang, J. Li, *et al.*, 2018a. Comparative Life Cycle GHG Emissions from Local

Electricity Generation Using Heavy Oil, Natural Gas, and MSW Incineration in Macau. *Renewable and Sustainable Energy Reviews*, 81.

Song, Q., Z. Wang, Y. Wu, *et al.*, 2018b. Could Urban Electric Public Bus Really Reduce the GHG Emissions: A Case Study in Macau? *Journal of Cleaner Production*, 172.

Song, Q., Y. Wu, J. Li, *et al.*, 2018c. Well-to-wheel GHG Emissions and Mitigation Potential from Light-duty Vehicles in Macau. *The International Journal of Life Cycle Assessment*, 23.

Sovacool, B. K., 2008. Valuing the Greenhouse Gas Emissions from Nuclear Power: A Critical Survey. *Energy Policy*, 36.

State Council, 2016. *Work Plan for Greenhouse Gas Mitigation in the 13th Five-Year Period.*

Stockholm Environment Institute, 2017. *Consumption-based Accounting Reveals Global Redistribution of Carbon Emissions.*

To, W. M., 2015. Greenhouse Gases Emissions from the Logistics Sector: The Case of Hong Kong, China. *Journal of Cleaner Production*, 103.

To, W. M., T. M. Lai, W. C. Lo, *et al.*, 2012. The Growth Pattern and Fuel Life Cycle Analysis of the Electricity Consumption of Hong Kong. *Environmental Pollution*, 165.

To, W. M., P. K. C. Lee, 2017a. A Triple Bottom Line Analysis of Hong Kong's Logistics Sector. *Sustainability*, 9(3).

To, W. M., P. K. C. Lee, 2017b. Energy Consumption and Economic Development in Hong Kong, China. *Energies*, 10.

To, W. M., P. K. C. Lee, 2017c. GHG Emissions from Electricity Consumption: A Case Study of Hong Kong from 2002 to 2015 and Trends to 2030. *Journal of Cleaner Production*, 165.

Wang, H., D. Zhou, P. Zhou, *et al.*, 2012. Direct Rebound Effect for Passenger Transport: Empirical Evidence from Hong Kong. *Applied Energy*, 92.

Wang, J., C. Yu, W. Pan, 2018. Life Cycle Energy of High-rise Office Buildings in Hong Kong. *Energy and Buildings*, 167.

Warner, E. S., G. A. Heath, 2012. Life Cycle Greenhouse Gas Emissions of Nuclear Electricity Generation: Systematic Review and Harmonization. *Journal of Industrial Ecology*, 16.

Wilburn, J. S., P. M. Roberts, 1995. Major Asian Subsea Pipeline to Start up This Year. *Oil and Gas Journal*, 93.

Woo, C. K., A. Shiu, Y. Liu, *et al.*, 2018. *Consumption Effects of an Electricity Decarbonization Policy: Hong Kong. Energy*, 144.

Xu, Y., G. J. Liu, 2015. Carbon Capture and Storage for Hong Kong's Fuel Mix. *Utilities Policy*, 36.

Yang, H., G. Zheng, C. Lou, *et al.*, 2004. Grid-connected Building-integrated Photovoltaics: A Hong Kong Case Study. *Solar Energy*, 76.

Yang, Z. Y., W. J. Dong, T. Wei, *et al.*, 2015. Constructing Long-term (1948-2011)

Consumption-based Emissions Inventories. *Journal of Cleaner Production*, 103.

Yim, S., S. Ng, M. Hossain, *et al.*, 2018. Comprehensive Evaluation of Carbon Emissions for the Development of High-rise Residential Building. *Buildings*, 8.

Zhang, C., D. Zhou, P. Li, *et al.*, 2014. CO_2 Storage Potential of the Qiongdongnan Basin, Northwestern South China Sea. *Greenhouse Gases: Science and Technology*, 4(6).

Zhao, S., Q. Song, H. Duan, *et al.*, 2019. Uncovering the Lifecycle GHG Emissions and Its Reduction Opportunities from the Urban Buildings: A Case Study of Macau. *Resources, Conservation and Recycling*, 147.

Chapter 4 Climate Change Policy and Regional Collaboration

Hong Kong and Macao are distinguished by their very high levels of population density, urbanization and economic outputs in China. More severe weather conditions in the South China region, brought about by climate change, present major challenges, as severe weather create many new risks that affect people's lives and livelihood. Aspects of Hong Kong's and Macao's current climate change policies in mitigation, adaptation and resilience started before reunification with China in 1997 and 1999 respectively, and some of them represent global best practices that provide useful reference for China as a whole and even elsewhere in the world. The climate policies of Hong Kong and Macao have continued to evolve after reunification, and there are many opportunities for active regional cooperation that could bring about multiple benefits to strengthen mitigation, adaptation and resilience.

This chapter starts with describing the overall climate change governance structure in Hong Kong and Macao and introduces their key climate policies, followed by describing and assessing their climate-related policies. This chapter ends with emphasizing those areas that are unique to Hong Kong and Macao, reviewing the regional cooperation framework, and proposing recommendations that could strengthen research and coping capacities, as well as regional cooperation.

4.1 Review of the Overall Climate Change Policies

4.1.1 Governance structure of the HKSAR and MSAR on climate change policies

Governance structures in Hong Kong and Macao are different from other regions in Mainland China. As the only special administration regions (SARs), they enjoy a very large measure of autonomy in local policy decision-making, including in climate change related policies.

Hong Kong and Macao are both SARs created under the *Constitution of the People's Republic of China*. They are referred to as the Hong Kong Special Administrative Region (HKSAR) and Macao Special Administrative Region (MSAR). Each has its own constitutional document, called the Basic Law, promulgated by the National People's Congress, which enshrines China's 'one country, two systems' policy for the SARs. The SARs enjoy a 'high degree of autonomy' in their local day-to-day affairs, including in decision-making related to climate change policies and action plans.

The HKSAR and MSAR are each headed by its own "chief executive", who is both the head of the SAR as well as the government. Their political system is referred to as an 'executive-led' system, which means the executive authorities, under the leadership of the chief executive, decide on government policies. "Principal officials" (i.e., ministers, who are referred to as "secretaries") have specific policy portfolio responsibilities. Each SAR has its own legislature to raise issues of public interest, scrutinize legislation and approve public funding. Hong Kong also has an Audit Commission to provide independent audit of government departments and public sector performance and accountability, and its reports are published. Some of the reports are relevant to aspects related to climate change, such as water resources[①].

While their constitutional arrangements are similar, there are organizational differences between the HKSAR and MSAR in their administrative structures, which are legacies of their respective colonial past[②].

Climate change decision-making authorities

According to the United Nations Framework Convention on Climate Change (UNFCCC), climate mitigation is referred as "efforts to reduce emissions and enhance sinks"[③], and adaptation is referred as "adjustments in ecological, social, or economic systems in response to actual or expected climatic stimuli and their effects or impacts"[④]. In this chapter, mitigation policies refer to the policies that

---

① The Director of Audit selects the topics and departments to audit. These reports are available at https://www.aud.gov.hk/eng/pubpr_arpt/rpt.htm. For example, on 1 April 2015, it published a report on the Water Supplies Department on the management of water supply and demand.

② Hong Kong became the HKSAR in 1997 and Macao became the MSAR in 1999 when they reunified with China respectively.

③ United Nations Framework Convention on Climate Change, "Introduction to Mitigation," United Nations Climate Change, accessed January 21, 2020, https://unfccc.int/topics/mitigation/the-big-picture/introduction-to-mitigation.

④ United Nations Framework Convention on Climate Change, "What Do Adaptation to Climate Change and Climate Resilience Mean?," United Nations Climate Change, n.d., https://unfccc.int/topics/adaptation-and-resilience/the-big-picture/what-do-adaptation-to-climate-change-and-climate-resilience-mean.

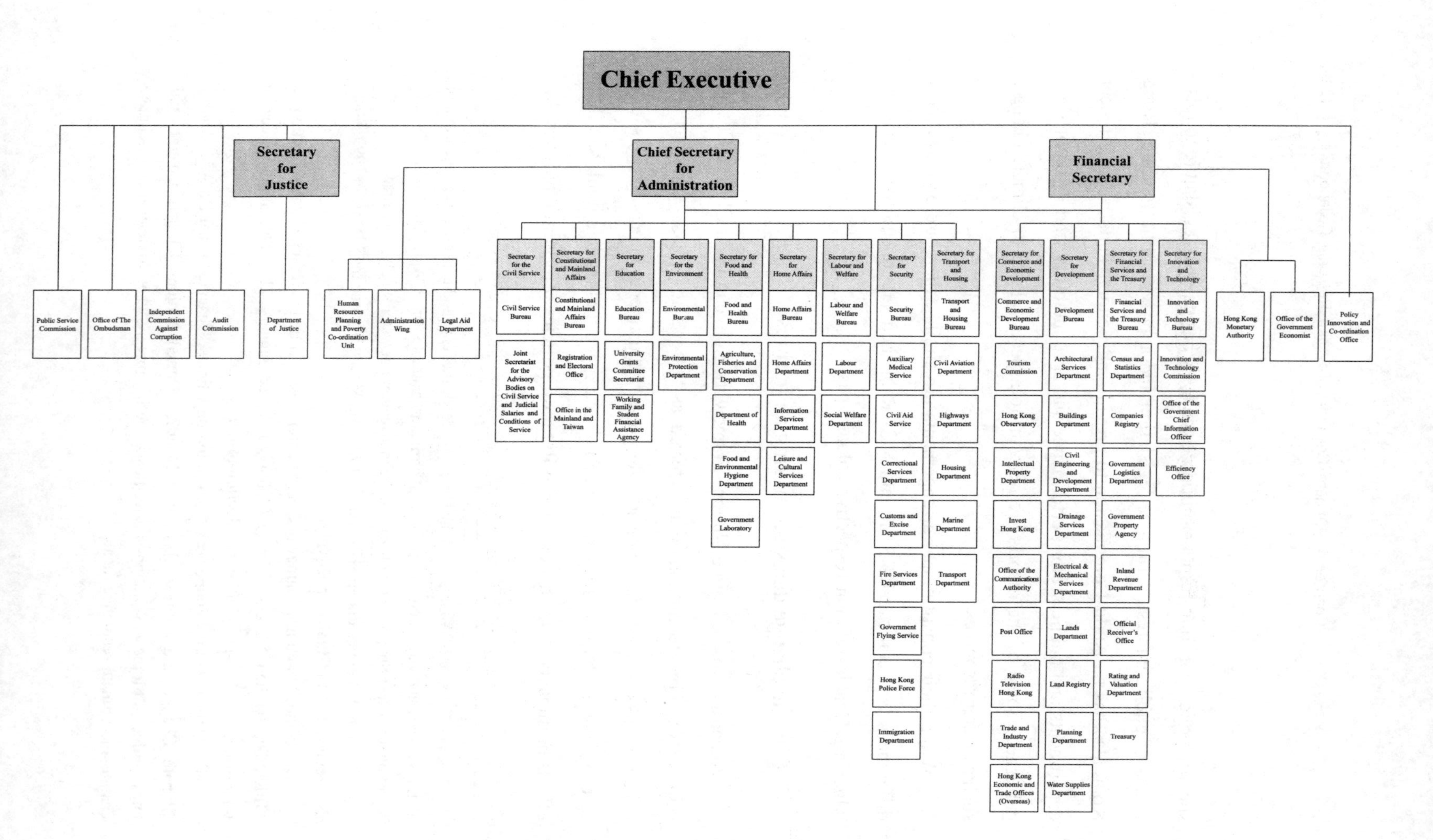

Figure 4–1 Government Organizational Chart of the HKSAR

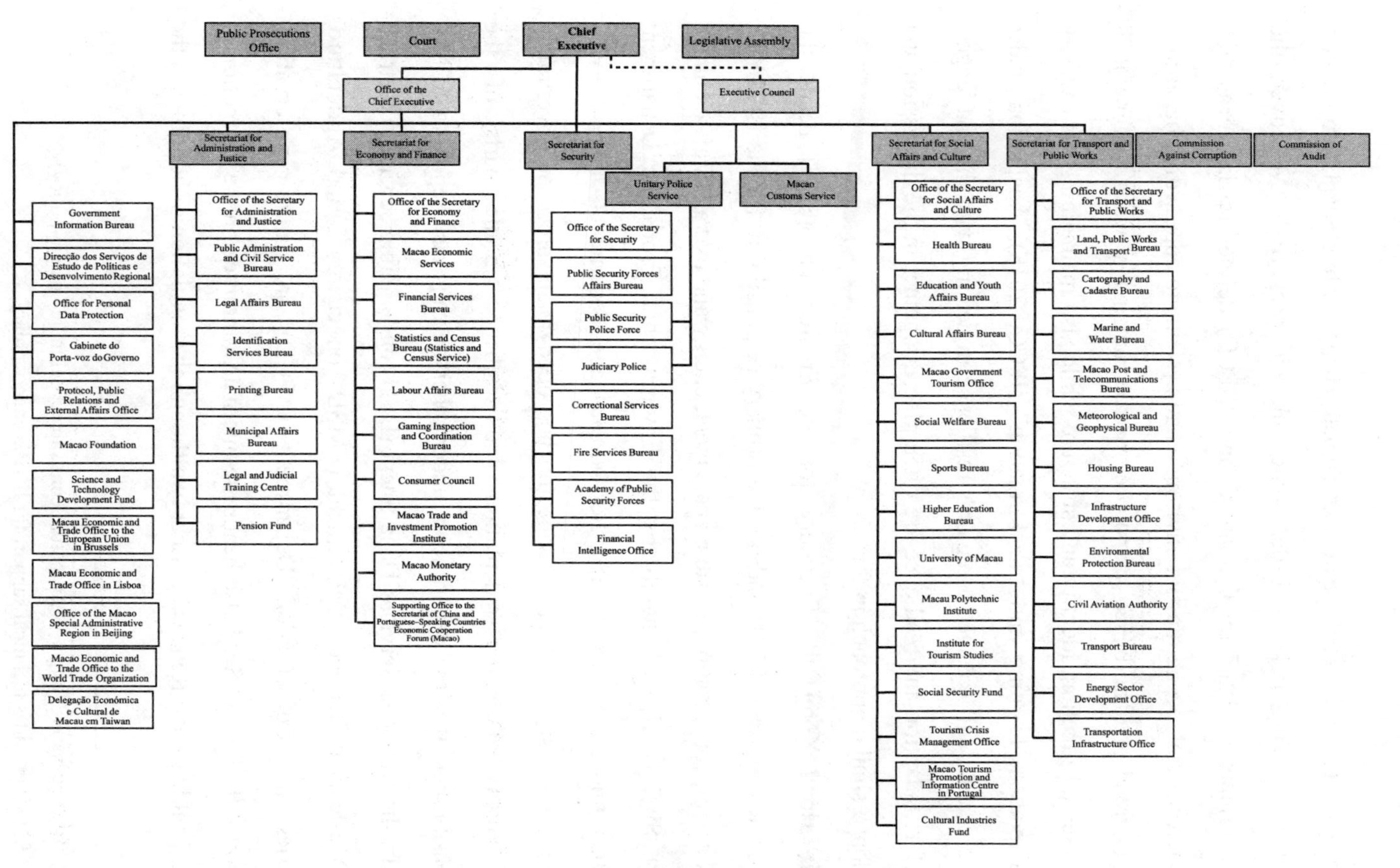

Figure 4–2 Government Organizational Chart of the MSAR

are designed to minimize the extent of climate change, while adaptation policies aim at minimizing risks and defending life and property[①]. Moreover, the Intergovernmental Panel on Climate Change (IPCC) defines resilience as "the capacity of social, economic and environmental systems to cope with a hazardous event or trend or disturbance, responding or reorganizing in ways that maintain their essential function, identity and structure while also maintaining the capacity for adaptation, learning and transformation" (IPCC, 2018). Adaptation under climate change will enhance a region's resilience in dealing with the challenges. Therefore, adaption and resilience are often placed together in the discussion and we adopt a similar approach in this chapter.

In Hong Kong, the Secretary for the Environment (SEN) heads the Environment Bureau (ENB) and has the primary responsibility for the HKSAR's environment, conservation, waste management, energy and overall climate change policy. SEN is assisted mainly by the Environmental Protection Department (EPD). Although a number of other bureaus and departments have relevant responsibilities since climate change touches upon many areas of government work across mitigation, adaptation and resilience. Some parts of the responsibilities of some departments under other bureaus also report to SEN – namely, the Drainage Services Department (DSD) on handling sewage, Electrical and Mechanical Services Department (EMSD) on energy[②], and Agriculture, Fisheries and Conservation Department (AFCD) on nature conservation. Moreover, the current SEN also heads the internal Committee on the Promotion of Green Buildings and Renewable Energy although the regulation of buildings is the

①Also see Nature (https://www.nature.com/subjects/climate-change-policy).

②EMSD provides technical support for ENB's energy policy.

responsibility of the Buildings Department under Development Bureau (DEVB)[①].

DEVBhas vast responsibilities dealing with urban planning, land use and building development, along with infrastructure development. As departments of DEVB, WSD has the mission to provide quality services of water supplies, DSD is responsible for flood protection, EMSD provides electrical and mechanical services, and CEDD is responsible for slope safety and development of coastal infrastructure.

Transport and housing policies are the responsibility of the Transport and Housing Bureau (THB). Emergency preparedness, including severe weather events, is the responsibility of Security Bureau (SB) and district relations is under the Home Affairs Bureau (HAB). The formulation and implementation of public health policies are the responsibility of the Food and Health Bureau (FHB), which works with the Department of Health (DOH). The HKSAR Government's climate change science authority is the Hong Kong Observatory (HKO), which is under the Commerce and Economic Development Bureau (CEDB).

Macao, being a much smaller jurisdiction, has a simpler governing structure at the secretary level. The Secretary for Transport and Public Works (STPW) heads all the sub-units relevant to climate mitigation and adaptation, such as climate science – the Macao Meteorological and Geophysical Bureau (SMG)[②]– land use, public works, transport, housing and environmental protection. Energy policy also comes under STPW, who supervises the Office for the Development of

---

①The Committee on the Promotion of Green Buildings was set-up in January 2013. Its scope was expanded in 2017 to cover also renewable energy. The current SEN was made the chair because of his professional qualification as an architect known for promoting green buildings. DEVB has oversight on the Buildings Department, which regulates building codes.

② See http://www.smg.gov.mo/smg/introduction/e_smg_intro.htm for reference.

the Energy Sector (ODES).

Both SARs have advisory bodies to assist the government in various policies, including issues that are relevant to climate change. These bodies are sounding boards for government ideas and avenues for collecting views from experts and non-government organizations (NGOs).

4.1.2 Climate policy implementation and coordination: the local, national and international perspectives

Local implementation and coordination

In light of the complexity of Hong Kong's governing structure split among many bureaus and departments, coordination is an issue for cross-cutting policy matters. In 2007, the HKSAR government set-up the Inter-departmental Working Group on Climate Change (IWGCC) chaired by the ENB to monitor the formulation and implementation of relevant policies and measures among bureaus and departments. The IWGCC served as the platform to monitor progress after Hong Kong's first climate change action plan was announced in 2010. The Steering Committee on Climate Change (SCCC) was created to replace the lower level IWGCC and is chaired by the Chief Secretary for Administration – the most senior principal official after the chief executive. It held its first meeting in April 2016 to steer and coordinate climate change-related actions within the bureaucracy. Members of the SCCC include 13 bureaus and three departments. Other departments are to attend meetings as and when needed.

The MSAR Government set-up the Interdepartmental Working Group on Climate Change coordinated by the Directorate of SMG, with the Environmental

Protection Bureau as one of its members in 2015. The first plenary meeting was held in March 2015 to introduce the responsibilities, structure, and operation methods of the group, and review the recent work of the MSAR government in addressing climate change.

National plans and international treaties on the local policy

The 'one country, two systems' principle gives ample autonomy to the two SARs to decide their own local climate policy implementation and coordination. They could have chosen not to be bound by the national carbon intensity reduction targets or adopt different targets. In the case of the HKSAR, its government adopted higher carbon emissions reduction target than the national government on the basis that it was an economically well-developed city, while the MSAR adopted the same target as the national target[①].

As for international agreements entered into by China, consultation is required for them to be applied to the two SARs. The UNFCCC and Kyoto Protocol initially did not apply to Hong Kong and Macao[②], but upon requests from the SARs, they were applied to Hong Kong on 5 May 2003, and to Macao on

① Hong Kong's Climate Change Strategy and Action Agenda – Consultation Document published in September 2010, and the 2017 Hong Kong Climate Action Plan 2030+ noted that the HKSAR would exceed China's national GHG emissions reduction targets.

② The respective Basic Laws of the HKSAR and MSAR provide that the application of international agreements to them that China enters into 'shall be decided by the Central People's Government, in accordance with the circumstances and needs of the Region, and after seeking the views of the government of the Region'. On 30 August 2002, China informed the United Nations that it would 'provisionally not apply' the Kyoto Protocol to Hong Kong and Macao.

14 January 2008[1]. Likewise, the Paris Agreement – the successor treaty to the Kyoto Protocol, was applied to Hong Kong and the Macao after consultation with the two SARs.

The Central People's Government is supportive of the SARs choosing to adopt such international agreements and be part of national effort to fulfil national commitments. The HKSAR government has regularly sent officials to be part of the Chinese delegation to attend UNFCCC's Conference of the Parties. In 2011, HKSAR and Guangdong government established the 'Hong Kong - Guangdong Joint Liaison Group on Combating Climate Change' on the existed Hong Kong - Guangdong Co-operation Joint Conference.More recently, the Greater Bay Area Outline Development Plan (GBA Plan)[2], promulgated in February 2019, is a national plan that encourages greater cooperation among Guangdong, Hong Kong and Macao, where climate change is an important substance of the cooperation (Section 5). In the same year, Hong Kong-Guangdong Joint Working Group on Sustainable Development and Environmental Protection Group and Hong Kong-Guangdong Joint Liaison Group on Combating Climate Change were merged into Hong Kong-Guangdong Joint Working Group on Environmental Protection and

---

① The UNFCCC entered into force in 1994 to which China was a signatory and the Kyoto Protocol became opened for countries to join in 1997. China approved the Kyoto Protocol in 2002. At the time, the national authorities did not apply it to the HKSAR and MSAR as it had yet to seek the views of the two SARs. Hong Kong requested to abide by the UNFCCC and it became applicable to the HKSAR on 5 May 2003. When the Kyoto Protocol became operational (which it did on 16 February 2005), it applied to the HKSAR as well. Macao requested to abide by the UNFCCC and Kyoto Protocol in 2007 and that became effective for the MSAR on 14 January 2008.

②The Greater Bay Area Outline Development Plan, see https://www.bayarea.gov.hk/tc/home/index.html.

Combating Climate Change, which advances and deepens cooperation and scientific research on mitigation and adaptation to climate change.

Paris Agreement 2015

On 4 November 2016, the Paris Agreement came into force succeeding the Kyoto Protocol. The key provisions call for global actions to: (1) achieve "peak" carbon emissions as soon as possible and achieve a balance between carbon sources and sinks in the second half of the 21st century (i.e. to reach "carbon neutrality" between 2051 and 2100); and (2) keep global average temperature increase well below 2℃ above pre-industrial levels and to pursue efforts to limit it to 1.5℃. Global actions are based on a "bottom-up" approach, where all the Parties must devise their own suitably ambitious "nationally determined contributions" (NDCs) with targets and timelines. Each signatory must prepare NDC every 5 years, and each successive NDC to represent a progression beyond the previous one.

Hong Kong and Macao both have carbon intensity reduction targets, as noted in Chapter 3. They should review their climate change efforts every 5 years and align them with the submission timelines under the Paris Agreement. The timeline for review up to 2030 for the two SARs is shown in Table 4–1. The Central People's Government is requiring the two SARs to submit inventories and reports to comply with UNFCCC requirement along a similar timing.

Table 4–1 Paris Agreement timeline for HKSAR and MSAR

Year	2017	2019	2020	2024	2025	2029	2030
Action	Set 2030 carbon target	Review its actions	Paris Agreement comes into operation (i.e., commit to further actions)	Review its actions	Next climate action plan (i.e., commit to further actions)	Review its actions	Next climate action plan (i.e., commit to further actions)

4.1.3 Key climate change policies since 2010

As noted in Chapter 1, HKO provided Hong Kong with its first study on climate change projection for the 21st century in 2004; and updates climate science regularly taking into account new research. As noted in Chapter 3, the HKSAR government commissioned a feasibility study to identify long-term measures to reduce carbon emissions in 2008. Since then, the HKSAR government has published several key policy documents related to climate change and energy saving, which together show the progression of policy:

Hong Kong's Climate Change Strategy and Action Agenda – Consultation Document (September 2010)[①];

Public Consultation on Future Fuel Mix for Electricity Generation (March 2014)[②];

Report on the progress and outcome of measures as set out in the Action Agenda 2010 (April 2014)[③];

Public Consultation on the Future Development of the Electricity Market (March 2015)[④];

---

① Hong Kong's Climate Change Strategy and Action Agenda – Consultation Document, September 2010, https://www.epd.gov.hk/epd/sites/default/files/epd/english/climate_change/files/Climate_Change_Booklet_E.pdf.

② Environment Bureau, Public Consultation on Future Fuel Mix for Electricity Generation, March 2014, https://www.enb.gov.hk/sites/default/files/en/node2605/Consultation%20Document.pdf.

③Report on the progress and outcome of measures as set out in the Action Agenda, 28 April 2014, https://www.legco.gov.hk/yr13-14/english/panels/ea/papers/ea0428cb1-1292-6-e.pdf.

④ Environment Bureau, Public Consultation on the Future Development of the Electricity Market, March 2015, https://www.enb.gov.hk/sites/default/files/en/node3428/EMR_condoc_ e.pdf.

Energy Saving Plan for Hong Kong's Built Environment 2015-2025+ (May 2015)①;

Hong Kong Climate Change Report 2015 (November 2015)②;

Hong Kong's Climate Action Plan 2030+ (January 2017)③; and

Deepening Energy Saving in Existing Buildings in Hong Kong Through '4Ts' Partnership (June 2017)④.

Macao promulgated the *Environmental Framework Law* in 1991, which established the general outline and principles of Macao's response to climate change. The authorities issued *Towards a Thematic Strategy for the Climate Change in Macao Special Administration Region* in 2008 after the Kyoto Protocol was extended to Macao. Macao has one key policy document that covers its overall environmental protection policy that includes climate change: *Environmental Protection Planning of Macao 2010-2020.*⑤

---

① Environment Bureau, Energy Saving Plan for Hong Kong's Built Environment 2015-2025+, May 2015, https://www.enb.gov.hk/sites/default/files/pdf/EnergySavingPlanEn.pdf.

② Environment Bureau, Hong Kong Climate Change Report 2015, November 2015, https://www.enb.gov.hk/sites/default/files/pdf/ClimateChangeEng.pdf.

③ Environment Bureau, Hong Kong's Climate Action Plan 2030+, January 2017, https://www.enb.gov.hk/sites/default/files/pdf/ClimateActionPlanEng.pdf.

④ Environment Bureau, Deepening Energy Saving in Existing Buildings in Hong Kong Through '4Ts' Partnership, June 2017, https://www.enb.gov.hk/sites/default/files/pdf/EnergySaving_EB_EN.pdf.

⑤ Macao Environmental Protection Bureau, 2012, https://www.dspa.gov.mo/Publications/EnvPlanningBook/201209-EnvPlanningBook_PB_EN.pdf.

4.2 Overview and Assessment of Climate Change Mitigation Policies in Hong Kong and Macao

4.2.1 Drivers for change and reduction targets

There were two sets of drivers to climate change mitigation policies between 1997 and 2018 for the HKSAR and MSAR, one is the local governmental initiatives to improve air quality and the other is the influence from the national and global efforts.

Chapter 3 provides the background to the two SARs' efforts to reduce air pollutant emissions through switching to cleaner fossil fuels, as well as the relocation of manufacturing to the mainland, both of which brought about substantial reductions in carbon emissions. Thus, carbon reduction was initially a co-benefit of meeting local air quality goals. The other driver was the growing willingness and pressure from national and global levels to set specific carbon reduction targets, which the two SARs provided policy responses.

Today, improving air quality remains important for Hong Kong and Macao. In the case of Hong Kong, local laws require the government to review air quality standards at least once in every 5 years and to consider if they could be tightened. The two SARs and the Guangdong environmental authorities collaborate on air quality improvement. In light of the national efforts to fight air pollution, carbon reduction will continue to be a co-benefit arising from the efforts of the three authorities. At the same time, national and global efforts to mitigate climate change will also drive continuing attention from the two SARs, which are both promoting 'low-carbon living' as a general theme.

4.2.2 Climate change mitigation policies in Hong Kong and Macao

In 2018, Hong Kong had a population of about 7.45 million people – a growth of about 25% since 1997 when it became an SAR; and Macao's current population is about 670 000 people – a growth of nearly 55% since 1999 when it became an SAR. In 2018, Hong Kong's GDP was approximately US$361.7 billion and Macao's GDP was about US$55 billion. Both SARs are active service economies. While Hong Kong has a more diversified economy, Macao's main business is gaming – it is the world's largest casino market with gaming accounting for 50% of its GDP. Both economies have a large tourism component.

While carbon emissions are calculated by sectors as shown in Chapter 3, mitigation policies in Hong Kong and Macao have two clear focuses: (1) revamping fuel mix to reduce air pollutants and carbon emissions, as this sector is by far the largest pollution and carbon emissions contributor; and (2) saving energy (see Table 4–2 for the summary). As Chapter 2 points out, energy saving in buildings is especially important in highly urbanized cities. For example, buildings take up 90% of the total carbon emissions from electricity in Hong Kong. The various programs to reduce vehicle-related emissions, including adopting electric vehicles (EVs), have resulted in reducing air pollutants at roadside but have not contributed much in reducing carbon, the reasons for which are explained in 4.2.3 below.

Table 4–2 Summary of the SARs main mitigation focuses

Mitigation in HKSAR
1. Revamping fuel mix
a. In electricity generation
i) Phase down coal and use more natural gas for local generation
ii) Increase the proportion of zero carbon energy (including RE) in the fuel mix
b. In transport
i) Support EVs adoption
ii) Build EVs charging infrastructure
iii) Government adopt EVs in its fleet
iv) Tax incentives for EVs buyers
v) Established Pilot Green Transport Fund to support the transport sector to test green innovative transport technologies, including EVs
vi) Support trials of electric franchised buses franchised buses
vii) Support rail expansion as backbone of passenger transport system
viii) Facilitating personal mobility means
2. Promoting energy saving
a. In buildings
i) Government to lead by example
ii) Promote green building practices
iii) Tighten Building Energy Codes and regulations
iv) Promote energy audits, retrofitting and retro-commissioning for existing buildings
v) Provide tax incentive to private building owners
vi) Foster exchanges with building owners through dialogue platforms
b. In other areas
i) Government to lead by example
ii) Strengthen and promote good practices
iii) Tighten mandatory energy efficiency labelling standards and cover more electrical appliances
iv) Promote water-cooled air-conditioning
v) Implement district-cooling systems
vi) Galvanize community action
Mitigation in MSAR
3. Changing fuel mix
a. In electricity generation
i) Use more natural gas for local generation
ii) Promote RE
b. In consumption
i) Support light rail
ii) Support hailing e-taxis
4. Promoting energy saving
a. In buildings
i) Techniques for Building-Energy Consumption Optimization
ii) Macao Green Hotel Awards
b. In other areas
i) Macao Energy Conservation Week
ii) Energy Saving Activities on Campus

Changing fuel mix

(1) Regulating the electricity and gas sector

As Chapter 2 also points out, the HKSAR and MSAR have very high population densities – 6 930/km^2 and 21 400/km^2 respectively – and require highly reliable electricity to keep their socio-economic activities functioning. One of Hong Kong's major competitive advantages is its highly reliable energy supply in the form of electricity and gas. As for Macao, in light of its policy in 2001 to open-up and greatly expand its gaming industry, it was essential for it to improve its energy reliability.

The electricity and gas businesses in Hong Kong have always been privately owned. The first utility company in Hong Kong was the Hong Kong and China Gas Company Ltd. (referred to as Towngas) founded in 1862. The two electricity companies also have long histories – CLP Power Hong Kong Limited (CLP) was founded in 1901 and The Hongkong Electric Company, Limited (HEC) was founded in 1889. In light of the land and investments needed for the electricity business, each of the companies is effectively the only service provider in its respective service areas[①]. All three companies are well-managed, profitable and either listed on the local stock exchange or wholly-owned subsidiaries of local listed entities[②].

---

①Towngas supplies territory wide. In 2017, Towngas has a 3 632 km network supplying gas to 1.88 million residential, commercial and industrial customers in Hong Kong. Towngas is also a supplier of LPG to Hong Kong's taxi fleet, which are powered by LPG; as well as supply aviation fuel to Hong Kong's airport. CLP is the larger of the two electricity providers covering Kowloon and the New Territories and HEC provides power to Hong Kong Island and Lamma Island.

② All three companies have extensive businesses outside Hong Kong. This chapter only deals with their operations in the HKSAR.

The HKSAR government regulates the electricity business through non-exclusive contracts, called Scheme of Control Agreements (SCAs), with CLP and HEC. The SCAs provide the framework for the government to monitor their financial affairs and operating performance. The SCAs entitle CLP and HEC to a permitted rate of return on average net fixed assets, which was designed to encourage them to continue to invest in plants and equipment to ensure supply reliability. The investment in power generation would be very substantial and for very long terms. As such, it had to make the investments commercially attractive for the private sector①. Apart from the oversight terms in the SCAs, the two companies mostly manage their businesses independently from the government.

There had not been any similar contractual agreement between Towngas and the Hong Kong authorities. It was not till April 1997 that the government and the gas supplier entered into a voluntary Information and Consultation Agreement (ICA) for Towngas to provide financial, operational, environmental and safety information to the public. More importantly, the ICA set-up a process for Towngas to consult the government when it wishes to change the tariff and have major system additions. Thus, the government was essentially concerned about the impact of tariff increases and wanted an opportunity to put its views to the company②.

All three energy companies in Hong Kong have set their own carbon

① For a detailed explanation about how the SCAs work, see Environment Bureau, Chapter 1, Public Consultation on the Future Development of the Electricity Market, 2015, https://www.enb.gov.hk/en/resources_publications/policy_consultation/public_consultation_future_development_electricity_market.html.

② See Monitoring of the Towngas Company, https://www.enb.gov.hk/en/about_us/policy_responsibilities/financial_monitoring.html#monitor2.

reduction targets. For CLP and HEC, they were set in accordance with their expected fuel mix change, and Towngas set its own target in 2008①.

As for Macao, a private company provided electricity from 1906 but dissatisfaction with its performance led to the then Macao authorities establishing Companhia de Electricidade de Macau (CEM) in 1972 to supply power and build a new plant. The government effectively took over CEM but has since divested much of its shareholding to mainland, overseas and local investors after restructuring the company. In 1985, CEM received an exclusive 25-year concession contract from the government to generate, transmit, distribute, sell, import, and export electricity. The concession also has a permitted rate of return and oversight terms similar to Hong Kong's SCAs. The concession was extended in 2010 for another 15 years but exclusivity on generation was removed, as the government wanted to leave the option open to bring in competition.

CLP and HEC will complete the rollout of smart meters to customers in their service areas by 2025. The HKSAR government has been in discussion with them and Towngas on the disclosure of electricity consumption and generation data to promote energy efficiency, open data and smart city. HKSAR government has reached agreements with three companies to disclose relevant data gradually and promote the development of smart city.

In the case of Macao, while the concession contract helps the MSAR to reserve a public service provider, it also helps the CEM to meet the fast-growing size of the electricity market and promote innovation. Except the principle of

① Towngas pledged to reduce its carbon intensity of its Hong Kong gas production operations by 30% by 2020, against that of 2005, and by 2017 it had achieved 23% reduction. The reduction can be attributed to the use of natural gas and landfill gas for town gas production, as well as energy efficiency enhancements implemented at its plant.

distributing electricity considering safety, environmental protection, efficiency and economy, the concession contract does not directly deal with carbon emissions reduction. CEM made its own efforts to reduce carbon emissions. It became the first utility company in Macao and the first electricity company in Hong Kong and Macao to obtain the ISO 14064 Greenhouse Gases Management System Certification by taking action against greenhouse gases emissions in 2010[①].

Renewable energy and adoption

Chapter 3 notes neither Hong Kong nor Macao has favorable conditions for large-scale commercialized RE generation, such as wind or solar farms, with both being small in land and sea areas.

The HKSAR authorities estimated in 2017 that Hong Kong had about 3%～4% of realizable RE potential arising from wind, solar and waste-to-energy (WTE) that could be exploited based on existing commercially available technologies. The power companies have been exploring the feasibility of the construction of two windfarms (300 MW in total) within Hong Kong's territorial waters that could contribute less than 1.5% of Hong Kong's total electricity need. In terms of WTE, Hong Kong's landfill gas, energy from co-digestion of sewage sludge and food waste generates certain amounts of energy. Towngas is taking landfill gas from government-owned landfills. A 3 000 tons/day municipal solid waste incineration plant is being built and will be commissioned around 2025. By then, WTE could contribute another 1%～2% of the city's electricity need. The 2017 RE potential estimate did not include options that had not yet been explored (such as co-digestion of sewage sludge and food waste currently being trialed by DSD and

① Asian Power, CEM first to obtain ISO 14064 in Macau, 2010, https://asian-power.com/environment/news/cem-first-obtain-iso-14064-in-macau.

EPD, wave energy etc.) and possible nearer term technological advancement (such as enhancement of efficiency of solar photovoltaic (PV) systems)[①].

The HKSAR government has been taking the lead in developing RE where technically and financially feasible and has been creating conditions that are conducive to encouraging the private sector to participate. For the public sector, the HKSAR government has earmarked HK$2 billion[②] in total since the 2017-18 financial year for the installation of small-scale RE facilities at government buildings, venues and facilities. In addition, government departments are actively considering the development of larger scale RE projects. For instance, the largest solar farm of 1.1 MW in Hong Kong was installed at the Siu Ho Wan Sewage Treatment Works in 2016, and the HKSAR government is considering the installation of larger-scale solar PV systems at suitable locations in reservoirs and landfills.

For the private sector, under the current SCAs which took effect in October 2018/January 2019, the HKSAR government and the electricity companies have introduced the Feed-in Tariff (FiT) Schemes, providing financial incentives to encourage the private sector to invest in distributed RE. The FiT rates were set at HK$3, HK$4 and HK$5 per unit of electricity depending on the capacity of the RE system with a view to shortening the payback periods of these systems to about ten years. At the same time, RE certificates are sold by CLP and HEC to

---

①Chapter 4, Hong Kong's Climate Action Plan 2030+.

②The Policy Address of January 2017 announced that, to support the development of RE, HK$200 million was earmarked from 2017-18 for government bureaux/departments to implement small-scale RE projects. A further HK$800 million and HK$1 billion was earmarked under the Financial Secretary's 2018-19 and 2019-20 Budgets respectively, making a total of HK$2 billion.

users who want their electricity to be from clean sources[①]. The government has also introduced a series of measures to facilitate and support members of the public in developing RE. Examples include relaxing the restrictions in relation to installation of PV systems at the rooftop of village houses and introducing a new scheme called "Solar Harvest" to install small-scale RE systems for eligible schools and welfare NGOs. From April 2018, the government also allows tax deduction to be claimed for capital expenditure by companies in procuring RE devices[②]. All these have contributed to the accelerated RE development in the past year or so.

Macao's waste incineration plant provides about 3% of the MSAR's electricity need. The incineration plant was built in 1992. In order to recycle the heat and waste produced during waste incineration, electricity-generating facilities were added to the plant in 2001[③]. In 2006, the MSAR government constructed three more incinerators as the first incinerator alone no longer met the incineration capacity that the city needed. By 2013, the electricity generated from incineration took up 23.6% of the total locally generated electricity (Song *et al.*, 2018).

The MSAR has also been facilitating the long-term supply of natural gas and CEM's new plant approved in 2018 is a gas-fired plant. The concession contract and government's partial ownership is an effective way to implement policies. As for energy saving, the Public Administration and Civil Service Bureau (SAFP) together with Gabinete para o Desenvolvimento do Sector Energético (GDSE),

---

① For details on the FiT and RE certificate scheme, see relevant webpage of EMSD's "HK RE Net" at https://re.emsd.gov.hk/english/fit/int/fit_int.html.

② Financial Secretary's annual budget 2018-19, paragraph 168.

③ DSEC, 2016. Yearbook of Statistics 1990-2015. Macao: Statistics and Census Service of Macao.

have been promoting various programs to government departments since 2010, and extending to the private sector through education campaigns. The GDSE launched the "Regulation for Safety and Installation of Solar Energy PV Interconnections" on 26 January 2015, standardizing the installation of solar power PV generation systems and related equipment in public or private buildings, and formulating safety specifications for connecting solar power PV generation systems to public electric networks of both Low Voltage and Medium Voltage directly or through power distribution systems.

(2) Promoting greener transport vehicles

EVs have no tailpipe emissions, replacing conventional fuel vehicles, especially commercial vehicles, with EVs will help improve roadside air quality and reduce carbon emissions. The HKSAR government's HK$300 million Pilot Green Transport Fund, in place since 2011, allows the transport sector (such as operators of taxis, light buses and non-franchised buses) and organizations to try out green innovative transport technologies, including cleaner commercial vehicles and ferries[①]. To further promote transportation industry to experiment green innovative transport technologies, HKSAR government reviewed the scope of funding by 2020 with HK$800 million of additional funding to expand the funding scope which cover 'Applications for Trial' and 'Applications for Use'. At the meanwhile, more new energy commercial transportation vehicles are categorized in the funding scope, such as boats, motorcycles, off-road vehicles and so on. PGTF was renamed as New Energy Transport Fund in September 2020. As

①For a discussion about the Pilot Green Fund, see LegCo Paper, "Progress on Improving Roadside Air Quality", Environment Bureau, 19 December 2018, https://www.legco.gov.hk/yr18-19/english/panels/ea/papers/ea20181219cb1-319-4-e.pdf.

at the end of November 2020, the NET Fund approved 196 trials, amounting to some HK$154 million, and included 163 electric commercial vehicles (e-CVs) (including single-deck buses, light buses, taxis and goods vehicles) and 103 hybrid commercial vehicles (including single-deck buses, light buses and goods vehicles).

The Macao Environmental Protection Bureau (DSPA) has also formulated policies to improve energy utilization in transport, including “Guidelines for Green Travel in the Public Sector”, introduced tax incentives for environmentally friendly vehicles with charging spaces in public parking spaces, and formulated “Environmental emission standards for new light vehicles” which tighten relevant standards in 2018 to achieve the greatest environmental outcomes.

1) First Registration Tax for EVs

Hong Kong grants first registration tax (FRT) concessions for the purchase of EVs, which reduces the cost to buyers substantially. The FRT is imposed on all types of newly registered vehicles. The FRT is fully waived for e-CVs, electric motorcycles and electric tricycles from 1 April 1994 to 31 March 2021. For electric private cars (e-PCs), there used to be a full FRT waiver till 31 March 2017. To strike a balance between the objectives of promoting the use of e-PCs and not increasing the overall number of private cars at the same time, the government imposed a cap on FRT concessions for e-PCs on 1 April 2017 and introduced the “One-for-One Replacement” scheme on 28 February 2018 to allow e-PC buyers who scrap an older car to enjoy a higher FRT concession. The current FRT concession arrangement lasts till 31 March 2024[①]. The government has also been installing EV chargers in the city. As at the end of September 2020, there

---

① Financial Secretary’s annual budget 2018-19, paragraphs 169-170.

were 3219 EV chargers for public use.

2) Trails for electric and hybrid franchised buses

Franchise buses operated by private companies provide about 33% of public transport patronage and is an essential part of Hong Kong's public transport system. There are about 6 200 franchised buses with 95% of them being double-deck buses. Single-deck buses make up only 5% of the total franchised bus fleet. The technology of single-deck electric buses is already used in places outside Hong Kong, including in Shenzhen and other cities in mainland China.

Nevertheless, the HKSAR government's general policy in this area is to provide subsidies to the franchise bus operators to acquire new technology buses for trial under very demand local conditions due to Hong Kong's hilly terrain and hot summer conditions, which require buses to be air-conditioned. The result of the trials will be discussed in the 4.3 of this chapter. As for double-deck electric bus, NET fund approved the experiment of 2 models of 4 double-deck electric buses, the expected trails will be carried on in the mid-2022.

3) Taxis and public light buses switch from diesel to LPG

Taxis in Hong Kong provide 7.4% and public light buses provide 15% of the public transport patronage respectively [①]. Between 2000～2003, Hong Kong switched its diesel taxi fleet to using liquified petroleum gas (LPG) through an incentive replacement scheme. A similar scheme was provided for Hong Kong's public light buses between 2002-05[②]. Currently, almost all the taxis are running on LPG (over 99%). For current franchised mini-bus fleet public light bus at present,

---

① https://www.td.gov.hk/filemanager/en/publication/ptss_final_report_eng.pdf (see Sections 3.6 and 3.8 on the webpage).

②https://www.info.gov.hk/gia/general/200311/05/1105158.htm.

about 80% of them are running on LPG[①]. The reason for the switch was to reduce air pollutants at roadside as LPG is a cleaner fuel than diesel. Studies show that the carbon emissions between LPG and diesel in automotive use is about the same[②].

While taxi owners could apply under the Pilot Green Transport Fund to trial e-taxis, which are FRT exempt, trials have been a failure in Hong Kong because battery life could not yet cope with the hilly terrain in Hong Kong, the need to have air-conditioning during summer, as well as the two-shifts-24-hour business model of taxis[③]. Hong Kong's taxi fleet is due for another major change as the existing LPG model is being phased out by the manufacturer in Japan. The new model is an LPG-electric hybrid, which has been introduced in Hong Kong and has so far received positive functionality feedback from the trade[④].

Macao's taxi fleet has about 1 600 vehicles and it begun to trial e-taxis in 2018, which is still continuing.

4) Diesel Commercial Vehicles (DCVs)

Hong Kong is a major container port and logistic hub with a high level of

---

①For information about the switch, see https://www.emsd.gov.hk/en/gas_safety/lpg_vehicle_scheme/.

②Atlantic Consulting, *LPG's Carbon Footprint Relative to Other Fuels: A Scientific Review* (2009), http://www.shvenergy.com/wp-content/uploads/2015/05/atlantic_consulting_scientific_review_carbon_footprint_june_2009.pdf.

③ LegCo Question, Promoting use of electric vehicles and hybrid vehicles, 7 February 2018, https://www.info.gov.hk/gia/general/201802/07/P2018020700429.htm; and The Standard, "End of the road for our only EV taxi", 30 May 2018, http://www.thestandard.com.hk/section-news.php?id=196234&sid=11.

④ South China Morning Post, "First hybrid Toyota taxis hit Hong Kong roads, but wider use will need incentives, industry says", 25 January 2019, https://www.scmp.com/news/hong-kong/transport/article/2183725/first-hybrid-toyota-taxis-hit-hong-kong-roads-wider-use.

road transport carrying goods within the city and to the mainland. From March 2014 to June 2020, the government provided an incentive-cum-regulatory program to phase out over 80 000 pre-Euro IV DCVs, a new scheme is under planning in October 2020 which aims at progressively phase out about 40 000 Euro IV DCVs by the end of 2027.

5) Environmental emission standards for vehicles

In 2012, the MSAR implemented the "Environmental emission standards for new light vehicles", stipulating an energy-efficiency rating for newly registered light vehicles. In November 2018, DSPA and the Transport Bureau (DSAT) upgraded the emissions standards for light duty vehicles. The qualifying standard for petrol/liquefied gas vehicles has been tightened from Euro 4 to Euro 6 limit, while the qualifying standard for diesel vehicles is Euro 5 instead of Euro 4. The standard for heavy-duty liquefied natural gas vehicles has also been advanced to Euro 5. The standards of new heavy- and light-duty motorcycles will mainly increase to the National 3 Emission Standard of China[①]. This measure is expected to be useful in controlling excessive transport energy consumption in Macao and there is associated effects in reducing carbon emissions.

Energy saving

As a member of the Asia-Pacific Economic Cooperation (APEC), Hong Kong endorsed the organization's very modest energy intensity reduction target in 2007 to achieve at least 25% reduction by 2030 compared with 2005 level. In 2011, APEC strengthened that goal to a 45% reduction of regional aggregate

①Macau Daily Times, New Vehicles' Emissions Requirements Tightened, November 21, 2018. https://macaudailytimes.com.mo/environment-new-vehicles-emissions-requirements-tightened.html.

energy intensity by 2035 compared with 2005. With the HKSAR government's publication of *The Energy Saving Plan for Hong Kong's Built Environment 2015 ～ 2025+* in May 2015, Hong Kong set its own more ambitious target to achieve a reduction of 40% in energy intensity by 2025 compared with 2005. Hong Kong would have to save about 6% energy by 2025 to meet that target. This plan, together with *Deepening Energy Saving in Existing Buildings in Hong Kong Through '4Ts' Partnership*, published in June 2017, set out the most important aspects of Hong Kong's energy saving efforts up to 2025 by focusing on achieving energy saving in buildings.

By 2018, Hong Kong had already achieved a reduction in energy intensity of above 30%. The HKSAR government have run many energy saving campaigns and programs with various quarters of the community and also the general public since 2012. The most prominent of these is rallying companies and organizations to sign energy saving charters that involved more and more facilities. The campaigns became more important as Hong Kong has an energy intensity target to achieve by 2025 requiring the community as a whole to take action.

Macao also has energy saving campaigns. In June 2006, the 'Macao Energy Conservation Week' was launched with a series of activities that encourage energy saving. Since then, the energy conservation week took place in every June and across all parts of Macao and it is the largest energy campaign in Macao. In 2019, activities included switching off lights for one hour; dressing casual clothes and reducing the use of air conditioner; lucky draw for 5% energy saving action for the public[①]. In addition, selected primary and middle schools hold annual 'Energy

① Activities during the Macao Energy Conservation Week 2019 are found: https://www.gdse.gov.mo/eco2019/index.html.

Saving Activities on Campus'[①], which is supplemented by the 'Energy Education: Affiliated Teaching Materials for Middle schools' started 2011 to embed energy-saving knowledge to the curriculum[②]. Macao generated efforts in the construction of energy-saving policy system. For instance, "Five-Year Development Plan of the MSAR (2016-2020)" and "Environmental Protection Planning of Macao (2010-2020)" listed out plans and actions for energy conservation and emission, along with further implementation and promotion.

(1) Energy saving in new and existing buildings

Chapter 3 notes the pattern of energy usage in buildings in the two SARs and the Mitigation Scenarios show the importance of achieving greater energy efficiency. An important initiative in Hong Kong was to rally major building owners to not only sign the voluntary energy saving charter noted above but also work together to help Hong Kong reach the 2025 target, as articulated in Deepening Energy Saving in Existing Buildings in Hong Kong Through '4Ts' Partnership. In 2016, ENB and EMSD made energy saving in existing buildings the target area for long-term stakeholder engagement. They believed that making energy use in buildings more transparent would help spur greater efforts. Based on EMSD's statistics of the electricity consumption by sector in Hong Kong, the top users among the buildings in the commercial sector (which include government and institutional buildings) were identified, as per Table 4–3. Since a relatively small number of major buildings used 20% of total electricity use in Hong Kong,

---

① ODES. Energy-saving Activities of Campus 2016. Office for Development of the Energy Sector. Macao; 2016. (http://www.gdse.gov.mo/public/chn/ecampus2016/2016/2016.html).

② ODES. Energy Education: The Affiliated Teaching Materials of Middle Schools. Office for Development of the Energy Sector. Macao, 2011.

energy saving efforts by them could have a sizable impact. As the government had committed to saving energy in its buildings by 5% between 2015 and 2020, it asked the building owners concerned to set targets of their own. In addition, the government allowed companies to further accelerate the tax deduction of capital expenditure for procuring energy efficient building installations from April 2018 to promote energy saving in buildings[①].

Table 4–3 Percentage of electricity use in commercial sector

Sector	Purpose	Usage percentage
Commercial existing buildings	Top 400 commercial buildings (common areas)	11%
	Hotel	4%
	MTR offices, depots and stations	1%
Public sector existing buildings	Government buildings	5%
	Schools & universities	4%
	Hospital authority	3%
	Housing authority	1%
	Airport authority	1%
Total	20% of total electricity use in Hong Kong	

Hong Kong and Macao have continued to add to their building stock in light of their expanding economies and demand for new commercial and residential buildings. Hong Kong's first attempt to implement energy saving through design was the Building (Energy Efficiency) Regulation enacted in 1995. It regulates the design and construction of external walls and roofs of certain types of buildings by

① Financial Secretary's annual budget 2018-19, Paragraph 168.

prescribing the maximum overall thermal transfer value. The next initiative was the voluntary Hong Kong Energy Efficiency Registration Scheme for Buildings (HKEERSB) launched in 1998 to promote the application of a Building Energy Code (BEC), that set the minimum energy performances standards of lighting, air conditioning, electrical installation, and lift and escalator. The Buildings Energy Efficiency Ordinance (BEEO) was then enacted in 2012 with the statutory BEC. BEEO requires compliance with the statutory BEC in the design of new prescribed buildings and major retrofitting works of prescribed existing buildings[①]. The BEC is reviewed every 3 years to reflect the development of international standards and latest technological advancement. Two reviews have been completed and the new standards have taken effect. By 2028, the implementation of the BEEO is expected to bring about energy saving of some 27 billion kWh from both new buildings and existing buildings in Hong Kong, equivalent to a reduction in carbon dioxide emissions of about 19 million tons. BEEO also requires owners of commercial buildings to carry out energy audit for central building services installation in accordance with the Energy Audit Code (EAC) every 10 years. The EAC is also reviewed and updated regularly.

Meanwhile, there has been a growing interest in green buildings, which are designed and built to meet sustainability standards as noted in Chapter 3 under Scope 2 emissions. The Hong Kong green building standard is BEAM Plus, which has been formulated having regard to the local context and environment

---

①Prescribed buildings are commercial buildings, hotels and guesthouses, residential buildings (common area only), industrial buildings (common area only), composite buildings (non-residential and non-industrial portion), composite buildings (common area of residential or industrial portion), educational buildings, community buildings, municipal buildings, hospitals and clinics, government buildings, airport passenger terminal building, and railway stations.

consideration[①]. Overseas standards, such as Leadership in Energy and Environmental Design (LEED), are also used. Since 2011, the government has required registration under BEAM Plus as one of the prerequisites for private developments to obtain gross floor area concessions for green features and amenities to encourage sustainable building design. Since 2010 and up until May 2019, over 1 160 projects with over 33 million square meters of space have been registered under BEAM Plus for new buildings. If other assessment standards were included, nearly 1 300 projects with over 40 million square meters space have been registered in relation to new buildings[②].

The HKSAR government has about 8 000 buildings and it begun to promote energy saving in buildings in 2003 by setting specific electricity reduction target for government buildings. Between 2003 and 2014, the government achieved a 15% reduction in energy and in 2015, it committed to an additional 5% saving by 2020 and has achieved about 4.9% saving by 2018. This target was set out in the *Energy Saving Plan for Hong Kong's Built Environment 2015 ～ 2025+*. The government also used its own new buildings, including public housing, to adopt green building features to reflect its commitment and to demonstrate the benefits of various green building features and technologies.

Macao is behind Hong Kong in this respect and the rated green buildings in

---

① The journey to promote green building certification in Hong Kong could be said to have begun in 1995 with BEAM, the forerunner to BEAM Plus, and the first standards were promulgated in 1996 with the founding of the Hong Kong Green Building Council (HKGBC) with the support of Development Bureau together with established professional bodies in the building and construction sector. The founding members were Construction Industry Council, Business Environment Council, BEAM Society Limited and Professional Green Building Council, see https://www.beamsociety.org.hk/en_about_us_2.php.

② Information update from BEAM Society, personal communication on 26 June 2019.

Macao are mostly gaming and resort buildings since it is in the operators' interest to save as much energy as possible[①]. In 2016, the ratio of carbon emissions from gaming industries in Macao takes up 62.46% while the emissions from residential buildings and commercial buildings only account for 20.51% and 17.03% respectively (Zhao *et al.*, 2019a). Nevertheless, the ODES issued the "Techniques for Building-Energy Consumption Optimization in Macao," in 2009, identifying six major areas of energy saving energy in buildings: (1) building materials; (2) lighting; (3) air conditioning and ventilation systems; (4) electrical power; (5) elevators; and (6) RE systems. Other policies issued by the ODES focus more on the use of energy-efficient electronic products and RE (e.g., solar power). Furthermore, focusing on the tourism industry in Macao, the DSPA started to issue Macao Green Hotel Awards from 2007[②].

(2) Energy saving in other areas

1) Energy saving appliances

The Energy Efficiency (Labelling of Products) Ordinance, enacted in 2008, provides the basis for implementation of the Mandatory Energy Efficiency Labelling Scheme in Hong Kong. This scheme requires that the energy labels be shown on prescribed products in order to inform consumers of the products' energy performances. The first phase, covering room air conditioners, refrigerating appliances, and compact fluorescent lamps, was implemented on

---

① Nelson Moura, Local green building sustainability mainly driven by gaming operators' initiatives-environmental advisor, Macau News Agency, 30 November 2017, http://www.macaubusiness.com/macau-local-green-building-sustainability-mainly-driven-gaming-operators-initiatives-environmental-advisor/.

② Macao Green Hotel Award. Environmental Protection Bureau of Macao, Macao; 2016 (http://www.dspa.gov.mo/h_index.aspx).

November 9, 2009. The second phase extended the coverage to washing machines and dehumidifiers on September 19, 2011. The energy efficiency grading standards of room air conditioners, refrigerating appliances, and washing machines were tightened in November 2015. The third phase covers both the heating and cooling functions of room air conditioners, washing machines with higher washing capacity, televisions, storage type electric water heaters, and induction cookers and commenced in December 2019. The appliances covered by all three phases of the scheme now represent about 70% of the electricity consumption in the residential sector. EMSD estimated that about 600 million kWh will be saved annually. There are a further 22 types of electrical and gas products under a voluntary energy labelling scheme for home and office appliances and equipment, as well as for vehicles (see below), in order to help the public to choose energy-efficient products. Macao does not have an energy labelling scheme for electricity appliances yet.

2) Energy saving in transport

Chapter 3 provides the details on energy usage and carbon emissions from transport. The transport sector energy end-use is aligned with their respective carbon emissions. The rail and tram systems are powered by electricity (about 43% of total public transport patronage), while road vehicles are mostly powered by diesel, petrol and LPG, and local ships are powered by marine diesel. As already noted earlier in this chapter, taxis use LPG and public light buses are powered by LPG or diesel. DCV, such as trucks, are powered by diesel; and private cars are mainly powered by petrol with a small number by diesel. This section focuses on public transport road vehicles as they carry the largest number of passengers and includes a short section on DCV.

The HKSAR government mitigate carbon emissions by saving energy in transport through the following two approaches:

First, ensuring public transport remains the preferred choice for daily mobility. This measure requires extending the rail network and improve interfaces with other public transport services (buses and light buses), as well as coordinating planning for housing and transport in new towns in the New Territories to reduce use of private vehicles. The HKSAR government has been expanding its rail network continuously for many years to bring 70% of the population within easy access of a railway station by 2021. It also aims to increase the share of rail in public transport trips from 40% to 43% by the same year. Further expansion plans up until 2031 are expected to serve areas inhabited by 75% of the population, taking up the share of rail in public transport trip to 50%[①].

Second, promoting zero-carbon transport modes. This measure requires improving walkability through the city so that people can walk shorter distances with relative ease, including building pedestrian bridges and elevators/escalators so people can navigate the city more easily on foot. In many respects, Hong Kong has been an innovator in light of its hilly terrain. By helping make it easier for pedestrians, walking is a daily activity for a large number of residents, and a survey found Hong Kong people is the most ambulatory population out of 46 territories surveyed[②]. Cycling paths in the New Territories and in new town areas have also been extended and are popular.

The Macao Transport Bureau developed the "General Road Traffic and Transport Policy of Macao (2010～2020)" to encourage the public to use public

---

① Railway Development Strategy, September 2014, https://www.thb.gov.hk/eng/psp/publications/transport/publications/rds2014.pdf.

② Stanford University, Activity Inequality: Large-scale physical activity date reveal worldwide activity inequality, 2017, http://activityinequality.stanford.edu/.

transport as their preferred choice[1]. The monthly trip volume by public bus increased from 13.3 million in 2012 to 15.5 million in 2014[2]. The policy aims to achieve the target that contribution of public transit (public bus, light rail, and taxi) reaching 50% of all trips by 2020.

3) Voluntary Energy-Labelling Scheme for Private Vehicles

Since 2002, Hong Kong's the voluntary energy labelling scheme noted above applies to private petrol vehicles carrying up to 8 passengers, the purpose of which is to give car buyers comparative information about the fuel consumption of different models under standard test conditions[3]. The scheme has not achieved its intended purpose – there are no models currently listed. The vehicle trade has been concerned about the fairness of comparison (with some justification) on fuel consumption figures generated by the different testing standards from Japan, the EU and the US as noted in an international review[4]. After consideration on how to harmonize the different testing methods, the EU developed a new test that went into effect in September 2018. The Hong Kong authorities will have to consider whether and how to move forward.

---

① Macau transportation policy on the road travel 2010-2020. Macau Transport Bureau. Macau, 2010.

② Annual summary 2011–2014 on Macau transportation policy on the road travel 2010–2020. Macau Transport Bureau. Macau, 2014.

③ Legislative Council Secretariat, Information Note: Electricity Market in Hong Kong, 24 May 2006, https://www.legco.gov.hk/yr05-06/english/sec/library/0506in26e.pdf.

④Legislative Council Secretariat, Information Note: Electricity Market in Hong Kong, 24 May 2006, https://www.legco.gov.hk/yr05-06/english/sec/library/0506in26e.pdf.

4.2.3 Assessment of climate change mitigation policies

Carbon reduction in the electricity and gas sector

It is vital to focus on electricity generation, as it emits the largest portion of carbon emissions in Hong Kong and Macao. In the case of the HKSAR, the decision to stop building more coal-fired generating capacity in 1997 and switch to natural gas was crucial. From commissioning its first gas-fired electricity plant in 1996, the pace of Hong Kong's switch from coal to natural gas in electricity generation is quite rapid. By 2015, there were 10 gas plants, and by 2020, natural gas will generate about 50% of Hong Kong's electricity, while coal will drop to about 25%. This was achievable because the SCAs provided a contractual framework under which CLP and HEC are assured of a reasonable return on the investment required to make such a change; and the HKSAR government scrutinizes the investment proposals to ensure that the transition would take place to meet the relevant environmental targets.

Indeed, the use of SCAs to regulate the electricity sector is a unique characteristic of the Hong Kong energy system that has provided Hong Kong with safe and highly reliable electricity – among the most reliable in the world – that is also reasonably priced for consumers, while minimizing the environmental impact of electricity generation. At the same time, it has enabled CLP and HEC to be profitable companies. New terms in the latest SCAs provided commercially attractive rewards for RE generation by both companies and their customers and also energy saving by consumers. It may be concluded that the use of contract between the government and the two electricity companies has been effective in implementing policies through negotiations. One oversight in the SCAs is the lack of provision to require the electricity companies to provide detailed relevant electricity production and consumption data to the authorities and for public disclosure, as these are now seen as essential for policymaking, particularly in demand-side management. The veracity of an argument that more fulsome disclosure would involve privacy issue needs to be debated as data transparency is essential for any jurisdiction to achieve better policies, including 'smart' city actualization.

The SCAs had been criticized on the levels of investment allowed and rates of return[①]. Strengthening mechanisms to reduce the reserve capacities became a focus for SCA renewal negotiations. Reserves capacities of CLP and HEC have been reduced over the course of the past two SCAs from over 50% to around 30%. There has also been pressure for the networks of the two companies to be linked so as to enable a measure of competition that could reduce Hong Kong's overall reserve capacity and improve tariff. Moreover, to import more natural gas in the future requires new infrastructure to receive gas. The two companies are currently planning to jointly build an offshore LNG terminal in Hong Kong waters for commissioning in 2022[②].

On the issue of competition, as long as more land and large investments are needed to provide electricity, it would be unrealistic to assume there would be other private investors interested to invest in electricity generation for both Hong Kong or Macao. In the case of Hong Kong, it may be possible to allow

① Legislative Council Secretariat, Information Note: Electricity Market in Hong Kong, 24 May 2006, https://www.legco.gov.hk/yr05-06/english/sec/library/0506in26e.pdf.

② Environment Bureau, 2018-23 Development Plans and 2019 Tariff Review of the Two Power Companies, https://www.legco.gov.hk/yr17-18/english/panels/edev/papers/ edev20180704-enbcr145760818pt28_enbcr245760818pt27-e.pdf.

competition should the government wish to pursue building the two off-shore wind farms noted above but the investors would still need to sell the power to CLP and HEC for distribution, assuming the projects are in themselves sufficiently attractive commercially. In fact, the government and public sector is currently Hong Kong's third biggest generator of energy with its various RE (including WTE) projects. If the government department generating the RE could not directly use all of the energy itself, it has to feed it to CLP, HEC or Towngas' distribution systems. Distributed power may one day mature to the extent that buildings could generate their own power through on-site PV installations, including on external walls. However, the public might still prefer the existing companies to invest in such new technologies since they have good track records and are trusted to provide good service. This is an area for further policy development.

Up until recently, there was no reward for energy saving by the two electricity companies. This is a general challenge for governments around the world to shift from rewarding sales to rewarding saving. Since the last SCA (2008/09-18), modest rewards were provided for saving energy and these were made more attractive in the current SCA (2018/19-33). Moreover, the previous effort in the last SCA to promote RE was too weak to drive uptake but the current SCA is generating greater interest in Hong Kong, on top of other contributing factors such as the advancement of PV technology, the downward trend of the cost of PV installation and operation, and the raising public awareness on RE. Combined with more determined effort by the HKSAR government since 2015 to promote energy saving and RE in public infrastructure and buildings, and together with extra funding and incentives for RE projects, this has generated much more interest by government departments, public sector institutions, schools and the private sector. Furthermore, to support the energy saving initiatives under the new

SCAs, both electricity companies will replace their electromechanical meters with smart meters with backend facilities. These smart meters will help individual customers achieve energy saving by providing them with near real-time power consumption information and enable implementation of demand response schemes, which dovetails Hong Kong's overall policy to become a 'smart city'. The Government of the HKSAR and electricity companies had agreement on gradually disclose relevant data to promote the development of smart cities.

CEM is installing grid-connected PV systems on rooftops of primary substations and to see how PV projects could be expanded in Macao①. In 2014, the MSAR government issued the "Regulation for Safety and Installation of Solar Energy PV Interconnections", which went into force in 2015, and published a new solar regulation and a Feed-in Tariff (FiT) scheme. Up until end-2018, there were three-grid-connected photovoltaic power generation systems with the installed capacity around 22.9 kW. Though the total electricity contribution from solar-generated power is modest, it is seen as an important part of the local electricity mix in Macao.

Local generation vs. import power

Hong Kong started to import nuclear electricity in 1994 from the Daya Bay Nuclear Power Station in Guangdong Province under a long-term supply contract. Currently, CLP imports around 80% of the total electricity output of the Guangdong Daya Bay Nuclear Power Station, 70% of which is purchased under a 20-year contract expiring in 2034, and an additional 10% under a separate contract

---

① ODES. Macau owns the large potential to develop the solar energy. Office for Development of the Energy Sector. Macau; 2014 (http://www.gdse.gov.mo/gdse_gb/ newsDetails.asp?newID=433).

that will expire in 2023. The nuclear energy imported accounts for about a quarter of the HKSAR's total fuel mix. CLP, through its wholly-owned subsidiary, Hong Kong Nuclear Investment Company Limited, has 25% shares in Guangdong Nuclear Power Joint Venture Company, which owns the Daya Bay Nuclear Power Station. CLP was an investor and the plant was financed through selling the bulk of the electricity to Hong Kong in hard currency. This arrangement has been a resounding success – it set-off China's development in commercial nuclear power and supplied a sizable source of safe and reliable electricity to Hong Kong at reasonable price over an extended period of time. Meanwhile, China has developed other nuclear power plants in Guangdong Province and elsewhere in the country.

Macao was able to secure electricity from CSPG when it needed to expand supply quickly from the early 2000s. In 2004, Macao was still supplying 93% of its own electricity and imported only 7% from the mainland. By 2014, it was importing 88% from the mainland, and in 2018, the rate came to 88.8% of the total power consumption in the city, while the remaining 3.1% was purchased from the Macao Refused Incineration Plant. In 2018, the MSAR government approved the building of a new gas-fired power plant in Macao, which is expected to be commissioned in 2024. CEM decided it is in Macao's interest to expand local generation capacity in order to increase stability and reliability.

For both Hong Kong and Macao, importing more electricity from the mainland is an option – it could be directly from a dedicated plant, such as in the case of CLP buying electricity from the Daya Bay Nuclear Power Station, where dedicated transmission capacities are built to deliver electricity to Hong Kong, or buying electricity directly from the mainland's electricity grid, as per the contract between CEM and CSPG. In the early 1990s when the Daya Bay Nuclear Power

Station was being built, it was not possible for Hong Kong to buy power from the mainland grid as the mainland did not have sufficient power itself. However, CSPG is able to supply more power today to the two SARs should they wish, as electricity supply has increased dramatically over the years and reliability continues to improve. Hong Kong's consumption would only be about 5% of the total consumption of CSPG's supply region that covers Guangdong, Guangxi, Yunnan, Guizhou and Hainan[①]. Macao's portion would be much smaller.

Nevertheless, the two SARs are understandably concerned about reliability where they are not in direct control of the generating plants, such as during severe weather events or some other system failure on the mainland. Since 2010, the HKSAR government has deliberated whether to increase electricity purchase from the mainland but due to public concern about supply reliability, it has not pursued it although studies would be carried out on the investment and infrastructure that would be needed to purchase more electricity from the mainland in the future. Meanwhile, CLP will enhance its Clean Energy Transmission System with CSPG and Daya Bay Nuclear Power Station, which provides CLP the flexibility to use more "zero-carbon energy" (including nuclear and RE). When completed in 2025, the enhancement project should enable Hong Kong to achieve its 2030 carbon intensity reduction target (i.e., reduction of carbon intensity by 65% to 70% as compared with the level in the base year of 2005) five years ahead of its target year. In addition, it could delay and/or reduce the capital investment for building new natural gas electricity generating units in Hong Kong to replace the coal units

① Public Consultation on the Future Development of the Electricity Market, 2015, pages 6-7, https://www.enb.gov.hk/en/resources_publications/policy_consultation/public_consultation_future_development_electricity_market.html.

due for retirement in 2025 and beyond[①].

In the case of Macao, the majority of its electricity need will continue to come from CSPG but in building a new gas-fired generation plant, CEM will be better able to ensure reliability, especially when there are severe storms. In addition, RE application is limited by environmental resources which further evaluations on its capability is required. Referring to the experience of other countries and regions, the components of RE in power supply combination may hard to have significant growth in a long term. In a long run, thinking of MSAR's specific geography conditions and economic volume, regional cooperation in the areas of RE and other emission technologies is essential to reach carbon peak and carbon neutrality.

A point to note here is the change in carbon emissions distribution depending on where power is generated. Thus, carbon emissions from non-nuclear generation where electricity is imported from the mainland is counted in the mainland's emissions inventory and not that of the SARs. Thus, the carbon emissions arising from non-nuclear electricity that is produced on the mainland and sold to CEM is counted on the mainland. What is important is to reduce electricity consumption overall and not to increase generation capacity – especially where fossil fuels are used – when existing capacity can be used.

Demand-side energy saving

The policy document, *Deepening Energy Saving in Existing Buildings in*

---

① Environment Bureau, *2018-23 Development Plans and 2019 Tariff Review of the Two Power Companies*, https://www.legco.gov.hk/yr17-18/english/panels/edev/papers/edev20180704-enbcr145760818pt28_enbcr245760818pt27-e.pdf.

Hong Kong Through "4Ts" Partnership, published in June 2017, is seminal for Hong Kong in that it shows the HKSAR government taking leadership to work with public and private sector building owners, property managers and property developers. ENB used its political power in convening to bring corporate leaders together to explain the HKSAR's mission to reduce electricity consumption and sought their cooperation. EMSD plays a critical role together with Hong Kong Green Building Council (HKGBC) to promote energy saving in building to management and engineering professionals working in companies. By reaching out to chairmen and directors of such companies, the government wanted to convince finance directors the benefit of energy saving.

Engagement with stakeholders and the public is arduous work that must be done continuously. It also takes specific skills to bring the right people together to co-learn and collaborate in a setting where the government depends on voluntary action, since if Hong Kong does not reach its targets, there are no penalties to the users of energy. With the rise of investor and public scrutiny in corporate environmental performance, the leading companies are preparing sustainability reports but there were many laggards. The local stock exchange adopted a new policy in 2013 for companies to prepare Environmental, Social and Governance (ESG) reports on a voluntary basis. In 2016, it took another step forward to require listed companies to "report-or-explain". In May 2019, it proposed to introduce mandatory disclosure as the next step in a consultation document, which will come into force in 2020[①]. The gradual tightening of disclosure requirement is

---

① Hong Kong Exchanges and Clearing Limited, Consultation Paper: Review of the Environmental, Social and Governance Reporting Guide and Related Listing Rules, May 2019, https://www.hkex.com.hk/-/media/HKEX-Market/News/Market-Consultations/2016-Present/May-2019-Review-of-ESG-Guide/Consultation-Paper/cp201905.pdf?la=en.

heightening companies to look at their use and management of natural resources, in particular energy and water.

Macao is in a transition phase to adapt diversification and economic transformation, along with the re-competition of gaming licenses will influence a wave of investments in entertainment projects. Not only other emerging industries will be affected, but also the demand of electricity. The future of energy utilization for economic development will change, thus Macao's sustainable development requires an equilibrium relationship between economic activity and electricity consumption.

Overall, policy development has been slow in the HKSAR and the MSAR. A justification may bc that in reality, it took considerable time for policymakers to prioritize energy saving when there was little real pressure from society, and also for an industry sector, such as the building sector, to truly develop the commitment and capacity to save energy. Now that the foundation has been laid, continuing policy focus and engagement with stakeholders remain critical.

Overall carbon reduction in transport

Hong Kong has a clear policy that favors public transport. The availability of an affordable and efficient public transport system has helped to reduce reliance on private cars. Private car ownership is very expensive in the HKSAR, where the FRT can be as high as 100%–110% of the price of the vehicle; and parking is also expensive. These policies have resulted in a low car ownership ratio relative to Hong Kong's economic development and a high public transport ridership rate when compared to cities, such as London and New York. It is very successful to have over 90% of daily passenger trips made by public transport. Hong Kong has been consistently ranked as the leader in public transport among cities[①].

Chapter 3 provides details on transport sector carbon emissions. As regards vehicle-related policies, it needs to be noted that they were mostly designed to reduce roadside air pollutants. Overall, these policies have not had much of an impact on reducing carbon emissions.

As noted earlier in Part II, the HKSAR government have provided generous subsidies to the franchised bus operators to trial new technology buses to improve roadside air quality. Specifically, 6 double-deck EURO 6 hybrid diesel-electric buses were acquired for two-year trials starting in 2014. The results of the much more expensive hybrid models showed no overall benefits in fuel usage and therefore pollutant emissions and carbon emissions when tested under local operation conditions of hilly terrain and heavy use of air-conditioning in the hotter months (Keramydas *et al.*, 2018)

The HKSAR government also fully subsidized 36 single-deck electric buses (including 28 battery-electric buses and 8 supercapacitor buses) also for conducting two-year trials to test performance, reliability as well as economic feasibility under local conditions. As of end-2020, 26 battery-electric buses and 7 supercapacitor buses have commenced operation and the remaining buses would follow shortly.

Early trail results indicate that whether single-deck battery-electric buses can be widely used in Hong Kong depends on the development battery capacity, enabling it to travel about 300 km a day after a full charge; and/or whether there is adequate space for installation of charging facilities at the depots or public

---

① Transport and Housing Bureau, *Public Transport Strategy Study: June 2017*, https://www.td.gov.hk/filemanager/en/publication/ptss_final_report_eng.pdf.

transport interchanges for top-up charging taking into account high operation frequency of buses in Hong Kong[①].

As for hybrid vehicles, the commercial hybrid light buses tried out under the Pilot Green Transport Fund have so far incurred fuel expense savings of only 4% or less as compared with their conventional counterparts. The government's approach is to continue to encourage vehicle suppliers to introduce e-CVs and hybrid commercial vehicles suitable for use by the local transport sectors, and invite the sectors to apply for trials of green innovative transport technologies under the fund.

Moreover, the EPD completed a pilot scheme at four government open car parks (located at the Electrical and Mechanical Services Department Headquarters, Hong Kong Wetland Park, Wai Tsuen Sports Centre and Shek Kip Mei Park) managed by contractors in 2018, where a total of 11 outdoor medium chargers have been installed to assess their reliability. The result indicates that the outdoor chargers could be operated satisfactorily.

The major problem is the total number of vehicles in Hong Kong has been rising. In 2014, the total number of vehicles was just under 700 000. By the end of 2018, it had grown to just over 784 000. Most worrying was the number of private cars had grown from just over 495 000 to over 565 200. Vehicle ownership change is the largest contributor to the increase of passenger transport energy use for Hong Kong, and the authorities have acknowledged its inability to control the rise

① LegCo Question, "Electric buses", 12 December 2018, https://www.info.gov.hk/gia/general/201812/12/P2018121200670.htm.

in private cars①. The growing energy use and externalities imposed by transport requires concerted policy action. While policymakers see private car ownership as a target for regulation, they have yet to find the right solutions.

4.3 Overview and Assessment of Climate Change Adaptation and Resilience Policies in Hong Kong and Macao

The degree of damage arising from severe weather events is a function of the physical threat (severe weather event), exposure to the threat (the presence of people, resources and assets in places that could be adversely affected) and vulnerability (propensity to be adversely affected)②. Based on the projection of climate change in Hong Kong and Macao presented in Chapter 1 and the adaptation that is needed in the areas of water management, floods and landslides prevention, sea-level rise, and public health discussed in Chapter 2, Hong Kong and Macao continue to face high degree of threats to the people and assets in the future.

In the case of Hong Kong, being a significant commercial and financial

---

①Transport Advisory Committee, Report on Study of Road Traffic Congestion in Hong Kong 2014, https://www.thb.gov.hk/eng/boards/transport/land/Full_Eng_C_cover.pdf; and Environment Bureau, Hong Kong's Climate Action Plan 2030+, January 2017, https://www.enb.gov.hk/sites/default/files/pdf/ClimateActionPlanEng.pdf.

② IPCC, *Assessing and Managing Risks of Climate Change*, https://www.bayarea.gov.hk/tc/home/index.html.

center, it has had to ensure there is adequate fresh water supply, and that the city's infrastructure could withstand severe weather during the summer typhoon season. Stronger and wetter typhoons present multiple challenges, in particular landslides, floods and storm surges – as Hong Kong is hilly and coastal. Likewise, Macao has to deal with landslides, floods and storm surges too although the risk for landslide is very much lower than Hong Kong. The experience of super typhoons in 2017 (Hato) and 2018 (Mangkhut) have been reviewed by the respective SAR governments. They provided useful reflection on the two SARs climate change adaptation and resilience systems. Lessons that were learnt in 2017 were applied in 2018. Nevertheless, to do better still, the authorities have to engage more widely in the future with the private sector and the community. Vulnerability can be reduced through various means, such as by good early warning systems, emergency preparedness, as well as appropriate defensive engineering works. These measures and public understanding of the risks are essential to improve emergency preparedness.

This section provides an overview, as well as an assessment of climate change adaptation and resilience in the HKSAR and MSAR, including the insights and lessons gained from the recent years' super typhoons. It is important to note that increased development increases both the likelihood and consequence of systems failure.

4.3.1 Managing excess water and flood prevention

Hong Kong's DSD deals with both wastewater treatment and stormwater drainage services. Municipal Affairs Bureau of Macao (IAM) previously known as the Civic and Municipal Affairs Bureau is responsible for the operation of the drainage network. Good drainage is essential in the two SARs, as both experience heavy rainfall. Even before climate change adaptation became an important policy narrative, Hong Kong had devoted a great deal of effort to prevent flooding with good results. Macao still needs to do more and its plans involve working closely with Zhuhai City in the mainland.

Chapter 2 notes that up until the 1990s, during particularly heavy rainstorms, flooding occurred in the rural low-lying areas and natural floodplains in the northern part of Hong Kong; and in parts of the older urban areas. Rapid development resulted in intensive development in the New Territories, including on floodplains. Development turned natural ground into hard paved areas and as a result, rainwater which formerly was retained, became surface flow. The expansion of built-up areas in close proximity to the major watercourses also reduced their flood-carrying capacities[①]. Problems in the old urban areas were due to expansion of development out-pacing the previous design standards and deterioration of the existing systems[②].

Between 1994 to 2010, DSD completed a series of studies, referred to as Drainage Master Plan (DMP) Studies and Drainage Studies, covering Hong Kong's catchment areas to provide comprehensive solutions to flooding. These studies examined the adequacy of the existing drainage systems and recommended

---

① For a historical discussion, see Ken KC Luk, Alex KF Kwan, John KY Leung, "Drainage Master Planning for Land Drainage Flood Control in the Northern New Territories", HKIE, Proceedings, Vol 3 No. 1, July 2001, https://www.dsd.gov.hk/EN/Files/Technical_Manual/technical_papers/LD0102.pdf.

② For a historical discussion, see KK Chan, KW Mak, Alan SK Tong, and KP Ip, "Flood relief solutions in metropolitan cities the Hong Kong experience theme: planning for sustainable water solution oral presentation", Water Practice & Technology Vol 6 No. 4 doi:10.2166/wpt.2011.081.

short to long-term improvement measures to meet higher flood protection standards and future development needs. The studies consider technical constraints, cost effectiveness and environmental consideration in formulating sustainable short-and- long-term improvement measures[①]. While new drainage systems and village flood protection schemes had to be constructed in the New Territories, various measures, such as stormwater interception, pipeline upgrading, as well as stormwater storage and pumping had to be built in the urban areas. Since 2008, the Review of these DMP Studies has progressively commenced and the review process is on-going. From the establishment of DSD in 1989 to 2018, the government spent about HK$30 billion in drains (~100 km), river channels (~110 km), four drainage tunnels, four stormwater storage schemes and 27 polder schemes in rural villages. From 1995, 126 flooding blackspots were eliminated which currently there are only 5 flooding blackspots left.

Over the years, DSD's thinking on planning and design of drainage infrastructure has gradually evolved from resistant to adaptive approach. It has adopted the "blue-green" infrastructure approach that uses nature and natural processes for greater resilience. While traditional engineering aims to defend assets by stopping, repelling or controlling natural forces, the new approach requires the design and planning of infrastructure that can both enhance flood resilience and improve utilization of water resource at the same time. This is similar to the idea of "sponge city" being implemented in mainland China. The various schemes that have been implemented have proven to be both successful,

---

① Categories of improvement measures can be found at https://www.dsd.gov.hk/EN/Flood_Prevention/Long_Term_Improvement_Measures/Categories_of_Long-Term_Improvement_Measures/index.html.

ecologically-sound and visually attractive[①]. They also provide the blending of a number of disciplines beyond engineering to include urban design and ecosystems resilience.

While flood risks have been much reduced in Hong Kong, it cannot be totally eliminated. Flooding could still occur under exceptional rainstorm condition when the rain intensity is more severe than the design standards. The drainage system may not function as designed if some sections, and in particular inlets such as catch-pits, are blocked by rubbish, fallen trees or landslide debris. Flooding at low-lying coastal areas may occur when there is exceptionally high-water level caused by storm surge coupling with wave action during the passage of a typhoon.

Going forward, DSD's approach is to pay more attention to aligning development, especially in the New Development Areas, with providing sewerage and stormwater infrastructure, and to design and build the relevant infrastructure that adopts the "blue-green" approach. Furthermore, flood proofing measures and non-structural measures shall also be explored in additional to large-scale flood prevention infrastructure so as to mitigate flood risk in a more effective manner in front of the extreme weather events.

Flooding has continued to occur in Macao as a result of storms and storm surges. While the MSAR government has invested in local equipment, such as more pumps, it has also been discussing with the Central People's Government to create a flood prevention system that would benefit Macao and the immediate neighborhood. On 23 August 2017, the passage of Typhoon Hato struck southern

---

① YK Ho, DSD, "Sustainable Drainage System in Hong Kong", 1 April 2017, https://www.hk2030plus.hk/document/KSS_PPT/5th_KSS/Sustainable_Drainage_System_in_HK.pdf. These include the Lam Tsuen River and Yuen Long Bypass Floodway. New projects being built include a retention lake at Development of Anderson Road Quarry Site.

China. Among the hardest hit cities, Macao experienced the worst flood since 1925. Half of the Macao Peninsula was inundated and the Inner Harbor area suffered the most, with a maximum inundation depth of 2.38 m at the coast[①]. Typhoon Hato expedited the decision at the Central People's Government to advance studies and plans for a flood prevention system involving a tidal barrier at Wanzai in Zhuhai that would help Macao, Zhuhai and Zhongshan[②].

Slope management

The threats arising from heavy rainfall is very high in Hong Kong because of its hilly condition, extensive development and dense population. In the 1960's and early 1970's, geotechnical input on site formation works was very limited. The rapid development of Hong Kong in this period brought about a large stock of potentially substandard man-made slopes. Coupled with many housing and other developments on and near hillsides, the risk of loss of life and damage to property was high. GEO/CEDD is the responsible department for slope management. Over the years, it has created one of the world's most impressive system in managing slopes.

The GEO was set up in 1977 to provide geotechnical control on new developments and redevelopments, and to develop a strategy in dealing with the large stock of potentially man-made slopes. The designs of new slopes which have been built since then have generally been checked by the GEO to ensure that they

---

①https://www.smg.gov.mo/zh/subpage/355/page/38.

② Studies by Guangdong's Pearl River Hydraulic Research Institute had been done in 2016-17. See also "Macau and Guangdong strengthen flood prevention cooperation", Macau News, 12 September 2017, https://macaunews.mo/macau-guangdong-strengthening-flood-prevention-cooperation/.

conform with the required safety standards. Up to 2010, substandard government-owned manmade slopes were upgraded systematically under the Landslip Preventive Measures Programme (LPMP). From 2010 onwards, the work was extended to include landslide risk mitigation for natural hillside under a Landslip Prevention and Mitigation Programme (LPMitP).

(1) Landslip Preventive Measures Programme (LPMP)

The LPMP was launched in 1976 to deal with the large stock of potentially substandard slopes affecting existing developments in accordance with a risk-based priority ranking system. The works and studies under the LPMP were initially undertaken by GEO in-house staff. In 1991, streamlined procedures were introduced to boost the output of slope studies. A program was commenced in 1992 to use private sector resources to improve the progress of the LPMP.

As part of the implementation of the Slope Safety Review Report in 1995, the GEO received increased resources to further accelerate the LPMP. On 1 April 1995, the GEO launched the 5-year Accelerated LPM Project to upgrade about 800 government-owned manmade slopes and to carry out safety-screening studies on about 1 500 private man-made slopes.

As part of the government's long-term strategy to improve slope safety in Hong Kong, the 10-year Extended LPM Project was launched in 2000 on completion of the 5-year Accelerated LPM Project. This 10-year Extended LPM Project boosted the level of LPM output in terms of upgrading of high priority substandard manmade slopes affecting developments and major roads. The 10-year Extended LPM Project upgraded about 3 100 substandard government-owned slopes and carried out safety-screening studies on about 3 300 privately-owned manmade slopes. This amounts to more than 4 times the output before the commencement of the 5-year Accelerated LPM Project.

(2) Landslip Prevention and Mitigation Programme (LPMitP)

Upon completion of the LPMP in 2010, the overall landslide risk from manmade slopes had been substantially reduced to less than 25% of what existed in 1977, reaching a reasonably low level that is commensurate with the international best practice in risk management. However, there were still landslide risks that pose a hazard to existing developments and important transport corridors. If investment in slope safety were not maintained, landslide risk would progressively increase with time due to slope degradation, population increase and encroachment of more urban development or redevelopment on steep hillsides, and potential impacts of extreme weather conditions will result in more frequent and more severe rainfall due to climate change. This will cause, in addition to risk to life, significant economic losses and social disruption consequence in road blockages and building evacuation due to landslides, which will compromise public safety, sustainable development and Hong Kong's reputation as a modern metropolitan city and tourist hub.

At the end of 2007, the Hong Kong authorities launched the LPMitP, which is a rolling program to dovetail with the LPMP in dealing with landslide risks arising from both substandard manmade slopes and vulnerable natural hillside catchments. The target annual outputs of the LPMitP are to:

a. upgrade 150 government-owned manmade slopes,

b. conduct safety-screening studies for 100 private man-made slopes, and

c. implement risk mitigation works for 30 natural hillside catchments.

The upgrading and mitigation work under the LPMitP commenced in 2010.

The government has spent some HK$24.6 billion on slope safety since

1977[①]. Notably today, Hong Kong's slope safety system has been recognized by the United Nations as a good practice for managing landslide risk[②]. Regular reports and updates are published by GEO/CEDD – a testament to the government's commitment[③].

Despite Hong Kong's achievements, slope degradation and climate change pose new challenges to the SAR's slope safety management system. Nevertheless, aging slopes, population growth and climate change could increase the risks of landsides. More frequent and extreme rainstorms increase both the likelihood and consequence of landsides.

To enhance community resilience and preparedness against landslide disasters, the GEO/CEDD also implements a year-round public education program and publicity campaign through a variety of events and activities (e.g., media briefings, exhibitions, school talks, TV programs) to raise public vigilance of the increasing landslide risk and to educate the public of the precautionary measures during heavy rain.

According to Macao's Lands, Public Works and Transport Bureau (LPWT), Macao has 214 dangerous slopes with only two that are considered high risk. The LPWT and IAM are responsible for maintaining roadside slopes and slopes in

---

① Report No.1/2020 on Landslip Prevention and Mitigation Studies and Works Carried Out by the Geotechnical Office.

② Report entitled "Landslide hazard and risk assessment" by United Nations Office for Disaster Risk Reduction, https://www.preventionweb.net/files/52828_03landslidehazardandrisk-assessment.pdf.

③Report No. 2/2018 on Landslip Prevention and Mitigation Studies and Works Carried Out by the Geotechnical Engineering Office, https://www.cedd.gov.hk/eng/projects/landslip/land_quar.html#intro for the September 2018 report.

public parks, and they have issued contracts to reduce landslide risks①.

Water resources

The three major water resources in Hong Kong mentioned in Chapter 2, namely Dongjiang water, local yield and seawater for toilet flushing are managed by the WSD under DEVB. While the Macao Water Supply Company (SAAM) has an exclusive concession to provide water services in Macao②, the Civic and Municipal Affairs Bureau Laboratory, under IAM, is responsible for the public water supply network. Since water supply is a key public service, private company is under government's supervision as well③.

Hong Kong and Macao are heavy water consumers due to their dense population and economic development. Hong Kong currently consumes ~1 010 million cubic meters (mcm) of fresh water per annum④; and Macao consumes ~88 mcm. Guangdong Province provides 70%~80% of Hong Kong's raw water (the exact figure changes every year depending on the local yield in each particular year) for fresh water supply and ~95% of Macao's raw water (101.2

---

① Macao Yearbook 2017, Page 396.

② In 1985, the Portuguese administration signed a 25-year contract with the Macao Water Supply Company, a joint venture between the New World Group of Hong Kong and Suez Environment of France (together holding an 85% stake and local investors hold 15%). The current concession contract in Macao runs till 2030.

③ For discussion on history of Macao's water concession contract, see Tao Fu, Miao Chang and Lijin Zhong, Chapter IX, "Water supply concession of Macao", *Reform of China Urban Water Sector*, IWA Publishing, Water 21, 2008.

④ Hong Kong uses seawater for toilet flushing for majority of its population to save fresh water resources. It currently consumes ~280 million cubic metres of seawater for toilet flushing per annum.

mcm[①]). The two SARs rely heavily on Guangdong for most of their fresh water supplies. In 2018, Hong Kong paid ～HK$ 4.8 billion for buying the raw water from Guangdong, and Macao's cost of water supplies in 2017 was MOP$184.6 million[②].

Chapter 2 already notes that Hong Kong and Macao started to import raw water from Guangdong in the early 1960s after suffering severe water shortage. While total rainfall is not predicted to be less in the future, there could still be periods of occasional drought. Indeed, this prediction of rainfall pattern affects South China as a whole. Thus, the long-term conservation of water and the efficiency of its use is vital for the whole of the GBA. This is an important area for regional collaboration (see below).

(1) Challenges for HKSAR and MSAR

The two SARs face similar water supply challenges.

a. Unstable annual yields: The annual local yields can fluctuate quite substantially, and there are no natural lakes, rivers or major underground water resources for water storage. Hong Kong's records show its local yield may fluctuate by more than 200 million cubic meters from year to year.

b. Rapid development: Both SARs have had increasing populations and continuing economic expansion, all of which requires more fresh water. They project higher consumption still for water in future years. The projected annual fresh water demand in Hong Kong is～1 100 mcm by 2040 in the absence of water demand management measures.

---

① Macao Water, Water Supply Statistics, 2019, https://www.macaowater.com/ frontend/web/index.php?r=statisticaldata%2Findex.

② Macao Water Annual Report 2017, https://www.macaowater.com/sites/default/files/report/annals/2017%20Macao%20Water%20Annual%20Report%20%28finalized%29.pdf.

c. Illusion of availability: With the ability to import water from Guangdong, even when Guangdong faced droughts, the two SARs were shielded from water shortage. The public did not feel the urgency to accept tough policies with targets and timelines to reduce water usage and the speeding-up of the implementation of technological and management means to save water.

Hong Kong's Total Water Management Strategy (TWMS), put forward in 2008 by WSD (TWMS 2008), illustrated the HKSAR's water challenges. TWMS 2008 was intended to reduce annual consumption by 236 mcm of fresh water by 2030 (equivalent to a fifth of projected consumption) through demand-and-supply side means. A review of TWMS, completed in 2019, has lowered that reduction target. The updated TWMS 2019 projected savings by 2030 to be only 110 mcm, and 120 mcm by 2040[①].

To achieve the abovementioned savings of 110 mcm and 120 mcm by 2030 and 2040 respectively, TWMS 2019 adopted a two-pronged approach with emphasis on containing fresh water demand growth and building resilience in the fresh water supply with diversified water resources. According to WSD, the review showed the current water supply will be able to meet the forecast water demand with demand management measures in water conservation, water loss management and expansion of the use of lower grade water (i.e., seawater and recycled water) for non-potable purposes. As for building resilience in Hong Kong's fresh water supply, a desalination plant is being built (see below); with back-up options as per Table 4–4 that could be considered in case there are

---

①WSD's dedicated webpage on TWMS 2019, see https://www.wsd.gov.hk/en/core-businesses/total-water-management-strategy/twm-review/containing-fresh-water-demand-growth/index.html.

deviations from water availability projects (such as higher-than-expected population, worse-than-projected impact of climate change on rainfall, or less-than-anticipated effect on containing demand growth)[①].

Table 4–4 Summary of TWMS 2019

Containing Fresh Water Demand Growth	Water loss management
	Expansion of use of lower grade water for non-potable uses
	Strengthening promotion of water conservation
Building Resilience in Water Supply	First stage of the Tseung Kwan O Desalination Plant
Additional Local Yield	Inter-Reservoirs Transfer Scheme (implemented by DSD)
Backup Options	Second stage of the Tseung Kwan O Desalination Plant
	Reactivation of mothballed water treatment works on Hong Kong Island
	Expansion of storage capacity by constructing a reservoir at high level in the Plover Cove Reservoir
	Expansion of the water gathering ground of Plover Cove Reservoir
	Expansion of the water gathering ground of High Island Reservoir
	Other desalination plants
	Increase of Dongjiang water supply

There are two longstanding aspects relating to Hong Kong's water management that are particularly noteworthy:

1) Hurdles to raise water tariffs

Various public campaigns have been launched to galvanize households to reduce water use but they have not been very successful. The HKSAR government

① WSD has a dedicated webpage on TWMS 2019, see https://www.wsd.gov.hk/en/core-businesses/total-water-management-strategy/index.html.

proposed to increase water tariff in 1996 and in 1999, but the Legislative Council did not support the proposal①. As a result, the HKSAR government has not raised water tariffs since 1995, making the HKSAR's water price the cheapest among developed economies, and water usage on a per capita per day basis is ~130 liters versus the world average of 110 liters②. The government pointed to a 2015 survey of cities around the world to justify that the effect of higher water tariff on lowering water consumption was not significant③, and the focus should be on public education. It should be noted, however, that water tariffs in many other developed cities, such as Singapore, Tokyo, Seoul, London, Paris, Geneva and New York, are much higher than the tariff in Hong Kong. Indeed, price is not unimportant in water demand management and non-government experts have continuously called upon the authorities to modernize and re-examine the water

---

① See the report by ADM Capital Foundation, Civic Exchange and WYNG Foundation, *The Illusion of Plenty: Hong Kong's Water Security, Working Towards Regional Water Harmony*, 2019, p.73.

② The water usage is much higher if seawater for toilet flushing is included. The reason for not raising the water tariff is political. There is concern that raising it would lead to political backlash from the Food & Beverage sector that has complained about higher trade effluent surcharge to treat wastewater, as well as from the Legislative Council members who could veto the legislation for raising the water tariff as they do not wish to see water charges being raised. See Frederick Lee, Hong Kong Water – Agenda and Goals, China Water Risk, 10 October 2013, http://www.chinawaterrisk.org/opinions/hong-kong-water-agenda-goals-2/.

③ In addition to the TWMS 2019, the government also referred to this survey here: https://www.wsd.gov.hk/en/faqs/index.html#41. According to the authors' correspondence with the WSD, the mentioned survey was part of the *Study on the Water Tariff Structure and Level in Other Major Cities* which was carried out by the University of Hong Kong in 2015 commissioned by the government.

tariff [①]. Moreover, public education needs to be better targeted to specifically communicate and motivate behavior change, especially for households, about the high fresh water consumption level of Hong Kong people, the inter-relationship between energy and water usage, and the fact that water tariff in Hong Kong is heavily subsidized (Zhao *et al.*, 2019b).

Under the waterworks operating account of the HKSAR government for 2017-18, only～26% of the cost of providing water is directly recovered through water tariff with～46% of the cost covered by the income of contribution from rates, ～10% covered by the income of contribution from government on free allowance to consumers and the rest is mainly subsidized from general revenue. Apart of the cost of water production (including desalination), the government needs to concern more issues, such as the affordability of citizens, financial status of water services, economic situation at the time, also the opinions of legislators when review water tariff. Thereby, the HKSAR government is not planning to raise water tariff in the near future.

Macao's total water consumption has increased in light of its vast expansion in gambling, hotel and hospitality establishments. In 2008, its water consumption was 67.46 mcm and by 2017, it had risen to 88.44 mcm. The concession contract to MSWC allows for the water tariff to be adjusted every year based on the actual cost of raw water, energy, labor, as well as repairs and maintenance, and

---

① Frederick Lee, Hong Kong Water – Agenda and Goals, China Water Risk, 10 October 2013, http://www.chinawaterrisk.org/opinions/hong-kong-water-agenda-goals-2/; "Extreme shortage of domestic freshwater", China Daily, 11 June 2017; Civic Exchange, "Modernising Hong Kong's Water Management Policy, Part I", June 2019, https://civic-exchange. org/wp-content/uploads/2019/06/Conservation-and-Consumption_ExSum-EN.pdf; and Ying Zhao, Yani Bo and Wai Ling Lee, "Barriers to Adoption of Water-Saving Habits in Residential Buildings in Hong Kong", Sustainability 2019, 11, 2036; doi:10.3390/su11072036.

investments needed. Cost has increased gradually and is still affordable and there are subsidies for low quantity users, who are likely to be lower income households.

2) Water loss from private water mains leakage

WSD has been tackling private water mains leakage through installing master meters for monitoring private water mains leakage of housing estates, providing property owners and building management agents with technical advice and support to detect leakage and subsequent repair to their private water mains. However, there is considerable administrative hassle of dealing with uncooperative private property owners. Current policy remains mild: WSD is assisting property owners rather than imposing penalties. WSD launched "Replacement and Rehabilitation Programme of Water Mains" in 2000 which replaced and repaired around 3km of aged water mains. The scheme was substantially completed in 2015 with significant improvement to water supply network.

Meanwhile, WSD has been implementing the Water Intelligent Network in phases as a more cost-effective way after completion of the large-scale replacement and rehabilitation of water mains for tackling water loss in the public water distribution mains. This new system enables WSD to collect and analyze data from the water distribution networks continuously so it can zero-in on problem areas and fix it as soon as practicable[①]. In tandem, WSD is exploring imposition of charges on property owners or building management agents who fail to rectify the leakage in their private communal mains according to the amount of

---

① For details of the Water Intelligent Network, see https://www.wsd.gov.hk/en/core-businesses/operation-and-maintenance-of-waterworks/reliable-distribution-network/index.html.

estimated water loss through legislative amendments.

China's average water loss is about 20%. The water loss rate of water in Macao is about 8.8%. While Hong Kong faces the challenges of high-water pressure in water mains due to its hilly topography coupled with vibration and disturbances to the water mains due to frequent roadworks, busy traffic and congested underground utilities which make the water mains more susceptible to leakage, Hong Kong needs to catch-up in this important area. It is noted that WSD targets to reduce the public water mains leakage to below 10% by 2030.

(2) Seawater for flushing

A unique way that Hong Kong has been able to limit fresh water usage is to use seawater for toilet flushing in areas that are close to the sea. Thus, a unique facet of the waterworks is the sea water supply systems with their separate networks of distribution mains, pumping stations and service reservoirs covering ～85% of the residents. In 2018, ～280 mcm of seawater was supplied for toilet flushing. Beyond saving fresh water, the effects of using seawater have other advantages but also a disadvantage. On the positive side, recent research showed that chlorinated saline effluent was generally less acutely toxic to the organisms than chlorinated freshwater (Yang *et al.*, 2015). It also helps to reduce the volume of sewage sludge through the SANI① process developed by the Hong Kong University of Science and Technology. On the negative side, seawater is more corrosive and requires the use of corrosion resistant plumbing materials.

(3) Desalination

TWMS 2008/2019 includes developing seawater desalination in Hong Kong

---

① SANI is the acronym of the technology of Sulphate Reduction, Autotrophic De-nitrification and Nitrification Integrated to reduce the volume of sewage sludge.

as a way to generate new water resources. Tseung Kwan O Area 137 was identified as a suitable site for a desalination plant for water production capacity of 50 mcm per annum expandable to 100 mcm to meet 5% to 10% of the overall fresh water demand of Hong Kong. The estimated cost for desalinated water including production, distribution and customer services in Hong Kong is ~ HK$13 per cubic meter, as compared with HK$10.5 per cubic meter for Dongjiang water and HK$4.7 per cubic meter for local yield. The plant is expected to commission by 2023.

(4) Recycled water

TWMS 2008/2019 includc thc use of recycled water, including water reclamation, grey water recycling and rainwater harvesting. WSD is implementing various projects, which are described in Chapter 2. WSD is also preparing the legislation for the supply of recycled water in Hong Kong after the completion of the public consultation in end-2018. In addition, WSD has been advocating the adoption of grey water reuse systems or rainwater harvesting systems for non-potable uses, such as toilet flushing and irrigation, in new government works projects. Regarding private buildings, bonus credits are awarded to buildings with a grey water reuse system or rainwater harvesting system under BEAM Plus to encourage developers to provide these facilities in lieu of using fresh water for non-potable uses.

4.3.2 Storms, storm surges and sea-level rise

As Chapter 1 projects, tropical cyclones will be more intense for the region as a whole. Moreover, based on the findings of the Fifth Assessment Report published by the IPCC and a recent study by CEDD, it is anticipated that under the scenarios of medium concentration of greenhouse gases, the likely range of sea level rise in Hong Kong by the end of this century would be 0.73 m – 1.28 m[①].

Coupled with rising sea level, the combined risks of storm surge and flooding is already greater than in the past and will intensify as sea level continues to rise. Super Typhoon Mangkhut in 2018 offered a glimpse into the ever-increasing risk. Storm surge (above astronomical tide) led the sea level in Hong Kong's Victoria Harbor to rise by 2.35 meters above the astronomical tide with maximum sea level of 3.88 meters (above Chart Datum) at Quarry Bay, which was considerable. Sea level could have been very much higher, possibly reaching 5 meters if Mangkhut had not crossed Luzon and weakened, and if its storm surge had coincided with the time of the astronomical high tide on that day (Choy *et al.*, 2020).

In the case of Macao, the maximum tide height and maximum flood height respectively reached 5.58 meters and 2.38 meters during Typhoon Hato and set a new historical record in Macao[②]. As noted in the section on flood prevention, the combination of severe storms, storm surges and sea-level rise, flooding will pose a much greater threat to the whole of the GBA. Hence it is prudent for the regional authorities to join hands to mitigate these risks, share climate data and work out commonly accepted design standards in tackling this problem.

Beyond making the standards for defensive structures more robust, such as for port works and sea walls, investment in studies to identify high-risk areas, as

---

①The above sea level projection has not accounted for possible downward vertical crustal movement which has been observed by the HKO at Tate's Cairn. Its effect might be incorporated into future sea level rise when there is sufficient data including geological evidence to substantiate the finding.

②Maximum tide height and flood height of Macao under the effects of storm surges, http://www.smg.gov.mo/smg/database/e_stormsurge_historicalRec.htm.

well as the non-linear and locally-specific nature of different types of risks (e.g., underground parking lots are a priority for flood prevention in Hong Kong and Macao) would be money well-spent. More controversial would-be studies to identify areas where development or further development should be limited or restricted, and whether certain assets should be relocated in the longer-term. Issues relating to limiting development and relocating assets are bound to be controversial and extensive consultation would be required. The conducting of studies, consultation, debate and decision-making will take time and will straddle administrations (five-year terms in the SARs). The issue today is how the current administrations intend to improve planning for the risks associated with climate change and sea level rise.

Moreover, in light of the very high socio-economic connectivity between Hong Kong, Macao and Guangdong (i.e., the GBA as a whole), the three authorities have a strong interest to cooperate and increase capabilities across the board.

4.3.3 Coastal waters and protecting biodiversity

Chapter 2 describes the increasing trend of eutrophication in the upper layer and hypoxia in the bottom layer of the coastal waters of Hong Kong, the severity of which will lead to acidification. The main cause of the problem is the increased nutrient loading of the Pearl River, as well as contributing factors from the changing climate, such as weakening wind speed and higher air temperatures.

It is important to note Hong Kong's exceptionally rich biodiversity arising from the interplay of climatic condition, geographical location and geology. For a small territory, the city has a myriad of terrestrial and marine habitats and niches, which are homes to a wealth of native flora and fauna. Some of these habitats and species are representative or particularly valuable. Hong Kong's exceptional biodiversity is described in the city's first Biodiversity Strategy and Action Plan (BSAP) published in December 2016. It represents a multifaceted five-year conservation plan prepared in light of the United Nations Convention on Biological Diversity[①]. As a signatory to the convention, China's National Biodiversity Conservation and Action Plan covers 2011-30[②].

The HKSAR government's Climate Action Plan 2030+ includes a chapter on protecting ecosystems, also shed light on the importance of BSAP[③]. For the next UN meeting on the Convention on Biological Diversity taking place in Kunming, Yunnan Province, on 2021, where a global post-2021 biodiversity framework is expected to be decided, it is likely that China will update its national plan, and Hong Kong too will develop a new plan following its first BSAP. An important area for GBA cooperation is whether Maca and Guangdong could also consider developing their own BSAP to fit in to the post-2021 biodiversity framework so that all parts of the GBA could improve conservation management as a part of strengthening resilience against climate change.

Designing and implementing BSAP requires the concerted efforts of many government departments. In the case of Hong Kong, SEN chairs an inter-departmental working group to co-ordinate the work among departments and to

---

① Environment Bureau, Hong Kong Biodiversity Strategy and Action Plan (2016-2021), December 2016, https://www.afcd.gov.hk/english/conservation/Con_hkbsap/files/HKBSAP_ENG_2.pdf.

② China National Biodiversity Conservation and Action Plan (2011-2030), https://www.cbd.int/doc/world/cn/cn-nbsap-v2-en.pdf.

③ Ibid, Chapter 8.

monitor progress. The HKSAR government earmarked HK$250 million for taking forward the initiatives under the BSAP in the five-year-plan. Thus, GBA regional cooperation on conservation management would also need cross-agencies coordination and additional resources.

4.3.4 Temperature changes and public health

Chapter 2 points out that climate change is altering many types of risks associated with weather conditions and that there are serious public health and safety issues for Hong Kong (and by implication, Macao, in light of their similar conditions). Moreover, the research of the impact of climate change on population health is still in its infancy, as is policymaking relating to climate change and public health. Development of climate policies should not be separated from scientific research since extreme temperatures and higher rainfall can lead to new health and safety risks, as healthcare professionals and social services need to be strengthened in order to prepare for greater surge capacities[①].

Very hot periods and cold spells

Chapter 1 and Chapter 2 show there will be longer periods of high temperatures in the summer and heatwaves will become more frequent and severe in future, and they will tend to start earlier in the year and occur later. The duration of individual episodes can also become longer. Heat is exacerbated by the dense, high-rise urban conditions of Hong Kong and Macao. Thus, urban planning and

---

①CARe 2018 Hong Kong Conference Summary Report and Policy Recommendations, December 2018, Chapter 6, Changing Risks and Public Health, available at: http://care2018.ust.hk/index.php/report01/.

design will need to take into account rising heat.

The HKSAR government uses its planning standards and guidelines to reduce urban heat island effect and improve micro-climate of the city's urban environment. The government follows these qualitative guidelines on urban design and air ventilation in new development areas. For existing built-up areas, project proponents are encouraged to take on board these design principles in planning and designing their development or redevelopment projects as to pursue incremental improvement of the urban wind environment. Since 2006, the government requires air ventilation assessments to be done for all major government projects so that the result can improve the design to facilitate wind penetration to their surrounding area, and the private sector is encouraged to follow this practice[①]. The HKSAR and MSAR would both need to consider what more they could do with respect to urban planning and design to cool their dense urban environment in light of the changing weather noted in Chapter 1 and health risks in Chapter 2.

Chapter 2 is clear that extreme heat can overpower the human body and cause dehydration, heatstroke or even failure. The population most at risk include children, people living in sub-standard conditions with poor ventilation, and the elderly. Hong Kong and Macao have also experienced very cold spells in recent winters and extreme cold can be deadly. As buildings and homes generally do not have heating in the SARs, low temperatures can lead to hypothermia, especially among older populations. While more research should be done, the work by Hong Kong experts should be useful to the GBA as a whole.

---

① Please see Hong Kong Climate Change Report 2015, p. 55, available at: https://www.enb.gov.hk/sites/default/files/pdf/ClimateChangeEng.pdf.

Weather warning, health services and awareness of health workforce

The effectiveness of the new warning system has been assessed in a survey conducted by the Chinese University of Hong Kong in 2017[①]. Among the respondents, 87% reported that they were aware of the heat warning issued by the HKO; 97.5% reported that they drank more water during the heatwave as a protestive measure, while 37.2% applied sunscreen. Nonetheless, 45.3% regarded high temperature would not affect their health at all and 28.8% of the over 65-year-old group were found to share the same neglect. There is room to improve public understanding of the risks by the health authorities, particularly awareness of heat-related health impacts (see Policy Recommendations in section 4.6).

4.4 Climate Change Policy Supported by Scientific Research and Cross-sectoral Cooperation

Scientific research and cross-sectoral cooperation are important elements of climate change policies for both SARs. Scientific research determines the direction and contents of decision-making, cross-sectoral cooperation generates a way to achieve the decision outcomes.

---

① The Chinese University of Hong Kong, Chan EYY, Man A, Lam HCY. Personal Heat Protective Measures During the 2017 Heatwave in Hong Kong: A Telephone Survey Study. Dec 2018 n=470. In that year, Hong Kong's recorded hottest temperature in 50 years reached 36.6℃, taken on 22 August 2017 at King's Park.

4.4.1 Scientific research on climate change

As noted in chapter 1 and chapter 2, Hong Kong and Macao have been delivering meteorological services to the public for a long time. During the development process, HKO has generated sufficient meteorological data which is valuable for the understanding and studying of climate change science. HKO established in 1883 and had extraordinary performance in developing and providing meteorological services, while the meteorological data that generated in Macao can be traced back to 1861. The meteorological service and study results generated by HKO and SMG offer a decision-making model on local climate change which gradually become a part of the adaptation policy. Tropical cyclone warning, heavy rain and landslide warning, extreme cold and hot weather warning.

Meanwhile, to achieve sustainable development, both SARs set Environmental Protection Funds to mobilize civilians to change personal behaviors and lifestyles; also support and promote the development of environmental protection industries by introducing innovative technologies and practical methods to improve environment and energy emissions. HKSAR built Environment and Conservation Fund (ECF) in June 1994, which operated in August 1994. ECF sets Environmental Research, Technology Demonstration and Conference Projects, prioritized research projects include 'Climate Change–Mitigation, Adaptation and Response', 'Biodiversity, Conservation and Geological Conservation' etc., which assist the development of projects relevant to climate change response. ECF approve approximately HK $29.5 million to fund relevant projects, which the budget come to be HK $30 million in 2020. HKSAR government grant HK $200 million to establish 'Low-Carbon Green Research

Fund' and application was available in December 2020, more sufficient and approachable funding was enhanced on carbon emissions and environmental protection. MSAR government also established Environmental Protection and Energy Conservation Fund (FPACE) with 'Environmental Protection, Energy-Saving Products and Equipment Subsidy Scheme', aiming at purchasing and replacing equipment that can promote environmental quality, energy efficiency and water conservation. The environmental protection funds that established by both SARs is an essential element for the transformation of local climate change studies.

Both SARs work closely with national and international climate change research institutions while developing its own climate adaptation policies, they exchange what is needed and feed back to the decision-making of climate change. Hong Kong and Macao joined World Meteorological Organization (WMO) in 1948 in 1996 respectively, which were represented by HKO and SMG. At the same time, HKO and SMG has maintained close working relations and developed joint projects with China Meteorological Administration, State Oceanic Administration and Guangdong (Provincial) Meteorological Service.

4.4.2 Cross-sectoral cooperation to implement climate change policies

While climate change policies are placed in practice, not only cross-sectoral cooperation within the government is essential, especially the cooperation between HKO and other sectors, effective communication and cooperation between government and the society is crucial as well. That is to say, cross-sectoral cooperation is an important element to ensure effective implementation of climate change policies.

Chapter 2 noticed the operation of warning system started with HKO's hot weather warning system in 2000. In 2014, HKO delivered 'special reminder for hot weather' service again along with precautions suggested to reduce risks from heatstroke, as it noticed the public to stay awareness even the hot weather is below the extreme heat level. Relevant departments, such as Home Affairs Department offered temporary shelters during the hot weather warning periods. HKO website listed out precautions under hot or extreme heat conditions, and advised outdoor workers, the elderly, people with chronic disease and other vulnerable groups to take precautions through maintain indoor air circulation. Centre for Health Protection (CHP) informed the latest development, precautions measures and health information to the public and stakeholders through various ways, so that citizens can take precautions to avoid climate risks after receiving the notifications.

Low-lying coastal areas of Hong Kong and Macao could be hit by levee waves under waves and storm surges generated by tropical cyclones. It is important for both SARs to reduce asset disclosure and identifying low-lying areas. In order to assess the goal, HKSAR completed the review on how climate change influence the design of coastal structures based on the IPCC Fifth Assessment Report and the Paris Agreement that the sea level rise and wind speed increase can be detected by the government. The design standards of sea level and rainfall parameters was updated in government infrastructure in 2018.

In terms of slope management, studies that aim to fill the knowledge gaps and guide policy and projects orientation were conducted by authorities and other experts through cooperation. Also, instead of general cooperation between HKO, CEDD and DSD, partnership between academics in engineering and professionals in other regions is essential since it is challenging to predict the location and time of potential landslide. Recently, such cross-sectoral cooperation come up with research thesis (Kwan *et al.*, 2018) and joint project, while a Smart Barrier System was designed to monitor the circumstances of debris dam and set alarming to trigger emergency actions when debris accumulation occurs.

4.5 Regional Collaboration

From Chapters 1, 2, and 3, it is clear that the integrity of the natural environment and resources require joint efforts from not only Hong Kong and Macao, but the GBA as a whole to work together on tackling climate change.

4.5.1 Existing framework of collaboration

Cooperation between Hong Kong and Guangdong in the environment evolved from the 1980s. Shenzhen and Hong Kong started to cooperate on pollution control across their borders in the early 1980s. Cooperation between the environmental protection agencies of Guangdong and Hong Kong began in the mid-1980s, and the Hong Kong-Guangdong Environmental Protection Liaison Group was set-up in 1990. In 1993, Guangdong, Hong Kong and Macao jointly conducted a seminar on environmental pollution control in the Pearl River Delta which opened the new era for three-way cooperation[①].

In 1995, Guangdong mentioned regional cooperation in its *Planning Outline for Modernization of the Pearl River Delta Economic Zone (1996-2010)* and

①See "Unity call over delta pollution" and "Joint bid to fight water pollution" reported by Kathy Griffin on 19 and 20 February 1993, *South China Morning Post*.

included many ambitious ideas. An outcome of the plan was the Hong Kong-Guangdong Cooperation Joint Conference (Joint Conference), created to foster communication between the HKSAR and Guangdong in 1998. The Hong Kong-Guangdong Environmental Protection Liaison Group was renamed the Hong Kong-Guangdong Joint Working Group on Sustainable Development and Environmental Protection (JWGSDEP) in 1999 as part of the Joint Conference to establish a framework and cooperation mechanism on environmental matters. JWGSDEP meets annually to review agreed work.

The JWGSDEP is made up of an expert group to discuss and coordinate work plans and make recommendations. There are currently seven special panels looking at specific areas on air quality, water quality, afforestation, marine resources and conservation, Mirs Bay and Deep Bay (Shenzhen Bay) protection, Dongjiang water, as well as marine environment management[①]. A Co-operation Agreement between Hong Kong and Guangdong on Combating Climate Change was signed between Hong Kong and Guangdong at the Joint Conference in 2011, which led to the formation of the Hong Kong-Guangdong Joint Liaison Group on Combating Climate Change[②] in the same year under the Joint Conference[③].

---

① For background, see ENB and EPD Guangdong, Full Co-operation by Hong Kong-Guangdong-Environmental Outcome for All to Share, 2017, https://www.enb.gov.hk/sites/default/files/en/booklet_jwgsdep.pdf.

② The Hong Kong-Guangdong Joint Liaison Group on Combating Climate Change, created in 2011, is co-chaired by SEN and the Director of the Guangdong Development and Reform Commission, and the purpose is to carry out collaboration and exchanges on mutually concerning topics of adaptation and mitigation in combating climate change.

③ The Hong Kong-Guangdong Cooperation Joint Conference, created in 1998, to strengthen horizontal coordination between Hong Kong and Guangdong across many policy areas covered by specific expert groups.

However, the more ambitious ideas in the *Outline of the Plan for the Reform of the Pearl River Delta 2008-2020*, such as for Guangdong, Hong Kong and Macao to gradually adopt unified standards on car fuel, vessel fuel, and emissions that would be the most advanced in the country, as well as for Guangdong and Hong Kong to conduct joint R&D of clean energy and renewable resources, cooperate on cleaner production, and construct profitable networks of energy supply and marketing, have yet to be done.

Regional environmental collaboration initially faced difficulties owing to lack of comparable data, the differences in their legal and administrative systems. Nevertheless, important innovations did come about, especially in air quality management, such as the Guangdong-Hong Kong-Macao Pearl River Delta Regional Air Quality Monitoring Network[①]. Another innovation is the Cleaner Production Partnership Programme (CP3) established in April 2008 to help Hong Kong-owned factories, in particular small-and-medium enterprises (SME). CP3 was launched by the EPD together with the then Economic and Information Commission of Guangdong Province (now renamed as the Department of Industry and Information Technology of Guangdong Province) to encourage and facilitate Hong Kong-owned factories in Guangdong and Hong Kong to adopt cleaner production technologies and practices, thereby contributing to improving the environment. Energy and water saving, pollutants emission reduction as well as production cost reduction, are key areas of focus. The HKSAR government subsidizes a part of the costs for on-site improvement assessments for factories in identifying and analyzing the problems they face and proposing practical

---

① Innovations are discussed in Christine Loh, "Hong Kong-Mainland Innovations in Environmental Protection since 1980", in *Asian Survey*, Vol 51, No. 4, July/August 2011.

improvement solutions, demonstration projects to show and evaluate the effectiveness of new cleaner production technologies, and trade specific promotion activities by industry associations for adoption of cleaner production technologies and practices by factories. The program has been seen as successful and extended three times. The current phase runs until March 2025[①].

4.5.2 Future perspective

The latest cooperation framework, *Outline Development Plan for the Guangdong-Hong Kong-Macao Greater Bay Area*, published on 18 February 2019, includes reference to climate change. It is more ambitious and more directional than the *Outline of the Plan for the Reform of the Pearl River Delta 2008-2020*. The new GBA plan is explicit that the region should pursue "green development and ecological conservation" including low-carbon production methods and lifestyles. Energy systems and protecting water resources, as well as improving flood prevention are included. The GBA Plan requires the authorities to cooperate in the short-term between 2019 and 2020, and to take a longer-term view towards 2035. At the time of the writing of this report, the authorities have yet to publish the structure and details of how they will strengthen and expand their cooperation beyond the existing mechanism.

It could be seen that over time, regional green development and cooperation became more and more important. The institutional set-up that has evolved

① The extension was announced in the HKSAR government's Policy Address 2019, see https://www.policyaddress.gov.hk/2019/eng/highlights.html for information. It was also written in the Policy Address's supplement Chapter VII, see here https://www.policyaddress.gov.hk/2019/eng/pdf/supplement_7.pdf.

provides a foundation for closer cooperation but it will have to continue to change to meet the socio-economic and multi-disciplinary policy needs for the region to achieve a much faster pace of change and to achieve wide-ranging green development outcomes.

4.6 Uniqueness of Hong Kong and Macao in Relation to Climate Change and Policy Recommendations

Hong Kong and Macao have long histories in building resilience against weather-related risks as they are highly-urbanized, coastal, sub-tropical service economies with high exposure to storms, heavy rains and other associated natural hazards that create high risks to the people and assets. They can do better still in facing climate change, where extreme weather events could be more frequent and/or more severe. Some of their weather-related risk management methods are unique and outstanding, and useful for other cities to consider. Sharing the experience also provides opportunities for the SARs to continue to make improvements. This chapter summarizes those unique aspects, based upon which recommendations for possible improvements are raised.

4.6.1 Unique experience in urban adaptation and resilience

Hong Kong's Tropical Cyclone Warning Signals System, and landslide and flood prevention systems are outstanding examples of climate adaptation and resilience, and Hong Kong's efforts are recognized internationally. The Tropical Cyclone Warning Signals System has a history of over 100 years and it serves as the first firewall that protects life and property. Measures and actions that should be taken by the authorities and the community under the issuance of each signal are clearly laid out and practiced. It sets out working norms under different emergency levels for the city. Hong Kong's slope safety management system classifies slopes with different levels of risks and tracks records for regular maintenance and management. Flood prevention strategy and plans deal with both short- and long-terms needs for handling and preventing flooding in highly populated areas to ensure minimum impact on daily life and work. In the GBA context, there should also be opportunities to co-design certain projects that affect the region as a whole. Collaboration between Hong Kong and other cities may include knowledge and expertise sharing.

The following are observations and recommendations:

1) Upgrading weather warning to include health warning – The already excellent weather warning system could be upgraded to include more targeted information to facilitate self-help and care behavior in the community during extreme weather days to reduce population risks. They could also review the climate readiness and surge capacities of their health services in order to reorganize and implement relevant public health, medical, education and social services at the community level to ensure services are relevant, appropriate, equitable and cost effective. Data on temperature-related mortality, morbidity/hospital admissions and extreme weather events should be thoroughly examined to ensure the health system can cope with increasingly frequent and intense extreme weather events of prolonged heat, occasional cold spells, and rainstorms that can cause casualties and increase risk of waterborne diseases. The

authorities also need to work with hospitals, medical schools and medical and nursing professions, as well as social welfare professionals to raise awareness and training about the health impacts of climate change. Training would need to cover technical skills to quantify population health risks and capacity building for health and climate change program development.

2) Prototyping slope management and landslides – Hong Kong has done excellent work on landslides warning, prevention and emergency preparedness, which could be widely shared. Hong Kong's experience is particularly useful to urbanized cities with hilly terrain on the mainland where landslides could threaten lives and properties. The GBA region could consider the following in the years ahead to increase resilience in their landslide and flood prevention systems:

• Supporting cross-disciplinary research and experimentation in dealing with excess water problems to be more climate-ready; Enabling the development of a science-based, data-drivenlandslide forecasting system that is based on better understanding of where and under what circumstances failures might occur; as well as the interaction between resisting barrier, excess water, soil and vegetation;

• Supporting "blue-green" approaches to designing infrastructure;

• Empowering communities to improve their disaster preparedness through well-designed communications of risks and actions they should take;

• Utilizing new tools for public education, such as Massive Open Online Courses aimed at unlimited participation and open access with appropriate visualization and interactive components; and

• Prototyping slope management and landslides to be used across the GBA on all the above fronts.

3) Protecting the shorelines and regional waters – The GBA has a long coastline and it would strengthen the region's adaptation capabilities to consider

how to protect its shorelines and regional waters. In addition, understanding in oceanography studies could guide GBA authorities to consider land-side developments, including reclamation, to ensure they do not affect the regional water movements that could result in negative effects, such as hypoxia. The GBA could also develop guiding design principles for eco-shorelines in reclamation projects by combining "hard" and "soft" engineering means to strengthen coastal defense structures. In addition, joint research on ocean-related issues to better understand how to control pollution loads and protect marine ecosystems would be highly desirable.

4) Enhancing climate resilience cooperation – Cooperation on a GBA-wide emergency awareness raising campaign on severe weather events would enhance community resilience and preparedness, as would promotion of green and low-carbon lifestyle of the GBA communities. The two SARs could consider the following as targeted aspects for cooperation enhancement in across the GBA:

• Improving urban planning and design principles in the GBA as a whole and explore whether it would make sense for the region to adopt similar design principles with respect to infrastructure to address potential hazards (i.e., wind, rain, sea level rise, etc.) since the region faces similar weather risks;

• Setting a timeline for greening the GBA's buildings so that they could commit to targets for achieving high energy and water saving;

• Encouraging cooperation on sponge city development, particularly on holistic flood prevention and resilience plans (blue-green infrastructure), as well as revitalization of rivers to restore their natural drainage and ecosystems;

• Expanding the sharing of data to areas beyond meteorology so that regional and national authorities can be more consciously built upon to strengthen the capacity of the relevant units to conduct studies and share knowledge in order to face extreme weather risks;

• Exchanging new technologies and business models to transform transport systems and services to achieve low or zero-carbon emissions;

• Consider whether there are GBA co-designing opportunities in sewage treatment so that the GBA authorities could improve the quality, efficiency and cost effectiveness and adopt the most appropriate level of sewage treatment;

• Strengthening energy supply by further integrating the energy supply systems between the mainland and the two SARs to achieve an optimal position from both climate mitigation and resilience perspectives;

• Making energy labelling compulsory in the GBA could mandate high energy requirements for electrical appliances and fuel efficiency for vehicles and continue to tighten them over time.

4.6.2 Unique advantages in governance, finance and research that enhance regional efficiency and competitiveness

The HKSAR and MSAR operate governance systems that are different from that of mainland China. This offers them flexibilities and opportunities to explore, experiment and adopt policies and practices that are more advanced. For example, Hong Kong has been a global leading financial center for decades. The quality of research at universities in Hong Kong are highly ranked worldwide. Comprehensive legal system attracts business and talents to Hong Kong and makes the city one of the most dynamic internationally. These existing advantages in should be retained.

The following are observations and recommendations:

1) Accelerating the development of green finance – Hong Kong is already seen as having the capabilities to be China's Green Finance Hub. Hong Kong is already active in raising green bonds for its own projects, as well as projects on the mainland and elsewhere. This role could be further expanded to other capital raising means through package green projects to raise funds in the GBA and the rest of the nation. With the increasing importance for listed companies in Hong Kong and on the mainland to provide ESG reports, the Hong Kong authorities can help to improve the quality of ESG reporting. Furthermore, the mainland and Hong Kong authorities should consider how Hong Kong could provide an international market for China's carbon emissions, as the mainland continues to evolve its emission trading system. Going further down the road, Hong Kong and Shenzhen could consider developing a system for the trading of emissions reduction from large buildings (cap-and-trade schemes), which Shenzhen has experimented on already, and which Tokyo has pioneered.

2) Innovating on technology and big data regulation – In the age of Big Data and Artificial Intelligence, each jurisdiction must update their policies from time to time. Furthermore, the GBA could consider a joint Big Environmental Data project, where relevant data from the region could be released.

3) Promoting professionalization development–The HKSAR authorities could consider ensuring the Director of HKEPD to be a qualified professional in the environmental field and for all deputy directors to also be professionally qualified. In the critical face of climate change and environmental issues, developing the advantages of professionals and formulating the frontier development and scientific designations are important for Hong Kong's own reformation and maintain competitiveness.

4) Cultivating open and collaborative culture across disciplines and sectors –

The complexity of dealing with climate change requires multi-disciplinary approaches and collaboration among government, universities, industry and professionals. Together, they could take both applied and cutting-edge research to implementation, including professional training. Healthcare and social services professionals should be included in strengthening adaptation and resilience capacities. Appropriate platforms and funding would be needed to support active and regular engagement. More importantly, since regional air quality management is an important GBA issue, the authorities could ensure carbon emissions outcomes are also factored in so that they could achieve optimal results for both.

5) Designing and developing regional cooperation mechanism – In light of the GBA Plan, appropriate cooperation arrangements are needed for the relevant authorities, as well as for them to work with universities and industry in order to achieve the plan's green development transformation.

References

Chan, E. Y. Y., A. Y. T. Man, H. C. Y. Lam, 2018. *Personal Heat Protective Measures During the 2017 Heatwave in Hong Kong: A Telephone Survey Study*. First Global Forum for Heat and Health, Hong Kong, December 2018.

Choy, C. W., D. S. Lau, Y. He., 2020. Super Typhoons Hato (1713) and Mangkhut (1822). Part II: Challenges in Forecasting and Early Warnings. *Weather*.

Keramydas, C., G. Papadopoulos, L. Ntziachristos, *et al.*, 2018. Real-World Measurements of Hybrid Buses' Fuel Consumption and Pollutant Emissions in a Metropolitan Urban Road Network. *Energies*,11(10).

Kwan, J. S. H., H. W. K. Lam, C. W. W. Ng, *et al.*, 2018. Recent Technical Advancement in Natural Terrain Landslide Risk Mitigation Measures in Hong Kong. *HKIE Transactions*, 25(2).

Song, Q. B., J. H. Li, H. B. Duan, *et al.*, 2017. Towards to Sustainable Energy-Efficient City: A

Case Study of Macau. *Renewable and Sustainable Energy Reviews*, 75.

Yang, M. T., J. Q. Liu, X. R. Zhang, *et al.*, 2015. Comparative Toxicity of Chlorinated Saline and Freshwater Wastewater Effluents to Marine Organisms. *Environmental Science & Technology*, 49(24).

Zhao, S. J., Q. B. Song, H. B. Duan, *et al.*, 2019. Uncovering the Lifecycle GHG Emissions and Its Reduction Opportunities from the Urban Buildings: A Case Study of Macau. *Resources, Conservation and Recycling,* 147.

Zhao, Y., Y. Bo, W. L. Lee, 2019. Barriers to Adoption of Water-saving Habbits in Residential Buildings in Hong Kong. *Sustainability*, 11(7).